This completely revised edition of *Critical Infrastructure Protection, Risk Management, and Resilience* brings a welcome update to the issues confronting today's Homeland Security Enterprise. These include the latest policies, increasing threats regarding cybersecurity, and the creation of new agencies such as CISA. Emerging threats including artificial intelligence are also discussed. This is an essential book both for security professionals and students.

Dr. Gregory Moore, *Director, Center for Intelligence & Security Studies, Notre Dame College*

As the threat landscape continues to rapidly evolve, staying versed in emerging concerns for critical infrastructure and the latest risk management strategies to address them is of the utmost importance. The second edition of *Critical Infrastructure Protection, Risk Management, and Resilience* enables students and practitioners to keep pace with these changes by offering valuable insight into evolving concepts like Artificial Intelligence (AI), cybersecurity, and FEMA's Threat and Hazard Identification and Risk Assessment (THIRA) process. This updated edition is a must-read for anyone with an interest in critical infrastructure protection.

Mark Christie, *CEM, former President of the Emergency Management Association of Ohio*

Critical Infrastructure Protection, Risk Management, and Resilience

This second edition of *Critical Infrastructure Protection, Risk Management, and Resilience* continues to be an essential resource for understanding and protecting critical infrastructure across the U.S. Revised and thoroughly updated throughout, the textbook reflects and addresses the many changes that have occurred in critical infrastructure protection and risk management since the publication of the first edition. This new edition retains the book's focus on understudied topics, while also continuing its unique, policy-based approach to topics, ensuring that material is presented in a neutral and unbiased manner. An accessible and up-to-date text, *Critical Infrastructure Protection, Risk Management, and Resilience* is a key textbook for upper-level undergraduate or graduate-level courses across Homeland Security, Critical Infrastructure, Cybersecurity, and Public Administration.

Kelley A. Pesch-Cronin is a Professor at Notre Dame College of Ohio, USA. Her research interests include homeland security and emergency management issues, especially as they pertain to policy and politics. Previously, she worked in municipal government and local law enforcement and has co-authored several books in the field.

Nancy E. Marion is a Chair and Professor of Political Science at the University of Akron, USA. Her research areas largely revolve around the intersection of politics and criminal justice. She is the author of numerous articles and books that examine how politics affects criminal justice policy.

Critical Infrastructure Protection, Risk Management, and Resilience

A Policy Perspective

Second Edition

Kelley A. Pesch-Cronin
and Nancy E. Marion

NEW YORK AND LONDON

Designed cover image: © Shutterstock

Second edition published 2024
by Routledge
605 Third Avenue, New York, NY 10158

and by Routledge
4 Park Square, Milton Park, Abingdon, Oxon, OX14 4RN

Routledge is an imprint of the Taylor & Francis Group, an informa business

First edition published by CRC Press 2017

ISBN: 978-1-032-56305-3 (hbk)
ISBN: 978-1-003-43488-7 (ebk)

DOI: 10.4324/9781003434887

Typeset in Times New Roman
by codeMantra

NEM: To Belilla cause she deserves it!
KPC: To Seth, Serena, and Leo XOXO

Contents

Figures

Tables

Boxes

Contents

Figures

Tables

Boxes

Foreword

The mission of protecting critical infrastructure continues to be a vital concern for the U.S. In recent years, there has been renewed urgency with newer and more complicated threats. These require greater attention to cybersecurity, understanding risk across all sectors, and protecting our critical infrastructure within a disparate set of threats. This second edition of *Critical Infrastructure Protection, Risk Management, and Resilience: A Policy Perspective* addresses these challenges and provides a robust update.

The second edition continues to frame the policy perspective giving students and practitioners alike a better understanding of the growing risk landscape. The federal government's role continues to expand along with its private sector partners. This updated version includes policy and law updates by the Trump and Biden administrations. New information on the creation of the Cyber and Infrastructure Security Agency (CISA) along with its mission for national risk management is discussed. In addition, the chapters include emerging cybersecurity threats to all 16 sectors and the updated Threat and Hazard Identification Risk Assessment (THIRA) process. The chapters include examples from recent events impacting critical infrastructure including the 2020 COVID-19 pandemic, the 2021 Texas Winter Storm Uri, and the 2023 East Palestine train derailment.

As we watch the threat spectrum shift, the challenges surrounding the management and protection of critical infrastructure continue to expand. Climate change, an aging infrastructure, the Internet of Things (IoT), Artificial Intelligence (AI), information sharing, and a renewed commitment to public–private partnerships are just some of the key concerns. Cybersecurity permeates each area, with much of our cyber infrastructure receiving attention like never before. Many of these vulnerabilities have been ignored for too long. It is our hope that this revised edition provides forward thinking and the insight needed to address these daunting challenges.

Kelley A. Pesch-Cronin, Ph.D.
Professor and Department Head, Public Service and Security Studies
Notre Dame College of Ohio

Abbreviations

ATF	Alcohol, Tobacco, and Firearms
C3	Cyber Crimes Center
C3VP	Critical Infrastructure Cyber Community Voluntary Program
CBP	Customs and Border Patrol
CCFI	Civil Cyber-Fraud Initiative
CCIPS	Computer Crime and Intellectual Property Section
CERT	Computer Emergency Response Team
CESER	Office of Cybersecurity, Energy, Security, and Emergency Response
CFAA	Computer Fraud and Abuse Act
CI	Critical Infrastructure
CIA	Critical Infrastructure Assurance
CIAO	Critical Infrastructure Assurance Office
CII	Critical Infrastructure Information
CIP	Critical Infrastructure Protection
CIS	Citizenship and Immigration Services
CISA	Cybersecurity and Infrastructure Security Agency
CISCP	Cyber Information Sharing and Collaboration Program
CIWN	Critical Infrastructure Warning Network
Co-SSA	Co-Sector Specific Agency
CSRIC	Communications Security, Reliability, and Interoperability Council
DHS	Department of Homeland Security
DOE	Department of Energy
DOJ	Department of Justice
DOT	Department of Transportation
EPA	Environmental Protection Agency
FAA	Federal Aviation Administration
FBI	Federal Bureau of Investigation
FCC	Federal Communications Commission
FCIRC	Federal Computer Intrusion Response Capability
FEMA	Federal Emergency Management Agency
FGDC	Federal Geographic Data Committee
FIDN	Federal Intrusion Detection Network
FirstNet	First Responder Network Authority
GAO	Government Accountability Office
GCC	Government Coordinating Council
GSA	General Services Administration

HIRT	Hunt and Incident Response Team
HSIN	Homeland Security Information Network
HSPD	Homeland Security Presidential Directive
IC3	Internet Crime Complaint Center
ICE	Immigration and Customs Enforcement
ICSC	Industrial Control System Cybersecurity
IJC3	Integrated Joint Cybersecurity Coordination Center
IOD	Integrated Operations Division
ISACs	Information Sharing and Analysis Centers
ISAO	Information Sharing and Analysis Organization
JTTF	Joint Terrorism Task Forces
NAC	National Advisory Council
NCIJTF	National Cyber Investigative Joint Task Force
NDPC	National Domestic Preparedness Consortium
NGAC	National Geospatial Advisory Committee
NGO	Nongovernmental Agency
NICC	National Infrastructure Coordinating Center
NICCS	National Initiative for Cybersecurity Careers and Studies
NIPC	National Infrastructure Protection Center
NIPP	National Infrastructure Protection Plan
NISC	National Institute of Standards and Technology
NSC	National Security Council
NSDI	National Spatial Data Infrastructure
NTIA	National Telecommunications and Information Administration
NTSB	National Transportation and Safety Board
OBP	Office of Bombing Prevention
OIP	Office of Infrastructure Protection
PCC	Policy Coordination Committees
PCCIP	President's Commission on Critical Infrastructure Protection
PCII	Protected Critical Infrastructure Information
PDD	Presidential Decision Directive
PKEMRA	Post-Katrina Emergency Management Reform Act
RFJ	Rewards for Justice
SFI	Strategic Foresight Initiative
SHIELD	Stopping Harmful Image Exploitation and Limiting Distribution
SLTTGCC	State, Local, Tribal, and Territorial Government Coordinating Council
SRNA	Strategic National Risk Assessment
SSA	Sector-Specific Agency
SSP	Sector-Specific Plans
THIRA	Threat and Hazard Identification Risk Assessment
TSA	Transportation Security Administration
USCG	United States Coast Guard
USSS	United States Secret Service
VAU	Virtual Assets Unit

Timeline

1947	National Security Act of 1947 passed by Congress
1950	Civil Defense and Disaster Compact
1972	FIRESCOPE initiated to aid in fighting fires. This stands for Firefighting Resources of Southern California Organized for Potential Emergencies
1978	FBI's National White Collar Crime Center created
March 31, 1979	President Carter created FEMA by signing Executive Order 12127
1984	State Department established Rewards for Justice program
September 1984	Rajneeshee cult poisoned salad bars in Oregon with *Salmonella*
December 21, 1988	Pan Am Flight explodes over Lockerbie, Scotland, killing 270 people
September 1, 1989	Hurricane Hugo hit the Carolinas
August 24, 1992	Hurricane Andrew
February 26, 1993	World Trade Center in New York bombed
January 17, 1994	Earthquake hits Los Angeles, CA
April 19, 1995	Oklahoma City bombing
1996	InfraGard started in Cleveland, Ohio: A public–private partnership between the FBI and businesses, academic institutions, law enforcement, security personnel, healthcare workers, and others to provide a secure communication system during an event
July 15, 1996	President Clinton signed Executive Order 13010: established the President's Commission on Critical Infrastructure Protection (PCCIP)
May 22, 1998	Presidential Decision Directive 63 (PDD-63): Signed by President Clinton to establish a national capability within five years that would provide for the protection of critical infrastructure
October 30, 2000	The Defense Authorization Act provided funds to protect the nation's critical infrastructure
September 11, 2001	Terrorist Attack on New York and Washington, D.C.
October 26, 2001	USA PATRIOT Act signed by President George W. Bush
October 8, 2001	Executive Order 13228: Established the Office of Homeland Security and the Homeland Security Council within the Executive Office of the President
October 29, 2001	President Bush issued the Homeland Security Presidential Directives: the first HSPD: concerned the organization and operation of the Homeland Security Council
October 29, 2001	HSPD-1: Organization and Operation of the Homeland Security Council
October 29, 2001	HSPD-2: Combating Terrorism Through Immigration Policies
November 19, 2001	Aviation and Transportation Security Act (PL 107-71) approved

March 11, 2002	HSPD-3: Homeland Security Advisory System (Color System)
September 28, 2022	Hurricane Ian hits Florida
November 25, 2002	Homeland Security Act passed by Congress that established the Department of Homeland Security
November 25, 2022	Critical Infrastructure Information Act signed by President George Bush
November 25, 2022	The Maritime Transportation Security At (PL 107-295) approved
December 2002	HSPD-4: National Strategy to Combat Weapons of Mass Destruction
February 1, 2003	National Strategy for the Physical Protection of Critical Infrastructure and Key Assets published
February 28, 2003	HSPD-5: Management of Domestic Incidents: created NIMS
September 16, 2003	HSPD-6: Integration of Use of Screening Information
December 17, 2003	HSPD-7: Critical Infrastructure Identification, Prioritization, and Protection
December 17, 2003	HSPD-8: National Preparedness
January 4, 2004	Tom Ridge begins as the first Secretary of Homeland Security
January 30, 2004	HSPD-9: Defense of U.S. Agriculture and Food (by President George W. Bush)
April 8, 2004	HSPD-10: Biodefense for the 21st Century issued by President George W. Bush
April 28, 2004	Intelligence Reform and Terrorism Prevention act of 2004 (Amended by PATRIOT Act Reauthorization Act of 2005)
August 27, 2004	HSPD-11: Comprehensive Terrorist-related Screening Procedures (by President George W. Bush)
August 27, 2004	HSPD-12: Policy for a Common Identification Standard for Federal Employees and Contractors (by President George W. Bush)
December 17, 2004	Intelligence Reform and Terrorism Prevention Act signed by President George W. Bush
December 21, 2004	HSPD-13: Maritime Security Policy (Also NSPD-41)
February 15, 2005	Michael Chertoff becomes Secretary of Homeland Security
April 15, 2005	HSPD-14: Domestic Nuclear Detection (Also NSPD-43)
August 2005	Hurricane Katrina hits the Southern U.S.
2006	Domestic Security Alliance Council, part of the FBI, was established to increase information sharing between public and private groups regarding criminal activity on the Internet.
2006	First National Infrastructure Protection Plan
March 9, 2006	USA PATRIOT Act Improvement and Reauthorization Act of 2005 signed
October 4, 2006	Post-Katrina Emergency Management Reform Act of 2006 signed by President George W. Bush
October 13, 2006	SAFE Port Act (Security and Accountability for Every Port Act) approved
October 26, 2006	Secure Fence Act of 2006 signed into law
December 19, 2006	Pandemic and All-Hazards Preparedness Act of 2006 passed
December 20, 2006	Tsunami Preparedness Act approved (PL 109-424)
2007	The State, Local, Tribal, and Territorial Government Coordinating Council (SLTTGCC) was formed
January 31, 2007	HSPD-18: Medical countermeasures against weapons of mass destruction (by President George W. Bush)
February 12, 2007	HSPD-19: Combating terrorist use of explosives in the U.S.

May 4, 2007	HSPD-20: National Continuity Policy published
June 2007	National Advisory Council established
October 18, 2007	HSPD-21: Public Health and Medical Preparedness
August 3, 2007	Implementing Recommendations of the 9/11 Commission Act approved
2008	National Cyber Investigative Joint Task Force created
January 2008	HSPD-23: National Cyber Security Initiative created by President George W. Bush (NSPD-54)
January 8, 2008	Homeland Security Presidential Directive 23: Cyber Security and Monitoring
June 5, 2008	HSPD-24: Biometrics for Identification and Screening to Enhance National Security (by President George W. Bush)
September 2008	A Guide to Critical Infrastructure and Key Resources Protection at the state, local, regional, local, tribal, and territorial levels published
September 13, 2008	Hurricane Ike hit Galveston, Texas
2009	National Infrastructure Protection Plan: Partnering to Enhance Protection and Resiliency (established 16 sectors)
January 20, 2009	Janet Napolitano becomes Secretary of Homeland Security
May 2009	President Obama creates the Cyberspace Policy Review committee
October 28, 2009	HR 2892, the Department of Homeland Security Appropriations Act was approved (PL 111-83)
February 2010	Quadrennial Homeland Security Review Report: A Strategic Framework for a Secure Homeland
July 15, 2010	Critical Infrastructure Protection: Key private and public cyber expectations need to be consistently addressed
September 10, 2010	DHS Risk Lexicon published
October 7, 2010	Reducing Classification Act passed (PL 111-258)
January 4, 2011	Congress passed a law to protect the food supply. Called Improving the Capacity to Prevent Food Safety Problems, this became Public Law 111-353
March 30, 2011	Obama issues PPD-8 (National Preparedness): with goals and 31 core capabilities
June 1, 2011	Community Resilience Task Force Recommendations
August 2011	Empowering local partners to prevent violent extremism in the U.S.
September 2011	National Preparedness Goal
December 2011	Strategic National Risk Assessment (SNRA): Secretary of Homeland Security identified the types of incidents that pose the greatest threat to the Nation's homeland
December 2011	A Whole Community Approach to Emergency Management: Principles, Themes, and Pathways for Action
2012	Congress passed the First Responder Network Authority (FirstNet) that established a wireless network solely for first responders
October 29, 2012	Hurricane Sandy hits New Jersey
December 20, 2012	The Coast Guard and Marine Transportation Act of 2012 approved
2013	National Infrastructure Protection Plan amended
February 12, 2013	Presidential Policy Directive/PPD-21: Critical Infrastructure Security and Resilience
February 12, 2013	Executive Order 13636: Improving Critical Infrastructure Cybersecurity (by President Obama)

February 25, 2013	Establishment of the Cyber Threat Intelligence Integration Center (by President Obama)
March 13, 2013	the Pandemic and All-Hazards Preparedness Reauthorization Act of 2013 approved and signed into law
April 15, 2013	Boston Marathon bombing
December 23, 2013	Jeh Johnson becomes Secretary of Homeland Security
February 12, 2014	Framework for Improving Critical Infrastructure Cybersecurity released by National Institute of Standards and Technology (Obama administration)
November 24, 2014	Sony Pictures Hack
December 2014	Cybersecurity Unit (part of Department of Justice) created
December 18, 2014	The Cybersecurity Workforce Assessment law approved (PL 113-246)
December 18, 2014	The National Cybersecurity Protection Act approved (PL 113-282)
December 18, 2014	The Cybersecurity Enhancement Act approved (PL 113-274)
December 18, 2014	Protecting and Securing Chemical Facilities from Terrorist Attacks Act signed by the president (PL 113-254)
February 13, 2015	Executive Order 13691: Promoting Private Sector Cybersecurity Information Sharing (by President Obama)
June 22, 2016	Public Law 114-183 approved; also called Protecting Our Infrastructure of Pipelines and Enhancing Safety Act
July 15, 2016	The FAA Extension, Safety and Security Act approved (PL 114-190)
January 2017	Elections systems declared a subsector of Government Facilities sector
May 11, 2017	Trump issues Executive Order 13800 called Strengthening the Cybersecurity of Federal Networks and Critical Infrastructure
May 11, 2017	Trump signs Executive Order 13799, Establishment of the Presidential Advisory Commission on Electoral Integrity. This was later rescinded in Executive Order 13820
2018	CISA formed to reduce the risk of cybercrimes against the U.S.
January 3, 2018	Trump signs Executive Order 13820, Executive Order on the Termination of Presidential Advisory Commission on Election Integrity
February 2018	U.S. Attorney General creates the Cyber-Digital Task Force
May 30, 2018	DHS and OMB publish the "Federal Cybersecurity Risk Determination Report and Action Plan to the President of the US" to help identify cyberattacks on federal civilian agencies, finding that most agencies were either "at risk" or "high risk" of an attack
August 14, 2018	NIST Small Business Cybersecurity Act approved (PL15-236) was signed into law
October 5, 2018	The Geospatial Data Act is signed by President Trump to protect geospatial data stored on networks
October 23, 2018	America's Water Infrastructure Act of 2018 approved (PL 115-270)
November 16, 2018	Trump signs the Cybersecurity and Infrastructure Security Agency Act of 2018 (PL 115-278). This transferred many responsibilities of protecting cyber networks to a new agency called the Cybersecurity and Infrastructure Security Agency (CISA)
December 5, 2018	The Emergency Information Improvement Act of 2015 signed into law (PL 114-111)
December 21, 2018	Strengthening and Enhancing Cyber-capabilities by Utilizing Risk Exposure Technology Act (SECURE Technology Act) signed (PL 115-390)

May 2, 2019	Trump signs Executive Order 13833 to advance America's cybersecurity workforce.
May 15, 2019	Trump signs Executive Order 13873 to block Chinese telecommunications companies from selling equipment in the U.S. that might compromise American computer networks
June 24, 2019	Pandemic and All-Hazards Preparedness and Advancing Innovation Act of 2019 approved (PL 116-22)
September 2019	Hackers broke into the SolarWinds computer network, placing malware on their software system. The hack went undetected for months, allowing hackers to glean information on thousands of users.
February 2020	Executive Order 13905 is signed by President Trump. Called "Strengthening National Resilience through Responsible Use of Positioning, Navigation and Timing Service," this was intended to protect critical infrastructure that relies on GPS data
March 11, 2020	COVID Pandemic begins
April 4, 2020	President Trump signs Executive Order 13913, called "Establishing the Committee for the Assessment of Foreign Participation in the United States Telecommunications Services Sector." This would help the FCC in reviewing the national security concerns raised by foreign governments that participate in the U.S. telecommunications services sector.
April 28, 2020	President Trump signed Executive Order 13917 to order established meat plants as critical infrastructure during the COVID pandemic
May 1, 2020	Trump signs Executive Order 13920, Securing the U.S. Bulk-Power System.
June 2020	President Trump opposes undersea Internet cable for national security concerns
August 2020	Trump signed Executive Order 13942, addressing the threat posed by TikTok and taking additional steps to address the national emergency with respect to the information and communications technology and services supply chain
January 1, 2021	Safeguarding tomorrow through Ongoing Risk Mitigation Act (STORM Act) became law (PL116-284)
January 5, 2021	CISA creates the Cyber Unified Coordination Group in response to the Solar Winds cyberattack
February 2021	Alejandro Myorkas becomes Secretary of Homeland Security
February 13–17, 2021	Winter Storm Uri hits the U.S.
May 2021	Colonial Pipeline ransomware attack
May 12, 2021	Biden issues Executive Order 14028, Improving the Nation's Cybersecurity
June 16, 2021	Biden issues Russian leader Putin a list of 16 critical infrastructure entities that were "off limits" to a cyberattack
July 2021	Biden signs a National Security Memorandum called Improving Cybersecurity for Critical Infrastructure Control Systems
October 29, 2021	Biden administration publishes a proclamation on Critical Infrastructure Security and Resilience Month
October 2021	Supply Chain Summit
November 15, 2021	Infrastructure and Jobs Act signed by President Biden (PL117-58)

May 12, 2022	the National Cybersecurity Preparedness Consortium Act signed by the president (PL 117-122)
September 2022	Hurricane Ian, a category 5 hurricane, hit Florida causing major damage and death
November 2022	Congress passes the Critical Infrastructure Information Act to create the Protected Critical Information program to ensure that shared information remains out of the public eye
December 20, 2022	the Community Disaster Resilience Zones Act passed into law (PL 117-255).
February 3, 2023	Norfolk Southern Train derailment in East Palestine, Ohio

Glossary

Adaptability includes designing risk management actions, strategies, and processes to remain dynamic and responsive to change.

Advisory Councils provide advice and recommendations to the government or private agencies about the best methods to protect critical assets. Councils are also able to develop better methods for information sharing between and among the public and private agencies. Two significant councils are the National Infrastructure Advisory Council (NIAC) and the Critical Infrastructure Partnership Advisory Council (CIPAC).

Automated Indicator Sharing (AIS) provides for sharing of machine-readable cyber threats and potential defensive measures.

Bureau of Counterterrorism an agency within the Department of State that coordinates national efforts to prevent acts of terrorism.

Bureau of Cyberspace and Digital Policy part of the Department of State that focuses on coordination of the department's work on cyberspace and digital diplomacy.

Capability Target defines success for each core capability and describes what the community wants to achieve by combining detailed impacts with basic and measurable desired outcomes based on the threat and hazard context statements developed in Step 2 of the THIRA process.

CARMA Cyber Assessment Risk Management Approach is used alongside the Cybersecurity Framework to assess risk. CARMA provides a methodology for sector stakeholders to define key business functions that must be protected and identifies risks posed to their functional viability.

CARVER Plus Shock Method used by the Food and Agriculture Sector to determine the vulnerabilities in its assets, systems, and networks. This is accomplished by encompassing the consequences and threats.

CISA Central shares technical information about cyber threats and other topics to public and private sectors and agencies.

Citizenship and Immigration Services a federal agency that oversees the process of immigration and naturalization services for those seeking to enter the U.S.

Coast Guard federal agency responsible for the security of the country's maritime territories.

Communications Sector-Specific Plan (CSSP) Risk Framework built on the goals of the CSSP, the plan attempts to increase the resilience of the Communications Sector. These basic risk assessment goals include a more resilient infrastructure, diversity, redundancy, and recoverability.

Computer Crime and Intellectual Property Section (of Department of Justice) investigates and prosecutes cyberattacks on critical assets and also participates in policy and legislation issues. The agency coordinates with other federal agencies to work on issues related to critical infrastructure protection.

Consequence the possible impact or result of an event, incident, or occurrence, such as the number of deaths, injuries, property loss or damage, or interruptions to necessary services. The economic impacts of an event are also critical consequences, as many events have both short and long-term economic consequences to communities or even to the nation.

Context a community-specific description of an incident, including location, timing, and other important circumstances.

Convergence the interconnected nature of critical infrastructure such that harm to one asset results in harm to other assets. At the same time, it means that if one sector is unable to provide a service, another asset may be able to provide that service.

Core Capability defined by the National Preparedness Goal, 31 activities that address the greatest risks to the Nation. Each of the core capabilities is tied to a capability target.

Countering Weapons of Mass Destruction Office part of DHS, this office seeks to prevent attacks against the U.S. that use weapons of mass destruction.

Critical Infrastructure the framework of man-made networks and systems that provide needed goods and services to the public. This includes any facility or structure, both physical and organizational that provides essential services to the residents of a community to ensure its continued operation. This term includes things like buildings, roads and transportation systems, telecommunications systems, water systems, energy systems, emergency services, banking and finance institutions, and power supplies. In addition to physical structures and assets, the term incorporates virtual (cyber) systems and people. Critical Infrastructure can be found in the local, state, and federal systems and can be owned and operated by private and/or public organizations.

Critical Infrastructure Assurance Office located in the Commerce Department, this office provided support to the sectors as they developed their protection plans, as well as providing assistance to National Coordinators as they integrated the sector plans into the National Plan.

Critical Infrastructure Information (CII) the data or information that pertains to an asset or critical infrastructure, such as knowledge about the daily operations of an asset, or a description of the asset's vulnerabilities and protection plans. CII can also include information generated by the asset such as patient health records or a person's banking and financial records, or evidence of future development plans related to the asset. It can also be information that describes pertinent geological or meteorological information about the location of an asset that may point out potential vulnerabilities of that facility (i.e. a dam at an earthquake-prone site).

Critical Infrastructure Partnership Advisory Council (CIPAC) established by DHS in 2006 to help foster effective communication between federal, state, local, tribal, territorial, and regional infrastructure protection programs. It provides a forum for these groups to discuss activities to support and coordinate resource protection. CIPAC membership plans, coordinates, and implements security programs related to critical infrastructure protection. The council also includes representatives from federal, state, local, and tribal governmental groups who are members of the Government Coordinating Councils for each sector.

Critical Infrastructure Protection (CIP) actions that are geared toward protecting critical infrastructures against physical attacks or hazards, or toward deterring attacks (or mitigating the effects of attacks) that are either man-made or natural. Most CIP activity includes preventative measures, but it usually refers to actions that are more reactive. Today, CIP focuses on an all-hazards approach. The primary responsibility for protecting critical infrastructure lies with the owners and operators, but the federal government and owners/operators work together to identify critical infrastructure.

Critical Infrastructure Warning Network an agency managed by the DHS to provide secure lines of communication between the federal government and other federal, state, and local agencies, the private sector, and international agencies.

Customs and Border Patrol an agency within the DHS that oversees securing American borders.

Cyber Unified Coordination Group created on January 5, 2021, by CISA in response to the Solar Winds cyberattack. The intent was to identify any victims and recommend further actions.

Cyber risk risks associated with cyberthreats, crimes, or attacks.

Cybersecurity any actions taken by the government or by private operators to prevent damage to, unauthorized use of, or exploitation of, information that is stored electronically. This includes any activity that is intended to protect or restore information networks and wirelines, wireless satellites, public safety answering points, and 911 communications systems and control systems.

Cybersecurity & Infrastructure Security Agency (CISA) a federal agency created in 2018 to coordinate security and resilience measures to protect against cyberattacks. As part of the Department of Homeland Security, the agency's goal is to reduce the risk of cyberattacks on public and private networks.

Cybersecurity Division of CISA an agency designed to increase the cybersecurity of critical networks, both public and private.

Cybersecurity Framework on February 12, 2013, President Obama issued Executive Order 13636 in which he described the development of a Cybersecurity Framework to increase protection to protect the nation's cybersecurity. This framework would establish standards, methodologies, procedures, and processes that the owners of critical infrastructure could use to reduce their cybersecurity risks.

Cybersecurity Framework Components the three components of the Cybersecurity Framework are the core, implementation tiers, and profile. The framework is a voluntary risk-based framework focusing on enhanced cybersecurity that can be used by organizations of any size in any of the 16 critical infrastructure sectors that either already have a mature cyber risk management and cybersecurity program, or even by those that do not have such programs.

Cybersecurity and Infrastructure Security Agency (CISA) Created by President Trump in 2018, the new agency was the primary federal organization to confront cyberthreats facing the U.S.

Cybersecurity Workforce Assessment Act required the Secretary of Homeland Security to conduct a yearly assessment of the cybersecurity workforce in DHS.

Cyberspace Policy Review Committee established in May 2009, by President Obama, this committee was asked to review the nation's cyberspace security policy and make recommendations to the president about ways to improve cybersecurity. The Committee recommended that the president appoint an Executive Branch Cybersecurity Coordinator and that the Executive Branch work more closely with all key players who are involved in U.S. cybersecurity policy.

Department of Homeland Security created in November 2002 when Congress passed the Homeland Security Act (PL 107-296). The DHS is a cabinet-level department that is responsible for developing a comprehensive national plan for securing the country's assets and for recommending policies to protect the nation's critical infrastructure and key resources.

Desired Outcome the standard to which incidents must be managed, including the timeframes for conducting operations or percentage-based standards for performing security activities.

DHS Risk Lexicon initiated in 2008, the DHS Risk Lexicon provides a common language to improve the capability of the DHS to assess and manage homeland security risk. At first, the document contained 23 terms that served as a tool to improve the capabilities of the DHS to assess and manage homeland security risk. In 2010, the second edition of the *DHS Risk Lexicon* was published with an additional 50 new terms and definitions.

DHS Risk Management Process encourages comparability and shared understanding of information and analysis in the decision-making process. This comprises seven planning and analysis efforts, including defining and framing the context of decisions and related goals and objectives; identifying the risks associated with the goals and objectives; analyzing and assessing the identified risks; developing alternative actions for managing the risks and creating opportunities, and analyzing the costs and benefits of those alternatives; choosing among alternatives and implementing that decision; and monitoring the implemented decision and comparing observed and expected effects to help influence subsequent risk management alternatives and decisions. Risk communications underlie the entire risk management process.

Domestic Security Alliance Council an agency developed to increase information sharing between public and private groups regarding criminal activity on the Internet.

Emergency Communications Division of CISA: promotes better communication between agencies and emergency responders to promote resiliency after an emergency.

Enhanced Cybersecurity Services Program allows classified information on cybersecurity threats and other technical information to be shared with infrastructure network service providers.

Executive Order 13010 a document signed by President Clinton in 1996 in which he revealed his plans to establish the President's Commission on Critical Infrastructure Protection (PCCIP) that would investigate the scope and nature of potential vulnerabilities and threats to the country's critical infrastructure, with a focus on cyber threats. Clinton also identified eight critical infrastructure sectors in the Executive Order, including telecommunications, transportation, electric power, banking and finance, gas and oil storage and delivery, water supply, emergency services, and government operations. The third change that Clinton identified was the expansion of the term "infrastructure."

Executive Order 13228 signed by President Bush on October 8, 2001, to create the Office of Homeland Security. The agency was to coordinate efforts to protect the United States and its critical infrastructure from another attack and maintain efforts at recovery.

Executive Order 13231 signed by President Bush on October 16, 2001, the document announced the president's intentions to create the President's Critical Infrastructure Protection Board. The board was responsible for recommending policies and programs to protect information systems for critical infrastructure.

Executive Order 13527 Medical Countermeasures Following a Biological Attack was signed by President Obama on December 30, 2009. This outlined plans for medical services if an attack occurred to mitigate illness and death.

Executive Order 13563 issued on January 18, 2011, by President Obama, this document was also entitled "Improving Regulation and Regulatory Review." In this document, Obama directed all federal agencies to develop a Preliminary Plan to review their regulations to determine whether any of the existing rules should be updated or altered in any way to make the agency's regulatory program more effective.

Executive Order 13636 Issued by President Obama on February 12, 2013, this document, also entitled "Improving Critical Infrastructure: Cybersecurity," stressed the need to protect the nation's Critical Infrastructure and cyber environment. Obama stressed the need for better communication and cooperation between the owners and operators of critical infrastructure and the federal government. In another part of the executive order, Obama sought to develop a Cybersecurity Framework to improve the nation's cybersecurity.

Executive Order 13691 issued by President Obama in February 2015, Executive Order 13691, also called Promoting Private Sector Cybersecurity Information Sharing, focused on the importance of sharing information pertaining to cybersecurity. Obama created Information Sharing and Analysis Organizations (ISAOs) to further this goal.

Executive Order 13799 establishment of the Presidential Advisory Commission on Election Integrity: Created an agency to ensure that the American electoral system was safe and secure. This was issued by President Trump and signed on May 11, 2017. It was later rescinded by President Trump on January 3, 2018.

Executive Order 13800 called "Strengthening the Cybersecurity of Federal Networks and Critical Infrastructure," this was signed by President Trump in May 2017. All federal agencies were asked to address their policies for managing cyber risks and standardize them with the NIST framework.

Executive Order 13820 executive order on the Termination of Presidential Advisory Commission on Election Integrity: Issued by Trump on January 3, 2018, it terminated the election commission after finding no evidence of voter fraud.

Executive Order 13833 signed by President Trump on May 2, 2019, to advance America's cybersecurity workforce.

Executive Order 13873 signed by President Trump in 2019. This was signed to prevent Chinese telecommunications companies from selling equipment in the U.S. that could compromise U.S. computer networks.

Executive Order 13905 Signed by President Trump in February 2020, this was called Strengthening National Resilience through Responsible Use of Positioning, Navigation and Timing Service. This was intended to protect critical infrastructure that relies on GPS data.

Executive Order 13913 signed by President Trump on April 4, 2020, and called "Establishing the Committee for the Assessment of Foreign Participation in the United States Telecommunications Services Sector" to assist the FCC is reviewing the national security concerns raised by foreign governments that participate in the U.S. telecommunications services sector.

Executive Order 13917 signed by President Trump on April 28, 2020, this executive order established meat plants as critical infrastructure during the COVID-19 pandemic.

Executive Order 13920 signed by President Trump on May 1, 2020, and called "Securing the United States Bulk-Power System."

Executive Order 13942 signed by President Trump in August 2020. This was called "Addressing the Threat Posed by TikTok and Taking Additional Steps to Address the National Emergency with Respect to the Information and Communications Technology and Services Supply Chain."

Executive Order 14028 issued by President Biden on May 12, 2021. This was called Improving the Nation's Cybersecurity.

Federal Emergency Management Agency (FEMA) the nation's federal agency that oversees the federal response to disasters, both natural and man-made. The agency provides support for citizens so they can protect against, respond to, recover from, and mitigate all hazards.

FEMA Grant Program an annual grant program overseen by the Grant Programs Directorate that provides financial assistance for programs that seek to increase the Nation's infrastructure protection and security of its assets. The focus is to promote communication with State, local, and tribal stakeholders and increase the nation's level of preparedness. The grant programs help to fund many activities related to homeland security and emergency preparedness, including planning, organization, equipment purchase, training, exercises, and management.

FEMA National Advisory Council established to increase coordination of policies for preparedness, protection, response, recovery, and mitigation efforts for events.

Financial Services Information Sharing and Analysis Center (FSISAC) a program to increase the financial sector's cybersecurity information sharing capabilities.

FirstNet First Responder Network Authority passed by Congress to establish a wireless network for first responders and public safety officials.

Fusion Centers one important way for federal, state, local, tribal, and territorial agencies to facilitate information sharing. Fusion centers have been established in many states and large cities to share intelligence within their own jurisdictions as well as with the Federal government. The fusion centers ensure that both classified and unclassified information can be shared among the group, with expertise at all levels sharing information.

Geospatial Data Act signed by President Trump on October 5, 2018, to protect geospatial data stored on networks

Hazard a source or cause of harm. There are different types of hazards, including natural hazards (caused by acts of nature such as hurricanes or wildfires); technological hazards (such as hazardous materials releases, dam or levee failure, an airplane crash, power failure, or radiological release); and human-caused hazards, which include incidents that are the result of intentional actions of an individual or group of individuals. Examples of this include acts of terrorism, active shooting, biological attacks, chemical attacks, cyber incidents, bomb attacks, or radiological attacks.

Hazard Effect the overall impacts to the community were an incident to occur.

Homeland Infrastructure Threat and Risk Analysis Center (HITRAC) found within DHS, HITRAC is the Department's Intelligence—Infrastructure Protection Fusion Center. The center's membership includes analysts from both the Office of Infrastructure Protection and the Office of Intelligence and Analysis who have the expertise to carry out infrastructure risk assessment responsibilities. HITRAC carries out Sector-Specific Threat Assessments, Sector-Specific Risk Assessments, Individual State Threat Assessments, and other assessments as needed. To assist in this process, they have a Critical Infrastructure Red Team that examines threats, vulnerabilities, and plans for mitigating risk to critical infrastructure.

Homeland Security Act of 2002 legislation passed by Congress after the terrorist attacks of 2002 that created the Department of Homeland Security.

Homeland Security Information Network (HSIN) a secure web-based system established by DHS to increase information sharing and collaboration efforts between government agencies and the private sector groups that have a concern with protecting critical infrastructure. HSIN comprises Communities of Interest (COI) that allows users in all 50 states to share information with others in their communities through a safe environment in real-time.

Homeland Security Presidential Directive 7 a document released on December 17, 2003, by President Bush that defined the responsibilities of various agencies that played a role in protecting critical infrastructure. The roles of sector-specific agencies were more clearly defined. The directive also reiterated the importance of establishing effective relationships between the federal government and agencies in other areas. Bush also asked for a comprehensive National Plan for Critical Infrastructure and Key Resources Protection.

Human-caused Hazard a potential incident resulting from the intentional actions of an adversary.

Immigration and Customs Enforcement an agency with the DHS that works to protect citizens from cross-border crimes and oversees immigration policies.

Impacts the impact of an event on an asset, such as the property damage caused or the disruption in services; how a threat or hazard might affect a core capability.

Information Sharing communication between agencies. This is essential between the private and public sectors to protect assets.

Information Sharing and Analysis Centers (ISAC) agencies that comprise representatives from both government and the private sector that would facilitate greater information. Information could also be shared after an incident to analyze why that event happened and how to make changes so prevent a similar event in the future.

InfraGard a cooperative, outreach effort between the federal government and private sector businesses, academic institutions, and state and local law enforcement agencies, who work cooperatively to increase the security of the nation's assets. The goal is to increase information between these groups so that assets are better protected.

Infrastructure Protection Executive Notification Service located within the DHS, the agency serves as a way for DHS to communicate more effectively with the Chief Executive Officers of major industrial firms to inform them of any incidents or threats that pertain to them.

Infrastructure Security Division of CISA cooperates with public and private agencies to conduct vulnerability assessments on networks to provide a better understanding of threats and risks.

Institutional Risks risk associated with an organization's ability to develop and maintain effective management practices, control systems, and flexibility and adaptability to meet organizational requirements. These risks are less obvious and typically come from within an organization. Institutional risks include factors that can threaten an organization's ability to organize, recruit, train, support, and integrate the organization to meet all specified operational and administrative requirements.

Integrated Operations Division of CISA prepares and plans operations to protect critical infrastructure and mitigate damage in the case of an attack.

Integrated Risk Management (IRM) established in May 2010, by then Secretary of Homeland Security, Janet Napolitano, the plan formalized many of the organizational aspects of the DHS risk effort. The policy supports the idea that security partners can most effectively manage risk by working together and that management capabilities must be built, sustained, and integrated with Federal, state, local, tribal, territorial, nongovernmental, and private sector homeland security partners.

Interdependencies many sectors are interconnected so that damage to one leads to damage to others. For example, a disruption in water service caused by a natural hazard or terrorist act may threaten other sectors such as emergency services, healthcare, and transportation.

Internet Crime Complaint Center part of the FBI, the agency processes complaints about criminal acts committed with the use of the Internet.

IT Sector Baseline Risk Assessment (ITSRA) serves as a foundation for the IT sector's national-level risk management activities.

Key Resources assets, either publicly or privately owned, that are critical for the nation's economy and the government.

Likelihood the chance of something happening, whether defined, measured, or estimated in terms of general descriptors, frequencies, or probabilities, or in terms of general descriptors (e.g. rare unlikely, likely, almost certain), frequencies, or probabilities.

Management Directorate part of the DHS, this organization is responsible for the budgets and dispersing funds earmarked for protecting critical infrastructure. Many grant programs that help fund protection plans originate or are managed by this office.

Maritime Transportation Security Act of 2002 a law that requires ports and facilities to carry out vulnerability assessments of infrastructure and then develop plans to keep them safe.

Mitigation any activities geared toward lessening the possible impact of an event. This can include increasing physical security measures, hiring additional security guards, or installing barriers around a building.

National Advisory Council (FEMA) established in June 2007 by the Post-Katrina Emergency Management Reform Act (PKEMRA) to develop more effective coordination of federal policies for preparedness, protection, response, recovery, and mitigation for all events. The council was given the responsibility to develop the National Preparedness Goal, the National Preparedness System, the National Incident Management System (NIMS), the National Response Framework (NRF), and other national plans.

National Coordinator for Security a position created in the PDD-63 who would serve as the chair of the Critical Infrastructure Coordination Group.

National Counterintelligence Executive established under Executive Order 14231 and PPD-75, the NCIX works with the President's Critical Infrastructure Protection Board to address potential threats from hostile foreign intelligence services. The agency helps to identify critical assets located throughout the nation, implement counterintelligence analyses, develop a national threat assessment, formulate a national counterintelligence strategy, create an integrated counterintelligence budget, and develop an agenda of program reviews and evaluations.

National Cyber Investigative Joint Task Force part of the FBI, this task force comprises over 30 agencies that share information and coordinate investigations of crimes carried out online.

National Infrastructure Advisory Council created by Executive Order 13231, the Council was an advisory agency to the president regarding the security of information systems for critical infrastructure. The Council was also to increase partnerships between public and private agencies.

National Infrastructure Assurance Council a position created in PDD-63 as an advisory group to the president that included private owners, representatives from state and local government, and representatives from relevant federal agencies.

National Infrastructure Protection Center created as part of PDD-63 and housed within the Department of Justice and the FBI, this agency was responsible for defending the nation's public and private computer systems from possible cyberattacks and responding to illegal acts carried out by the use of computers and information technologies.

National Infrastructure Simulation and Analysis Center (NISAC) DHS's modeling, simulation, and analysis program that seeks to analyze critical infrastructure and key resources, along with their interdependencies, consequences, and other complexities. NISAC provides three types of products: preplanned long-term analyses, preplanned short-term analyses, and unplanned priority analytical projects. The reports produce provide information for mitigation design and policy planning and address the cascading consequences of infrastructure disruptions that could occur across all 18 CIKR sectors at national, regional, and local levels.

National Initiative for Cybersecurity Careers and Studies (NICCS) a program to increase education and skills in cybersecurity.

National Plan for Critical Infrastructure and Key Resources outlined by President Bush in HSPD-7, the document was to include (a) a strategy to identify, prioritize, and coordinate the protection of critical infrastructure and key resources, including how the department will work with other stakeholders; (b) a summary of activities to be undertaken to carry out the strategy; (c) a summary of initiatives for sharing critical infrastructure information and threat warnings with other stakeholders; and (d) coordination with other federal emergency management activities.

National Plan for Information Systems Protection issued by the Clinton administration in January 2000, the plan focused on protecting the nation's cyber infrastructure.

National Protection and Programs Directorate under PPD-21, the National Protection and Programs Directorate (NPPD) develops ways to identify the nation's critical infrastructure and prioritize them so that funds can be distributed accordingly. Officials in the agency also seek to reduce possible risks to critical infrastructure, including both physical and virtual threats. One way this directorate achieves its goal is by training others (owners and operators) about identifying risks and mitigating them.

National Risk Management Center work with public and private owners of critical infrastructure to identify risks and oversee resiliency measures.

National Security Staff created when President Obama merged the Homeland Security Council and the National Security Council into one agency.

National Strategy for Homeland Security a document that sets out government efforts to protect the nation against terrorist threats of all kinds. It also added public health, the chemical industry and hazardous materials, postal and shipping, the defense industrial base, and agriculture and food to the list of sectors that have critical infrastructure in them. It also combined emergency fire service, emergency law enforcement, and emergency medicine as emergency services and eliminated those functions that belong primarily to the federal government. The report also introduced key assets.

National Strategy for the Physical Protection of Critical Infrastructures and Key Assets a document published in February 2003, by the Office of Homeland Security that defined "key assets." It also identified the roles and responsibilities of agencies and people, actions that needed to be taken, and guiding principles for protecting the nation's key assets.

National Strategy to Secure Cyberspace a 2003 document that stressed methods to protect critical information and data stored electronically or available on the Internet.

Natural Hazard a potential incident resulting from acts of nature.

NIMS-typed Resource a resource categorized, by capability, the resources requested, deployed, and used in incidents.

NIPP or the National Infrastructure Protection Plan, was first published in 2006. This document outlined a national plan for managing risk for the country's critical infrastructure. The updated plan, published in 2013, emphasized the goals of critical infrastructure security and resilience, including the identification, deterrence, detection, and disruption of threats, along with reducing vulnerabilities.

Office of Cybersecurity and Communications (CS&C) the sector-specific agency for both the Communications and Information Technology (IT) Sectors. The CS&C coordinates national-level reporting that is consistent with the National Response Framework and works to prevent or minimize disruptions to critical information infrastructure to protect the public, the economy, and government services.

Office of Homeland Security an agency created after the terrorist attacks of 2001 that would develop and coordinate a comprehensive strategy to keep the United States safe from terrorist threats and attacks.

Office of Intelligence and Analysis (I&A) this Office seeks to improve information sources to assist other agencies in protecting critical infrastructure. The IAIPD carries out many tasks such as gathering and disseminating information from other sources that will assist them in identifying and assessing the risk of a possible terrorist threat and assessing the vulnerabilities of critical infrastructure to determine the possible risks posed by attacks.

Operational Risks risk that has the potential to impede the successful execution of operations with existing resources, capabilities, and strategies. Operational risks include those that impact personnel, time, materials, equipment, tactics, techniques, information, technology, and procedures that enable an organization to achieve its mission objectives.

Partnership Engagement collaborating with sector partners and encouraging continuous growth and improvement of these partnerships.

Partnership for Critical Infrastructure Security an agency established in December 1999 to share information and strategies for infrastructure protection and to identify potential interdependencies across sectors.

Physical Security both cyber and physical security measures are needed to protect critical infrastructures against potential threats. Physical security measures prevent unauthorized access to servers and other technologies that carry sensitive information.

Pipeline and Hazardous Materials Safety Administration seeks to protect the environment and citizens by ensuring safe operations of pipelines, storage tanks, and shipping of hazardous materials.

Post-Katrina Emergency Management Reform Act (PKEMRA) of 2006 the law passed by Congress after Hurricane Katrina that reformed the Federal Emergency Management Agency to make it more effective in responding to national emergencies. Congress opted to keep FEMA as part of DHS but clarified FEMA's mission, which included to lead the nation's efforts to prepare for, respond to, recover from, and mitigate the risks of, any natural and man-made disaster, including catastrophic incidents; Implement a risk-based, all-hazards strategy for preparedness; and Promote and plan for the protection, security, resiliency, and post-disaster restoration of critical infrastructure and key resources, including cyber and communications assets.

Preparedness activities that are planned to prepare for an event. This can include establishing guidelines, protocols, and standards for training responders, or purchasing needed equipment.

Presidential Decision Directive 63 (PDD-63) stressed the importance of partnerships between the government and private ownership to protect critical infrastructure. The document described critical infrastructure as composed of five essential domains: banking and finance, energy, transportation, telecommunications, and government services. Specific critical infrastructure assets that required protection were referred to as "sectors" and included information and communications; banking and finance; water supply; aviation, highways, mass transit, pipelines, rail, and waterborne commerce; emergency and law enforcement services; emergency, fire, and continuity of government services; public health services; electric power, oil and gas production, and storage.

Presidential Policy Directive-8 signed by President Obama on March 30, 2011, this document, also called National Preparedness, was a way to increase the nation's security and resilience by focusing on preparation. Obama concentrated on an all-hazards approach to security that included planning for possible terrorist acts, including cyberattacks, technological events, and natural disasters. Additionally, the president recognized that national preparedness and security must involve government, the private sector, and individual citizens.

Presidential Policy Directive-21 announced in February 2013, by President Obama, this document, entitled "Critical Infrastructure Security and Resilience," established a policy to strengthen critical infrastructure that is also resilient to attacks. Both physical and cyber infrastructure was included. Throughout the document, the need for cooperation between private sectors and the government was stressed. The number of sectors and how they were changed as well.

President's Commission on Critical Infrastructure Protection established by President Clinton, the commission was to assess the vulnerabilities of the country's critical infrastructures, and then create a new plan to protect them. The final report indicated the need for greater exchange of information by all participants in critical infrastructure protection to predict or prevent an attack.

President's Critical Infrastructure Protection Board established by President Bush, the board was responsible for recommending policies and coordinating programs geared to protecting information systems for critical infrastructure.

President's Cup Cybersecurity Competition created by Executive Order 13870, this is a competition overseen by CISA to identify individuals who may be interested in, and capable, of a career in cybersecurity.

Private Sector any unit that is not operated by the state or federal government, such as companies, corporations, private banks, television or radio stations, or nongovernmental organizations. Since the goal of these entities is to make a profit, their actions are geared toward minimizing any financial risk.

Private Sector-Preparedness Program initiated by FEMA in 2009, the program initiated ways to improve the preparedness of private sector and nonprofit organizations for an event. The program involves establishing guidelines, best practices, regulations, and codes of practice.

Private Sector Resources Catalog a list of DHS publications aimed specifically toward the needs of private sector owners. This including resources on training, guidance, alerts, newsletters, programs, and other services. They have also created the "Critical Infrastructure Protection and Resilience Toolkit" to assist the owners and operators of critical infrastructure assets at both the local and regional levels to help them prepare for, protect against, respond to, mitigate against, and recover from any possible threats or hazards.

Protected Critical Infrastructure Information (PCII) a program developed by CISA to help share information about threats while maintaining privacy and confidentiality.

Protective Security Advisors Program (PSA) originally developed in 2004, the PSA program is part of the National Protection and Programs Directorate (NPPD) of the DHS. The PSAs are experts who have been trained in critical infrastructure protection and mitigation procedures for infrastructure protection. These analysts ensure that the owners and operators of critical infrastructure in the private sector are also given essential information. The PSA program focuses on three areas: enhancing infrastructure protection; assisting with incident management; and facilitating information sharing.

Public Sector agencies that are owned or operated by the government, such as federal, state, or local departments and agencies, water treatment plants, and power plants. These groups often get financial support from the government through taxes.

Public–Private Partnerships an agreement between a public agency and a private sector agency. The goal is to draw on the skills and resources of each separate group so that a task can be accomplished efficiently.

RAM-W a risk assessment tool that is widely used across the Water and Wastewater sector.

Recovery the ability to adapt and withstand any disruption that may occur after an emergency or event. It refers to the ability of a community to recover rapidly and bounce back, or regroup, after a disruption.

Resiliency See Recovery

Resilient Infrastructure critical infrastructure and its communications capabilities should be able to withstand natural or man-made hazards with minimal interruption or failure.

Resource Requirement an estimate of the number of resources needed to achieve a community's capability target. A list of resource requirements for each core capability is an output of the THIRA process.

Resourcefulness the capacity to mobilize needed resources and services in emergencies.

Risk the probability that an asset will be the object of an attack or another adverse outcome. Risk is the likelihood that an adverse event will occur, and is related to consequences (C), vulnerabilities (V), and threats (T), in the following relationship:

Risk = (function of) (CVT)

Risk Analysis an attempt to identify the probability, and possible consequences, of an attack. A risk assessment asks, "What can go wrong? What is the likelihood that it will go wrong? What are the possible consequences if it does go wrong?" This way, the probability of an incident occurring, and the impact of that incident will be better understood. The analysis can also be used to determine what assets are more critical and how should money be spent to protect them.

Risk Assessment analyses of critical infrastructure carried out to identify potential risks and possible actions to mitigate or prevent harm.

Risk-based Decision Making determination of a course of action predicated primarily on the assessment of risk and the expected impact of that course of action on that risk.

Risk Communications a key element of the risk management process is effective communication with stakeholders, partners, and customers. According to DHS, consistent, two-way communication will ensure that decision makers, analysts, and officials are able to implement any decision and share a common understanding of what the risk is and what factors may contribute to managing it.

Risk Fundamentals a doctrine to support homeland security practitioners. This doctrine includes promoting a common understanding of, and approach to, risk management; establishing organizational practices that should be followed by DHS components; providing a foundation for conducting risk assessments and evaluating risk management options; setting the doctrinal underpinnings for institutionalizing a risk management culture through consistent application and training on risk management principles and practices; and educating and informing homeland security stakeholders in risk management applications, including the assessment of capability, program, and operational performance, and the use of such assessments for resource and policy decisions.

Risk Management efforts to decide which protective measures should be implemented to reduce the risk of an event occurring.

Risk Transfer Products occurs when insurance companies offer customers the ability to transfer financial risks under a multitude of circumstances.

Science and Technology Directorate as part of DHS, this agency supports research and development regarding critical infrastructure protection. The agency carries out research in explosive detection, blast protection, and safe cargo containers. They monitor threats and develop ways to prevent those threats. There are three directors within the Science and Technology Directorate, including the Director of Support to the Homeland Security Enterprise and First Responders, the Director of Homeland Security Advance Research Projects Agency, and the Director of Research and Development Partnership.

Secret Service a federal agency whose mission is to protect the nation's elected officials, foreign heads of state, and diplomats. They also safeguard the U.S. financial infrastructure.

Sector Coordinator a person appointed by the private sector organizations in each sector who cooperated with others to develop plans for protecting assets.

Sector Liaison Official a sector official who communicates with others in private sector organizations to build methods to protect assets.

Sectors Categories of assets that helps to organize the country's critical infrastructure and protection plans.

Sector-Specific Plan (SSP) a way for each sector to tailor their response plan depending upon the unique operating conditions and risk landscape of its particular sector.

SEMS Security and Environmental Management Systems.

Shields Up a program conducted by CISA to help companies and individuals identify and respond to a potential cyberattack.

Stakeholder Engagement Division of CISA helps to create strong partnerships with all stakeholders on federal, state, local, tribal, and territorial levels.

State, Local, Tribal, and Territorial Government Coordinating Council (SLTTGCC) the SLTTGCC, under the NIPP, helps to ensure that state, local, tribal, and territorial homeland security officials or their designated representatives are integrated fully as active participants in national CIKR protection efforts. The SLTTGCC provides the organizational structure to coordinate across jurisdictions on state and local-level CIKR protection guidance, strategies, and programs.

Stop. Think. Connect a program in CISA to increase public awareness of cyber threats and attacks with the goal of increasing the security of computer networks from cybercrimes.

Strategic Foresight Initiative facilitated by FEMA, this organization assists members of the emergency management community in understanding changes to the field and their impact.

Strategic Homeland Security Infrastructure Risk Analysis (SHIRA) provides a common framework that sectors can use to assess the economic, loss of life, and psychological consequences resulting from terrorist incidents as well as natural hazards and domestic threats. It is a threat-based approach and is the result of an integrated "fusion" effort between the infrastructure protection and intelligence communities. Typically, intelligence initiates the planning and all functional areas participate in the entire process.

Strategic Risks risk that affects an organization's vital interests or execution of a chosen strategy, whether imposed by external threats or arising from flawed or poorly implemented strategy. These risks threaten an organization's ability to achieve its strategy, as well as position itself to recognize, anticipate, and respond to future trends, conditions, and challenges. Strategic risks include those factors that may impact the organization's overall objectives and long-term goals.

Technical Resource for Incident Prevention (TRIPwire) an online information sharing network for groups including bomb squads, law enforcement personnel, and other emergency services personnel that informs them about current terrorist tactics, techniques, and procedures. The agency was developed by the DHS Office for Bombing Prevention (OBP), which continues to maintain it. The group relies on expert analyses and reports alongside relevant documents, images, and videos that were gathered directly from terrorist sources to assist law enforcement to anticipate, identify, and prevent incidents.

Technological Hazard a potential incident resulting from accidents or failures of systems or structures.

Threat a natural or man-made event, person, entity, or action that has the potential to harm life, information, operations, the environment, and/or property. Threats can stem from humans, natural hazards, or technology.

Threat and Hazard Identification Risk Assessment a three-step risk assessment process that helps communities identify their greatest threats, hazards, and risks.

Traffic Light Protocol a program in CISA that identifies the sensitivity of information based on a color scheme similar to a traffic light. It labels information with different colors to indicate if the information can be shared and with whom.

Training and Exercise Support initiated by DHS, there are many programs to help local communities train their first responders and others for an event. Three levels of training are provided: awareness level training, performance level training, and management level training.

Transportation Security Administration became part of DHS when it was formed in 2002. The (TSA) oversees the security of the nation's transportation sectors. Officials screen airline passengers and their baggage to ensure that no dangerous material is brought onboard an aircraft. TSA also regulates the installation and maintenance of equipment to detect explosives. Agents provide security for airport perimeters. They also oversee the Air Marshals.

Unity of Effort reiterates that homeland security risk management is an enterprise-wide process and should promote integration and synchronization with entities that share responsibility for managing risks.

U.S. Computer Emergency Readiness Team (U.S.-CERT) makes information related to computer-related vulnerabilities and threats available to others and provides information about responses to incidents. U.S.-CERT collects incident reports from others around the country and analyzes that information to look for patterns and trends in computer-based crime. Officials here manage the National Cyber Alert System, which provides general information to any organization or individual who subscribes.

USA PATRIOT ACT legislation signed by President Bush after the terrorist attacks of September 11, 2001. The law defined critical infrastructure and added new terms to the homeland security lexicon.

Virtual Assets Unit part of the FBI, the agency focuses on the use of cryptocurrency to commit crimes.

VSAT Vulnerability Self-Assessment Tool

Vulnerability a physical feature or attribute of critical infrastructure that leaves it vulnerable to an attack or natural event. This could be a weakness or flaw in an asset that may cause it to be a target for an attack. In most cases, an attacker will identify a vulnerability in an asset and plan their attack on that using that liability to strike the asset.

Vulnerability Assessment assessments completed to identify risks to operations, assets, and individuals. These can be used to implement mitigation techniques to reduce possible harm in the case of an event.

Whole Community an approach to emergency management that reinforces the fact that FEMA is only one part of our Nation's emergency management team. We must leverage all of the resources of our collective team in preparing for, protecting against, responding to, recovering from, and mitigating all hazards, and collectively we must meet the needs of the entire community in each of these areas.

1 Introduction

Critical Infrastructure and Risk Assessment

Chapter Outline

Introduction

Many catastrophes and disasters have impacted the U.S. throughout history, resulting in devastating damage to property and injuries or deaths to residents. One disaster that most people think of is the terrorist attacks of September 11, 2001, in which suicide hijackers were able to fly two passenger jets into the Twin Towers in New York and one into the Pentagon in Washington, D.C. A fourth plane, possibly intended to hit the U.S. Capitol building, crashed into a field in rural Shanksville, Pennsylvania. In all, it is estimated that there were 2,977 people killed as a result of the attack[1] with an undetermined amount of critical infrastructure destroyed or impacted in some way.

DOI: 10.4324/9781003434887-1

This man-made attack is not the only type of misfortune that impacts critical infrastructure. Some catastrophes are natural or environmental. In 2005, Hurricane Katrina struck the city of New Orleans. A Category 5 hurricane, it caused winds of 140 miles per hour and flooding in over 80% of the city. The resulting damage totaled over $108 billion[2] and 1,392 residents lost their lives.[3] Another Category 5 hurricane, Ian, hit the coast of Florida in 2022. After this catastrophe, 125 people were dead[4] with approximately $112.9 billion in damages.[5]

Natural disasters are not always hurricanes. A major winter ice storm affected parts of the U.S. in 2021. Often referred to as Winter Storm Uri, many states, but particularly Texas, were blanketed with snow and ice, followed by extreme low temperatures. Two hundred and ten people were killed, mostly because of hypothermia and vehicle crashes, and there was between $80 billion to $130 billion worth of damages to infrastructure, largely due to the loss of power.[6]

Cyberattacks, another type of event that causes great damage to both people and infrastructure, have been on the rise in recent years. One example of this is the 2014 cyberattack on the entertainment industry. A hacker group referred to as the Guardians of the Peace was able to break into the SONY Pictures network and access confidential information including employee information, emails, salaries, and copies of unreleased films. To prevent the information from being made public, the group demanded that SONY executives pull a film from the market that revolved around the assassination of North Korean leader Kim Jong-un. The group threatened violence to movie theaters and moviegoers that chose to show the film. U.S. officials determined that the attack was likely supported by the North Korean government.

In 2021, an oil pipeline system that carried gasoline throughout the Southeastern U.S., Colonial Pipeline, was the victim of a ransomware cyberattack. The hack shut down company operations, leading to widespread gas shortages. President Biden declared a state of emergency that suspended limits on the transportation of petroleum products by road and rail in an effort to maintain the availability of gas to retail outlets. Executives in the company chose to pay the ransom ($4.4 million), much of which was later recovered.

Some catastrophes are related to the public's health and well-being. In late 2019, the coronavirus was first detected in Wuhan, China, and quickly spread throughout the world, becoming a global pandemic on March 11, 2020. The virus caused social and economic changes, including a global recession and deaths. It is estimated that well over one million people died due to the global pandemic.[7] Many businesses were permanently closed, and thousands of people lost their jobs and incomes.

Catastrophes like these demonstrate how vulnerable the nation's critical assets and systems can be to an attack or natural disaster. If they are damaged or incapacitated, even for a short time, there can be a debilitating effect on the nation's security, economic system, or public health.[8] Disasters can cause a disruption of vital government services that people rely on each day, and residents living in an area affected by a calamity often do not have the critical services they need to survive in the period after an event. Citizens may find themselves without access to water, food, or shelter. People may be prevented from traveling from one place to another easily, and needed goods and products may not be accessible. There may not be effective and reliable communication systems, financial services, power, or medical services. If these disruptions become prolonged, they could have a major impact on the country's health and welfare. If severe enough, the disruption can pose a serious risk to society—there is a risk that even more citizens will be harmed or killed or additional property will be damaged in the time after an event. If not prepared, officials may find it difficult to provide basic services that are needed in a community and to monitor and if necessary respond to hostile acts or natural disasters.

There are many incidents that could lead to services being interrupted, as noted earlier. They can include natural events, such as a hurricanes, earthquakes, or floods, or man-made events such as terror attacks, either domestic or international, that have the intent to harm people or disrupt services. Conflicts (war) can lead to an interruption of services or prohibit access to them. Health pandemics such as the COVID virus can affect people locally or globally, halt daily activities, and have a major effect on the economy. Cyberattacks on government computers and private networks are becoming common, leading to loss of data and privacy breaches. These can all put a halt to activities and services, disturbing supply chains to prevent people from having what they need.

The damages caused by recent events have made it clear that there is a need to constantly re-examine how the country protects its assets and seeks to ensure that critical services are available to citizens in the days and weeks following an event. Both victims and non-victims have called on government officials to enact policies that will protect the nation's critical infrastructure so they are better able to withstand events or, if damaged, can recover quickly. Previous disasters have made public officials realize that the government needed to put more emphasis on the security of the nation's infrastructure during a disaster or terrorist act, to ensure that basic services are available to citizens. It has become unmistakable that protecting the nation's critical infrastructure is essential to the public health and safety of residents, to the economic strength, to the way of life, and to national security.[9] Thus, one of the goals for government officials in recent years was to ensure the protection of the country's critical infrastructure to keep the country safer and more secure and even more resilient in the aftermath of an event.

Protecting the nation's critical infrastructure can be a very demanding task that becomes more difficult each day. Each critical asset is different, with different characteristics, different access points, and different needs. The U.S. society has become very open, transparent, and accessible to most people within and outside of its boundaries. The development of a global Internet, which most people, companies, and government agencies rely on each day, has also led to increased security challenges. A breach in, or attack on, a computer system can result in harm to individuals, businesses, or governments as private information can be stolen and made public. Methods for protecting critical infrastructure must constantly evolve and become more sophisticated as new challenges arise.

Adding to this, the aging infrastructure in the U.S. makes it more difficult to protect. Many of our assets are getting older and not being replaced at an adequate speed, making them more prone to failure. About one-third of the dams across the nation are at least 50 years old, and the number that have been rated as deficient has tripled. The Federal Highway Administration rated 600,000 bridges as structurally deficient.[10]

Today, the Department of Homeland Security (DHS) spends billions of dollars annually to prevent (or mitigate), prepare for, respond to, and recover from an incident. The government's goal has become National Preparedness, which it defines as:

> The actions taken to plan, organize, equip, train, and exercise to build and sustain the capabilities necessary to prevent, protect against, mitigate the effects of, respond to, and recover from those threats that pose the greatest risk to the security of the Nation.[11]

To fulfill the National Preparedness goal, the focus of the DHS shifted from focusing primarily on terrorism threats to all hazard threats. This shift has been significant and continues to be debated as to how best to balance the approach to prevention, response, and recovery.

It is critical that government officials and private individuals alike understand and support government policies that attempt to identify and protect the country's critical infrastructure as they seek to ensure that essential services and goods are available to residents in the aftermath of a disruptive event. The proper identification of structures deemed to be critical infrastructure and the strategies to protect them have become a priority in today's world. While essential, these steps have also become sometimes controversial. It is important to begin our analysis of critical infrastructure protection by defining essential terms that are used frequently by those who seek to protect the nation's critical infrastructure. Over time, the meanings of some terms have changed and become muddled. In some cases, the meanings of some terms vary regionally across the country. For that reason, it is important to define the terminology that will be used throughout the remainder of the book.

What is Critical Infrastructure?

What is meant by the term "critical infrastructure" has changed over time, and, because of that, the term is sometimes ambiguous or blurred. Prior to the September 11, 2001, terrorist attacks in the U.S., the term "infrastructure" referred primarily to public works (facilities that were publicly owned and operated) such as roadways, bridges, water and sewer, airports, seaports, and public buildings. The main concern at that time was how functional these services were for the public. This began to change during the early 1990s after the nation witnessed major disasters such as the bombings of the World Trade Center (1993) and the Oklahoma City building (1995). About this time, the threat of terrorism was also emerging in the U.S., and, consequently, the definition of what is meant by "critical infrastructure" has become much broader.

It is clear that in the post-9/11 era, the term "critical infrastructure" refers to the framework of man-made networks and systems that provide needed goods and services to the public. In other words, it is the facilities and structures, both physical and organizational, that provide essential services to the residents of a community that ensure its continued operation. This term includes things like buildings, roads and transportation systems, telecommunications systems, water systems, schools, hospitals, energy systems, roads and bridges, emergency services, banks, dams, and power supplies. In addition to physical structures and assets, the term incorporates virtual (cyber) systems and people.

According to the DHS, infrastructure can be defined as:

> The framework of interdependent networks and systems comprising identifiable industries, institutions (including people and procedures), and distribution capabilities that provide a reliable flow of products and services essential to the defense and economic security of the U.S., the smooth functioning of government at all levels, and society as a whole; consistent with the definition in the Homeland Security Act, infrastructure includes physical, cyber, and/or human elements.[12]

A more precise definition of critical infrastructure is provided in the U.S.A. Patriot Act of 2001. Here, critical infrastructure is defined as:

> Systems and assets, whether physical or virtual, so vital to the United States that the incapacity or destruction of such systems and assets would have a debilitating impact on security, national economic security, national public health or safety, or any combination of those matters.[13]

In general, "critical infrastructure" is all of the systems that are indispensable for the smooth functioning of the government at all levels. It is the assets that are vitally important or even essential to a community or to the nation that improve living conditions and commerce and, if disrupted, harmed or destroyed, or in some way unable to function, could have a debilitating impact on the security, economics, national health, safety, or welfare of citizens and businesses.[14] It forms the basis of the country's economy and prosperity as well as the health of the citizens. There could also be a significant loss of life if these services are not provided.

Critical infrastructure across the country is identified by the National Critical Infrastructure Prioritization Program that is located within the Cybersecurity and Infrastructure Security Agency. This agency identifies assets, found both in the U.S. and overseas, that are critical to the nation's ability to prosper.

A more recent addition to the field of critical infrastructure protection is cybersecurity. In recent years, people and businesses are much more reliant on technology, which increases the risk of cybercrimes such as theft and fraud. According to the DHS, cybersecurity is the

> prevention of damage to, unauthorized use of, or exploitation of, and, if needed, the restoration of electronic information and communications systems and the information contained therein to ensure confidentiality, integrity, and availability; includes protection and restoration, when needed, of information networks and wireline, wireless, satellite, public safety answering points, and 911 communications systems and control systems.[15]

President Biden made cybersecurity a top priority for his administration, and federal agencies including the DHS have to be more engaged in developing policies that improve, modernize, and protect computer networks.

Local Critical Infrastructure

Each community has assets that provide a service to its residents and need to be protected from both natural and man-made events. What assets defined or labeled as critical infrastructure can be different in different cities or regions of the country—critical infrastructure assets are different in Cleveland as compared to Los Angeles, or even Tampa or Denver, because they have different weather conditions, different needs, and different assets. In considering a community's critical infrastructure, it is essential to know how valuable an asset is to that community and whether or, to what extent, it needs to be protected. Community leaders must rank assets by placing some kind of value on them. In some cases, a community's critical infrastructure can be one major structure that is very costly to build, maintain, and operate, such as a water purification plant. Clearly, a community relies heavily on this service, but, because of the enormous cost, a community can only afford one of them. Protecting this structure would be vital to the community. This asset provides a needed service to residents, and there would be serious impacts on the health of the community should that plant be harmed in some way. Officials need to know if an asset is vulnerable to a natural disaster or if it would be an attractive target for an attack. It is also important to know if there is a backup method or secondary method for providing the service to residents.

Federal Critical Infrastructure

On the federal level, there are thousands of assets that are considered to be critical infrastructure. The Homeland Security Act of 2002 and the Homeland Security Presidential Directive-7 (2003) require officials in the DHS to identify the nation's critical assets and networks (the national infrastructure). This list is found in a document called the National Asset Database, maintained by the Office of Infrastructure Protection. There are 77,000 national assets on the list that are located across the country, with about 5% of those assets (only 1,700) labeled as critical.[16] This would include assets such as power plants, dams, or hazardous materials sites.[17] The critical infrastructure assets in the U.S. include a power grid that is essential for daily life that is interconnected with other national systems. There are four million miles of paved roadways with 600,000 bridges. There is also a complex rail system in the U.S. that includes 500 freight railroads and 300,000 miles of rail track. There are 500 commercial service airports along with 14,000 general aviation airports. In addition, there are two million miles of oil and gas transmission pipelines; 2,800 electric plants; 80,000 dams; and 1,000 harbor channels and 25,000 miles of inland, intracoastal, and coastal waterways servicing more than 300 ports and 3,700 terminals. Clearly, if any of these facilities were to be attacked and damage, communities and residents may be seriously impacted.[18]

The list of critical assets is sometimes controversial, as officials in the federal, state, and local governments, and the private sector owners, often disagree about what should be included in the directory. For example, the list includes many assets that are considered to be local assets, such as festivals and zoos, that some officials argue shouldn't be included. However, the DHS includes all assets in an attempt to create a comprehensive inventory of critical infrastructure around the country. Thus, identifying a comprehensive list of nationally critical assets continues to be an ongoing debate for the DHS.

Private Critical Infrastructure

In addition to having local assets and federal assets, there are privately owned critical infrastructure assets. Most people have the perception that critical infrastructure is owned and operated by the government, but in reality 80% of the critical infrastructure in the U.S. is owned and operated by the private sector.[19] Recent estimates find that the federal government owns 13% of the nation's infrastructure, whereas private organizations own and operate 87%. State and local governments own the majority of roads and interstates, schools, and water systems, and the federal government is more likely to own research facilities, dams, and parks.[20]

Because many assets are owned by private entities, the private sector must be involved in planning for protecting those valuable assets. Many federal government documents, including the National Strategy, the Homeland Security Act, and Homeland Security Presidential Directive 7, address the importance of including all partners in coordinating protection efforts. These documents make it clear that protecting the infrastructure cannot be accomplished effectively simply by the government and the public sectors. Instead, they must work jointly with private sector owners and operators. The government can assist owners and operators of critical infrastructure in many ways, such as providing timely and accurate information on possible threats, including owners in the development of initiatives and policies for protecting assets, helping corporate leaders develop and implement security measures, and/or helping to provide incentives for companies whose officials opt to adopt sound security practices.[21] It is imperative that the government work to reduce potential risk to critical infrastructure in partnership with the private sector that bear the responsibility of providing these services to the public.

Sectors

Critical infrastructure is divided into sectors. A sector can be thought of as a "logical collection of assets, systems, or networks that provide a common function to the economy, government, or society."[22] The DHS places critical infrastructure into sectors based on their function as a way to organize them and manage their resources and activities. Critical infrastructure that is related in some way or performs similar functions, or maybe has similar needs, may be grouped together in the same sector.

There are currently 16 critical infrastructure sectors, as identified in PPD-21. They are listed in Table 1.1

Table 1.1 Critical Infrastructure Sectors

Chemical sector
Communications sector
Dams sector
Emergency services sector
Financial services sector
Government facilities sector
Information technology sector
Transportation systems sector
Commercial facilities sector
Critical manufacturing sector
Defense industrial base sector
Energy sector
Food and agriculture sector
Healthcare and public health sector
Nuclear reactors, materials, and waste sector
Water and wastewater systems sector

These 16 sectors are often interdependent and rely on each other to provide community services. For this reason, the federal government has designated four of these as having "lifeline functions," which means that they provide such essential or critical services that if they were damaged or destroyed, it would have an immediate effect on the security of the critical infrastructure across or within other sectors. These sectors are transportation, water, energy, and communications.

Many of the sectors are divided into subsectors. One example of this is the Government Facilities sector, which houses the National Monuments and Icons, Education Facilities, and the newly created Elections subsector. While these are all considered to be government entities and therefore part of the Government Facilities sector, they are also different enough to warrant their own plan. Often, there is additional infrastructure included in the subsector that does not fit into that subsector. This structure helps facilitate information sharing in the case of an event.

These sectors and the subsectors are static and can change. This happened in January 2017, when the DHS officials elevated state-run election systems as a Critical Infrastructure as a subsector of the Government Facilities sector. This was done after the recognition of the electoral system as critical to the democratic system and way of life in the country. Because of this change, officials who manage elections in state and local governments could request assistance with cybersecurity concerns from federal agencies.

Key Resources

A related term is key resources. As defined in the Homeland Security Act of 2002 and the 2003 National Strategy, key resources are the assets that are either publicly or privately controlled that are essential to the minimal operations of the economy and the government. These documents identified five key resources: national monuments and icons, nuclear power plants, dams, government facilities, and commercial key assets. By 2009 the number of sectors and key resources expanded to 18 and were called critical infrastructure and key resources. Since then, the concept of critical infrastructure and key resources has evolved to encompass the sectors and resources. For the most part, key resources are not separated from critical infrastructure in today's nomenclature, and the terms are used interchangeably.

National Infrastructure Protection Plan

The activities of the sectors are guided in part by the National Infrastructure Protection Plan (NIPP) (2013). Stakeholders from each of the sectors and from all levels of government (federal, state, and local), as well as representatives from industry, were instrumental in creating the document. The plan is a guide to implement the nation's efforts to prevent, mitigate, and manage risks to critical infrastructure. It also establishes partnerships and encourages innovations to support risk management. The plan updated the government's approach to protecting critical infrastructure and its resilience. It also included cybersecurity efforts.[23]

SSA/CO-SSA/GCC

To help the sectors and subsectors succeed, many agencies have been created. First is the Sector Specific Agency (SSA). These are federal departments that are responsible for providing sector members with specific expertise in that area. They are given the task of leading and supporting the critical infrastructure sectors as they develop and revise security policies. These planning documents are called the Sector-Specific Plans. Each sector has an SSA associated with it, and each subsector has a Co-Sector Specific Agency (Co-SSA) associated with it. The Co-SSA works to understand the issues pertaining to that subsector and helps to expedite information between them.[24] A list of sectors, SSAs, and Co-SSAs is provided in Table 1.2.

A Government Coordinating Council (GCC) is a government office that is assigned to each sector that helps to increase communication and coordination between the private owners and operators of privately owned critical infrastructure and government agencies. It comprises sector stakeholder representatives from different levels of the government (federal and state, local, tribal, and territorial) as needed depending on the sector and its risk level. The private sector counterpart to the GCC is called the Sector Coordinating Council. The members meet to discuss best practices to mitigate threats.[25] An example of this is the Election Infrastructure sector that connects election officials from all levels of government.

Table 1.2 Sectors, SSAs and Co-SSAs

Sector	*SSA*	*CO-SSA*
Chemical	DHS	
Commercial facilities	DHS	
Communications	DHS	
Critical manufacturing	DHS	
Dams	DHS	
Defense industrial base	Department of Defense	
Emergency services	DHS	
Energy	Department of Energy	
Financial services	Department of Treasury	
Food and agriculture	Department of Agriculture	Department of Health and Human Services
Government facilities	DHS	General Services Administration; Department of Education; Department of Interior
Healthcare and public health	Department of Health and Human Services	
Information technology	DHS	
Nuclear reactors, materials, and waste	DHS	
Transportation systems	DHS	Department of Transportation
Water and wastewater systems	EPA	

Information Sharing and Analysis Centers

Information Sharing and Analysis Centers (ISACs) are . ....organizations created by the owners and operators of critical infrastructure facilities that help to keep their buildings, operations, and personnel safe and operating smoothly. The ISACs provide means for public and private stakeholders to share information about credible threats on a 24/7 basis, as well as possible solutions to keep everyone safe.[26] Closely related to this is the Information Sharing and Analysis Organization (ISAO). These organizations are groups of agencies that were created within either the public or private sector. They collect and analyze information pertaining to critical infrastructure threats as a way to have a greater understanding of any security problems or concerns related to protected systems and communicate those with others. This helps to detect threats and then prevent, mitigate, or recover from these events.[27]

Similarly, the National Infrastructure Coordinating Center (NICC) is also an information-gathering organization that helps to analyze pertinent information and share that with others who could benefit. When an event occurs that could impact critical infrastructure, that information is disseminated so that the DHS and any owners and operators of the infrastructure can coordinate any actions as needed.[28]

Critical Infrastructure Information

In addition to critical infrastructure assets, there is something called Critical Infrastructure Information (CII). This is the data or information that pertains to an asset or critical infrastructure and is sensitive but not always classified (secret). An example of CII is knowledge about the daily operations of an asset or a description of the asset's vulnerabilities and protection plans. CII can also include information generated by the asset such as patient health records or a person's banking and financial records. CII could also be any evidence of future development plans related to the asset or information that describes pertinent geological or meteorological information about the location of an asset that may point out potential vulnerabilities of that facility (i.e. a dam at an earthquake-prone site). In general, CII refers to any information that could be used by a perpetrator to destroy or otherwise harm the asset or its ability to function.

The importance of protecting CII was first identified in the Critical Infrastructure Information Act of 2002, passed by Congress. It was noted that when a private organization chose to share information with government officials, that information then became a public record and could be accessed by the public through public disclosure laws. Many companies did not want to make that information public, so they were reluctant to work with government agencies and officials. As a way to protect that information and encourage more cooperation, Congress created a new category of information called CII. According to the law, any federal official who knowingly discloses any CII to an unauthorized person may face criminal charges. They could be removed from their position and may face a term of imprisonment of up to one year and fines. The information may be disclosed to other state or local officials if it is used only for the protection of critical infrastructures. The law was passed to ensure that only trained and authorized individuals who need to know the information can access it and use it only for homeland security purposes.

Critical Infrastructure Protection

To protect the critical infrastructure and CII and to maintain services if an event occurs, it is essential that officials from the federal, state, and local governments, as well as private owners of the nation's critical infrastructure, develop plans to protect not only their assets from possible harm but also action plans to respond to an attack or other harm. These plans must be reviewed regularly and updated as potential threats continue to change. The term "Critical Infrastructure Protection" (CIP) refers to those actions that are geared toward protecting critical infrastructures against physical attacks or hazards. These actions may also be directed toward deterring attacks (or mitigating the effects of attacks) that are either man-made or natural. While CIP includes some preventative measures, it usually refers to actions that are more reactive in scope. Today, CIP focuses on an all-hazards approach.

The primary responsibility for protecting critical infrastructure, and for responding if it is harmed, lies with the owners and operators, but the federal government and owners/operators work together to identify critical infrastructure and then assess the level of risk associated with that asset. The assets' potential vulnerabilities are determined, and possible methods for reducing the risk are identified. If owners and operators are unwilling or unable to participate in this process, the federal government can intervene, assess the protection level, and devise a response.[29] While most critical infrastructure protection is carried out at the federal, state, and local levels, there is also a global perspective on protecting critical infrastructure as the world becomes more interconnected.

A related term is Critical Infrastructure Assurance. This revolves around the process by which arrangements are made in the event of an attack or if an asset is disrupted to shift services either within one network, or among multiple networks, so that demand is met. In other words, it has to do with detecting any disruptions and then shifting responsibilities so that services can continue to be met. This can often be done without the consumer's knowledge.

Risk

The probability that an asset will be the object of an attack or another adverse outcome is its risk.[30] Risk is the likelihood that an adverse event will occur[31] and is related to consequences (C), vulnerabilities (V), and threats (T), as described in the following formula. The NIPP expresses this relationship as:

Risk = (function of) (CVT)

It is essential that critical infrastructure owners and operators assess the potential risk to their assets using these three elements. This way they can make policies to protect the critical infrastructure and plans to respond if that were to occur. Each element is described below.

Consequence

A consequence is the effect or result of an event, incident, or occurrence. This may include the number of deaths, injuries, and other human health impacts; property loss or damage; and interruptions to necessary services. The economic impacts of an event are also critical consequences, as many events have both short- and long-term economic consequences to communities or even to the nation.[32] It is important that there is business continuity, which is the ability of an organization to continue to function before, during, and after a disaster.[33]

Vulnerability

A vulnerability is "a physical feature or operational attribute that renders an entity open to exploitation or susceptible to a given hazard."[34] It is easy to think of it as a weakness or flaw in an asset that may cause it to be a target for an attack. An aggressor may seek out a vulnerability and use that to strike the asset. In most cases, the major vulnerability is access control whereby unauthorized people can enter the asset (such as a building or open area) to gather information to plan an attack or even to carry out an attack. To reduce this possibility, it has become common practice to prohibit unauthorized people from entering these types of areas.[35]

Structural vulnerabilities need to be addressed and maintained over an extended time rather than relying on a temporary solution or a "quick fix." This extended approach is referred to as Long-Term Vulnerability Reduction. The National Preparedness Goal defines the Long-Term Vulnerability Reduction Core Capability as to

> build and sustain resilient systems, communities, and critical infrastructure and key resources lifelines so as to reduce their vulnerability to natural, technological, and human-caused incidents by lessening the likelihood, severity, and duration of the adverse consequences related to these incidents.[36]

According to the DHS, the initial national capability target is to "achieve a measurable decrease in the long-term vulnerability of the Nation against current baselines amid a growing population based and expanding infrastructure base."[37]

Threat

A threat is a "natural or man-made occurrence, individual, entity, or action that has or indicates the potential to harm life, information, operations, the environment, and/or property."[38] This term has also been more simply defined as "an intent to hurt us."[39] Threat has to do with the potential harm that can originate from any source, including humans (terrorists or active shooter), natural hazards (different threats for different parts of the country), or technology (a cyberattack). Those charged with protecting critical assets seek to identify possible threats to resources to mitigate harm that could result. It is much easier to identify natural threats such as storms and earthquakes. To a great extent, these threats can be predicted, and the possible impact is easier to judge. Plans can be established so that a community is prepared to respond. On the other hand, man-made threats are far less predictable and can occur at any time with unknown consequences, making mitigation and response planning much more difficult.

Cyber Risk/Cybersecurity

A specific risk of growing concern is cybersecurity and cyber risk. The term "cybersecurity" refers to actions that are taken by the government or by private owners and operators to prevent damage to, unauthorized use of, or exploitation of, information and communications held electronically. This also includes all actions geared toward restoring these systems after an attack or other harm. The goal of this sector is to ensure the confidentiality, integrity, and availability of online information and data. Activities regarded as cybersecurity include those that are intended to protect and restore information networks and wirelines, wireless satellites, public safety answering points, and 911 communications systems and control systems.[40]

Cyber risk refers to the risks associated with cyber threats, crimes, or attacks. It includes an attacker who seeks to damage an agency's or individual's computer network. This can include malware such as ransomware. When considering cyber risk, things like vulnerability and the likelihood of an attack must be considered, along with potential consequences of the attack and possible loss of revenue or profit (or other financial consequences). In some cases, other national security concerns may be raised.[41]

Risk Assessment

Risk assessments of critical assets are carried out as a way to identify potential risks that may exist surrounding an asset, which can then lead to developing courses of action to prevent or respond to an attack. Through data collection and analysis, a risk analysis is an attempt to understand and identify not only threats but also possible consequences of an attack. In general, a risk assessment asks, "What can go wrong? What is the likelihood that it will go wrong? What are the possible consequences if it does go wrong?"[42] This way, the probability of an incident occurring and the severity (consequence) of that incident will be better understood.[43] The analysis can also be used to determine priorities or what assets are more critical and how should money be spent to protect them. It can also help officials create plans to mitigate the possible effects of an attack to protect residents and keep their property safe.

Since 9/11 and Hurricane Katrina, public interest in risk analysis has grown dramatically. Risk analysis has become an effective and comprehensive procedure to reduce the possibility of an attack and subsequent damages, and they have become complex.[44] Government officials at the federal, state, and local levels, heads of agencies, and even legislators now incorporate risk analysis into their decision-making processes and address risk more explicitly at all levels.[45]

Risk assessments are completed on an asset, a network, or a system. They typically consider three components of risk as noted above and rely on a variety of methods, principles, or rules to analyze the potential for harm. Some risk assessments are heavily quantitative and rely heavily on statistics and probabilities, whereas others are less quantitative.[46] In general, a risk assessment report typically includes five elements. They are:

1 Identification of assets and a ranking of their importance

The first step in a risk analysis is to determine which infrastructure assets can be "critical." Since all assets vary as to how important they are, assets can be, and need to be, ranked. Officials must determine what properties are needed in a community to ensure services are required. Examples include buildings, water treatment plants, or power plants. They may also decide that certain people are critical, such as medical professionals, police officials, or government officials. Another possibility is to include information such as financial data or business strategies. Risk assessments are then done on those assets that are identified as the most critical. The time and resources that would be needed to replace the lost asset must also be part of the analysis. If that asset were lost, how quickly could it be replaced? Are there other assets that could substitute for that one? If the asset was lost, how would services be provided? What cascading effects might occur if one asset were lost or damaged?[47]

2 Identify, characterize, and assess threats

All potential threats to an asset need to be identified. Details about potential threats that should be considered include the type of threat (e.g. insider, terrorist, or natural threat); the attacker's motivation; potential trigger events; the capability of a person to carry out an attack; and possible methods of attack (e.g. suicide bombers, truck bombs, cyberattacks). An analyst can gather information on these topics from the intelligence community, law enforcement officials, specialists and experts in the field, news reports in the media, previous analysis reports, previously received threats, or "red teams" who have been trained to "think" like a terrorist.[48]

3 Assess the vulnerability of critical assets to specific threats

An asset's vulnerability can be analyzed in many ways. One is physical. Here, an analyst would determine things like an outsider's accessibility to an asset. The second is technical, which refers to an asset's likelihood of being the victim of a cyberattack or other type of electronic attack. The third type of vulnerability is operational or the policies and operating procedures used by the organization. The fourth is organizational or the effects that may occur if a company's headquarters is attacked.[49]

4 Determine the risk
Risk is the chance that a disruptive event may occur, as described above. Assets are usually rated on their risk, and resources can be allocated to reduce an asset's risk.

5 Identify and characterize ways to reduce those risks
An important part of a risk assessment is to determine ways to mitigate or eliminate the risk of an attack. This could be something as simple as banning unauthorized people from entering areas or reducing traffic around an asset. Of course, other ways to eliminate the risk of an attack may be more complicated such as building physical barriers or relocating assets.

Risk Management

In risk management, officials ask, "What can be done? What options are available and what are the associated tradeoffs in terms of cost, risks, and benefits? What are the impacts of current management decisions on future options?"[50] These are efforts to decide which protective measures to take based on an agreed-upon risk reduction strategy.

Convergence

Many of the critical infrastructure assets are connected to each other in some way. This integration of infrastructure is called "convergence." This means that if one asset is harmed and unable to serve people, the other assets linked to may also be unable to perform.[51] This is referred to as "cascading" or "escalating" effects. The interconnected nature of critical infrastructure could lead to even more harm to a community than if the assets were independent. In some cases, the interdependencies can be global since many of our assets are linked to those around the world.

An obvious example of convergence can be seen with cyber assets, which are linked to all other assets both in the U.S. and elsewhere. Computers have become an essential part of our society, and every other sector relies, at least in part, on information technology. A cyberattack may affect the power grid, water, financial services, and healthcare, causing great damage in both the short and the long term. It would also affect transportation and financial outlets, having an impact on the economy. Computer systems control equipment in the chemical, nuclear, and oil industries. Companies rely on IT for easy communications, personnel management, research, and online commerce. The computer network is so essential that, in the Comprehensive National Cybersecurity Initiative, the sector was identified as one of the most serious economic and national security challenges facing the U.S. Those assets that are interconnected to other assets and networks may be an attractive target for enemies because of the broad harm it may cause.

On the other hand, interconnected assets could be a benefit for communities. In the case that one sector is unable to provide a service, another asset may be able to step in so that there is minimal disruption, and the desired level of service can be provided. So clearly, the interconnected nature of critical infrastructure has both positive and negative components.

Recovery/Resiliency

If an attack or other disaster does occur, a community must take steps to return to "normal" or the conditions that existed prior to the event and subsequent disruption of services. This process is called recovery and is part of the emergency management all-hazards response cycle. Recovery has been defined as the ability to adapt and withstand the disruption that occurs after an emergency or event.[52] It is the ability to recover rapidly and bounce back, or regroup, after a disruption, which could be natural, technological, or human-caused.[53] In most cases, community agencies and facilities are able to return to their full capabilities in a reasonable amount of time after an event. However, in many cases, the costs of rebuilding are too high, and it becomes impractical to return to pre-event standards.[54]

A similar concept is resiliency, which refers to the ability of a community to resist, absorb, recover from, or adapt to a change in conditions. As part of the risk management process, resiliency is "the capacity of an organization to recognize threats and hazards and make adjustments that will improve future protection efforts and risk reduction measures." This has to do with a community making changes to reduce the risk of an event or the consequences of that event.[55] For example, communities may take steps to ensure that buildings are constructed so that they are able to withstand damage to, or the loss of, a supporting beam or column.[56]

Resiliency is made more difficult because, in many cases, when one infrastructure is impacted, others may be impacted alongside (convergence). Each system is interconnected to many other infrastructures, whether it be cyber, physical, or organizational, making them interdependent. These relationships are constantly changing. A risk to one subset becomes a risk to all.[57]

At the same time, however, if one infrastructure is damaged or lost, it can be offset by other infrastructure. If one is damaged, another infrastructure may be able to reallocate its services in a way to fill in the gap and reduce the impact caused by the event. For example, if the water supply is damaged, people in that community are less concerned with whether the water is coming through a central pipe or some peripheral parts of the system. Instead, residents are concerned if the water supply fails to provide water to their homes at all.

Beyond allowing a community to continue to provide services, resiliency also has a deterrent value or a protective value. If a community is well protected and is prepared to bounce back quickly, an attacker, whose goal is to disrupt services, may be deterred from attacking. An offender may look at the target's protection when considering a target, and if that target is one that will not fall prey to an attack, the offender may go elsewhere.

A community's resiliency is made up of robustness (strength), resourcefulness (innovation, ability to adapt), and recovery. Robustness refers to the inherent strength of a structure or system or its ability to withstand external damage without the loss of functionality.[58] Resourcefulness is the capacity to mobilize needed resources and services in emergencies, and recovery is the ability to return to a "normal" condition. This can be portrayed in the following:

$$R1 + R2 + R3 = \text{Resilience}$$

Some officials have indicated a fourth factor that should be included in this equation, which is rapidity, or the speed with which disruption can be overcome and safety, services, and financial stability restored.[59] Certainly, residents want essential services such as power and water restored as quickly as possible to return to a "normal" state.

Resourcefulness

Resourcefulness refers to the ability of a community to gather and coordinate any necessary resources, services, equipment, and personnel in the event of a damaging event. Those communities that are resourceful can recover more quickly than others. Some essential parts of this include identifying personnel and equipment that might be critical to a recovery operation; cross training so that first responders can respond quickly to more than one type of event; mutual aid agreements that allow agencies to share resources and ask for help under particular circumstances; purchasing of spare equipment so that there is never a gap in available resources; and maintaining a supply of personnel and equipment that could quickly respond when needed.[60]

Hazard

A hazard is a source or cause of harm.[61] There are different types of hazards. A natural hazard is a potential incident resulting from acts of nature or a weather phenomenon.[62] These would be incidents that are caused by acts of nature such as hurricanes, wildfires, avalanches, earthquakes, winter storms, tornadoes, disease outbreaks, or epidemics.[63] Another type of hazard is a technological hazard. These are potential incidents that are the result of accidents or failures of systems or structures.[64] Examples of these include hazardous materials releases, dam or levee failures, airplane crashes, power failures, or radiological releases.[65] These may be caused by human error or a failure of technology. The final type of hazard is human-caused hazards, which include incidents that are the result of intentional actions of an individual or group of individuals. Examples of this type of hazard include acts of terrorism, active shooting, biological attacks, chemical attacks, cyber incidents, bomb attacks, or radiological attacks.[66]

The "all-hazards" approach is a way to analyze and prepare for a full range of threats and hazards, including domestic terrorist attacks, natural and man-made disasters, accidental disruptions, and other emergencies.[67] This is a "grouping classification encompassing all conditions, environmental or man-made, that have the potential to cause injury, illness, or death; damage to or loss of equipment, infrastructure services, or property; or alternatively causing functional degradation to social, economic, or environmental aspects."[68]

Impacts

Impacts describe how an event might affect an asset or the impact it has on the provision of services to residents. An impact could be the damage caused by an event or the consequences that occur as the result of an event. Impacts are clearly linked to the size and complexity of an event—a more serious event will result in more serious impacts. In a risk analysis, the possible impacts identified should be specific to allow officials to have a better understanding of how to manage the risk.[69]

Preparedness

Preparedness has been defined as those activities that are "necessary to build, sustain, and improve readiness capabilities to prevent, protect against, respond to, and recover from natural and man-made incidents." Preparedness is a continuous process whereby vulnerabilities are continually being assessed and response plans are continually being updated and revised. Preparedness can be completed by officials at all levels of government and between government and private sector and non-governmental organizations. As described in the National Incident Management System, preparedness has to do with establishing guidelines, protocols, and standards for planning, training, and exercises, personnel qualification and certification, equipment certification, and publication management.[70]

Mitigation

Mitigation refers to lessening the impact of an event. All communities should develop plans that have the goal of reducing the potential impact of a natural or man-made event. Once a community completes an assessment report that identifies risks and vulnerabilities, officials should devise a mitigation plan. All members of a community should be invited to participate in making the plan, as well as private stakeholders. The plan should define the roles and responsibilities of all interested organizations and individuals. It may also include mutual aid agreements with other jurisdictions or memorandums of understanding that will help ensure that the plan is carried out when needed.[71] Some examples of mitigation measures that communities have taken to improve the safety of a facility include increasing physical security measures, hiring additional security guards, and installing barriers around a building. Examples of mitigating cybersecurity measures include enhancing firewalls and updating passwords.

Training is essential to mitigation efforts. Personnel can train on equipment, become familiar with policies, learn to communicate, and work with other agencies. Training exercises that simulate emergencies are important as agencies can assess how well they have planned. Since threats can change, exercises will keep people ready to react.[72] This way, when an event occurs, people will be ready and able to assist.

Conclusion

This book examines the government's role in identifying and protecting the nation's critical infrastructure as officials seek to protect the country and its citizens from harm and ensure that essential services and goods are available in the aftermath of a disruptive event. It will focus on the risk assessment of assets and the development of plans to protect the nation's infrastructure from damage resulting from both natural disasters and attacks. The purpose is to introduce these ideas to readers in a way that is easy to understand rather than with the use of complicated formulas.

A history of risk assessment and programs for critical infrastructure protection is given in Chapter 2. This helps to give readers a background into early government policies that form the basis of today's asset protection programs. The status of today's protection plans is the focus of Chapter 3. The role of the DHS in critical infrastructure protection is described in Chapter 4, and Chapter 5 provides information on other agencies that help the nation in these efforts. The importance and status of public/private partnerships is presented in Chapter 6. This is of particular importance because a great portion of our country's assets are privately owned. The information in Chapter 7 summarizes the laws pertaining to critical infrastructure protection that have been passed by Congress. Chapter 8 presents the DHS perspective on risk and details three key documents that were created to define the principles, processes, and operational practices of risk management. Chapter 9 provides an overview of earlier risk assessment methods, federal guidelines for risk, and application of the Threat and Hazard Identification and Risk Assessment process. Chapters 10–12 summarize the 16 critical infrastructures and their SSAs. In this section each chapter provides a sector profile, goals and priorities, and the various methods and approaches each sector takes to assess risk. The text concludes with Chapter 13 and a discussion of the issues that continue to challenge and shape our responses to critical infrastructure protection, risk management, and resilience efforts.

Key Terms

Critical Infrastructure (Local, Federal, Public, and Private)
Sectors
Key Resources
National Infrastructure Protection Plan
SSA/Co-SSA
GCC
ISAC/ISAO
NICC
Critical Infrastructure Information

Critical Infrastructure Protection
Risk/Risk Analysis/Risk Assessment
Consequence
Vulnerability
Threat
Cyber Risk
Risk Assessment
Convergence
Recovery/Resiliency
Resourcefulness
Hazards
Impacts
Preparedness
Mitigation

Review Questions

1 Why is it important for the U.S. to protect its critical infrastructure and key resources?
2 What is critical infrastructure?
3 Why would a government or agency carry out a risk assessment? How is a risk defined?
4 Describe the five steps to a risk assessment.
5 Why would a community be interested in recovery and resiliency?

Notes

1 *CNN* (September 11, 2013). "Terror Attacks Fast Facts. Deaths in World Trade Center Terrorist Attacks." https://www.cnn.com/2013/07/27/us/september-11-anniversary-fast-facts/index.htm
2 Hurricanes Katrina and Rita, Office of Response and Restoration, National Oceanic and Atmospheric Administration. https://response.restoration.noaa.gov
3 CNN Editorial Research, "Hurricane Katrina Statistics Fast Facts." https://www.cnn.com/2013/03/23/us/hurricane-katrina-statistics-fast-facts/index.htm
4 Zdanowicz, Christina, Willians, David, and Alvarado, Caroll. "Hurricane Ian Killed at least 125 People. Here Are Some of the Victims." *CNN*. https://www.cnn.com/2022/10//14/us/victims-hurricane-ian/index.htm
5 National Oceanic and Atmospheric Administration, "U.S. Billion-Dollar Weather and Climate Disasters in Historical Contexts," 2022. https://www.climate.gov/news-features/blogs/2022-us-billion-dollar-weather-and-climate-disasters-historical-context
6 Donald, Jess. "Winter Storm Uri 2021." https://comptroller.texas.gov/economy/fiscal-notes/2021/oct/winter-storm-impact.php#
7 U.S. Center for Disease Control and Prevention, "COVID Data Tracker." https://covid.cdc.gov/covid-data-tracker/#datatracker-home
8 US Department of Homeland Security, and US Department of Justice, Global Justice Information Sharing Initiative, "Critical Infrastructure and Key Resources, Protection Capabilities for Fusion Centers," December 2008. https://it.ojp.gov/documents/d/CIKR%20protection%20capabilities%20for%20fusion%20centers%20s.pdf
9 DHS, FEMA, "CIKR Awareness AWR-213, Participant Guide," September 2010, pp. 1:14–15.
10 Strategic Foresight Initiative, "Critical Infrastructure Long-term Trends and Drivers and Their Implications for Emergency Management," June 2011. https://www.fema.gov/pdf/about/programs/oppa/critical_infrastructure_paper.pdf
11 US DHS, "National Preparedness Goal," 2011.
12 Department of Homeland Security, Lexicon (2010).
13 §1016(e) of the USA Patriot Act of 2001 (42 U.S.C. §5195c(e).
14 US Department of Homeland Security, "National Preparedness Goal," 2011, p. A-1; US DHS, "National Infrastructure Protection Plan," 2009, p. 7; US DHS, "National Infrastructure Protection Plan," 2013, p. 29.
15 Department of Homeland Security. "Blueprint for a Secure Cyber Future," November 2011. https://www.dhs.gov/xlibrary/assets/nppd/blueprint-for-a-secure-cyber-future.pdf
16 Liscouski, Robert. Assistant Secretary. Infrastructure Protection, Department of Homeland Security, testimony before the House Select Committee on Homeland Security, Infrastructure and Border Security Subcommittee, April 21, 2004.

17 Moteff, John (2007). *Critical Infrastructure: The Critical Asset Database*. Washington, DC: Congressional Research Service, RL 33648. http://fas.org/sgp/crs/homesec/RL33648.pdf
18 Collins, Pamela, and Baggett, Ryan (2009). *Homeland Security and Critical Infrastructure Protection* Westport, CT: Praeger.
19 Federal Emergency Management Agency, "Strategic Foresight Initiative, Critical Infrastructure," June 2021. https://www.fema.gov/pdf/about/programs/oppa/critical_infrastructure_paper.pdf
20 Cato Institute, "Who Owns U.S. Infrastructure?" June 1, 2017. https://www.cato.org/tax-budget-bulletin/who-owns-us-infrastructure
21 DHS, FEMA, "CIKR Awareness AWR-213, Participant Guide," September 2010, p. 1-14-15.
22 NIPP, "Partnering for Critical Infrastructure Security and Resilience," 2013.
23 Department of Homeland Security, "National Infrastructure Protection Plan and Fact Sheet." https://www.cisa.gov/sites/default/files/2022-11/NIPP-Fact-Sheet-508.pdf; see also https://www.cisa.gov/sites/default/files/2022-11/national-infrastructure-protection-plan-2013-508.pdf
24 Source: PPD-21, 2013; 2009 NIPP.
25 2009 NIPP.
26 CRS Report for Congress, "Critical Infrastructures: Background and Early Implementation of PDD-63," June 19, 2001. https://www.everycrsreport.com/files/20010619_RL30153_733b6d58fef64d01369593e298619a5cf0555672.pdf
27 Homeland Security Act of 2002.
28 DHS.gov/national-infrastructure-coordinating-center
29 Moteff, John D. (2015). "Critical Infrastructures: Background, Policy and Implementation." Congressional Research Service, 7-5700, June 10. www.crs.gov
30 Lowrance, W. (1976). *Of Acceptable Risk*. Los Altos, CA: William Kaufmann.
31 Haimes, Yacov Y. (2004). *Risk Modeling, Assessment, and Management*, 2nd Ed. Hoboken, NJ: John Wiley and Sons, p. xii; US DHS (2010); DHS Risk Lexicon, p. 27.
32 US DHS (2013); NIPP, p. 109.
33 DHS, FEMA, "CIKR Awareness AWR-213, Participant Guide," September 2010, p. A-2.
34 US DHS, "National Infrastructure Protection Plan," 2013, p. 33.
35 US DHS, "National Infrastructure Protection Plan," 2013, p. 33.
36 US DHS, "National Preparedness Goal," 2011, p. 11.
37 US DHS, "National Preparedness Goal," 2011, p. 11.
38 US DHS, "DHS Risk Lexicon," 2010, p. 36; US DHS, "National Infrastructure Protection Plan," 2013, p. 33.
39 US DHS, "DHS Risk Lexicon," 2010, p. 36; US DHS, "National Infrastructure Protection Plan," 2013, p. 33.
40 DHS, FEMA, "CIKR Awareness AWR-213, Participant Guide," September 2010, p. A-2.
41 Rostick, Paul. "Risk vs. Cyber-risk." https://www.techtarget.com/searchsecrity/tip/Top-6-critical-infrastructure-cyber-risks
42 DHS, "DHS Risk Lexicon," 2010, pp. 27–28.
43 DHS, FEMA, "CIKR Awareness AWR-213, Participant Guide," September 2010, pp. 3–7.
44 See Ezell, Barry C., Farr, John V., and Wiese, Ian (September 2000). "Infrastructure Risk Analysis Model." *Journal of Infrastructure Systems*, (6)3, pp. 114–117; Ten, CheWooi, Manimaran, Govindarasu, and Liu, Chen-Ching (2010). "Cybersecurity for Critical Infrastructures: Attack and Defense Modeling." *IEEE Transactions on Systems, Man and Cybernetics*, Vol. 40, pp. 853–865.
45 Haimes, Yacov Y. (2004). *Risk Modeling, Assessment, and Management*, 2nd Ed. Hoboken, NJ: John Wiley and Sons.
46 Motef, John (2005). "Risk Management and Critical Infrastructure Protection: Assessing, Integrating, and Managing Threats, Vulnerabilities and Consequences." CRS Report for Congress; Congressional Research Service, February 4. https://www.fas.org/sgp/crs/homesec/RL32561.pdf
47 Motef, John (2005). "Risk Management and Critical Infrastructure Protection: Assessing, Integrating, and Managing Threats, Vulnerabilities and Consequences." CRS Report for Congress; Congressional Research Service, February 4. https://www.fas.org/sgp/crs/homesec/RL32561.pdf
48 Motef, John (2005). "Risk Management and Critical Infrastructure Protection: Assessing, Integrating, and Managing Threats, Vulnerabilities and Consequences." CRS Report for Congress; Congressional Research Service, February 4. https://www.fas.org/sgp/crs/homesec/RL32561.pdf

49 Motef, John (2005). "Risk Management and Critical Infrastructure Protection: Assessing, Integrating, and Managing Threats, Vulnerabilities and Consequences." CRS Report for Congress; Congressional Research Service, February 4. https://www.fas.org/sgp/crs/homesec/RL32561.pdf
50 Ezell, Barry C., Farr, John V., and Wiese, Ian (2000). "Infrastructure Risk Analysis Model." *Journal of Infrastructure Systems*, September, pp. 114–117.
51 US DHS, "National Infrastructure Protection Plan," 2013, p. 31.
52 US Department of Homeland Security, "National Preparedness Goal," 2011, p. A-2.
53 DHS, FEMA, "Advanced Critical Infrastruction Protection, MGT-414, Participant Guide," April 2013.
54 DHS, FEMA, "Critical Asset Risk Management, Participant Guide," September 2014, p. 6-15.
55 US DHS, "DHS Risk Lexicon," 2010, pp. 26, 46; DHS, FEMA, "CIKR Awareness AWR-213, Participant Guide," September 2010, p. 1–11.
56 Mueller, John, and Stewart, Mark G. (2011). *Terror, Security and Money.* New York: Oxford University Press.
57 Haimes, Yacov Y. (2004). *Risk Modeling, Assessment, and Management*, 2nd Ed. Hoboken, NJ: John Wiley and Sons, p. 684.
58 O'Rourke (2009). "Setting Performance Goals for Infrastructure," p. 2.
59 O'Rourke, T.D. (2007). "Critical Infrastructure, Interdependencies and Resilience." *The Bridge*, Vol. 37, Issue 1, pp. 22–29. http://www.nae.edu/File.aspx?id=7405
60 DHS, FEMA, "Critical Asset Risk Management, Participant Guide," September 2014, p. 6–15.
61 US DHS, "DHS Risk Lexicon," 2010, p. 17.
62 US DHS, "Threat and Hazard Identification and Risk Assessment Guide (CPG-201)," 2nd Ed, 2013, p. B-1.
63 US DHS, "Threat and Hazard Identification and Risk Assessment Guide (CPG-201)," 2nd Ed, 2013, p. 5.
64 US DHS, "Threat and Hazard Identification and Risk Assessment Guide (CPG-201)," 2nd Ed, 2013, p. B-1.
65 US DHS, "Threat and Hazard Identification and Risk Assessment Guide Comprehensive Preparedness Guide (CPG 201)," 2nd Ed, 2013, p. 6.
66 US DHS, "Threat and Hazard Identification and Risk Assessment Guide (CPG-201)," 2nd Ed, 2013, pp. B-1, 6.
67 DHS, FEMA, "Critical Asset Risk Management, Participant Guide," September 2014, p. 2–5.
68 DHS, FEMA, "CIKR Awareness AWR-213, Participant Guide," September 2010, p. A-2.
69 US DHS, "Threat and Hazard Identification and Risk Assessment Guide (CPG-201)," 2nd Ed, 2013, p. 11.
70 DHS, FEMA, "CIKR Awareness AWR-213, Participant Guide," September 2010, p. 1–11.
71 DHS, FEMA, "Critical Asset Risk Management, Participant Guide," September 2014, p. 6–16.
72 DHS, FEMA, "Critical Asset Risk Management, Participant Guide," September 2014, pp. 6-16–6-17.

2 Early History

Chapter Outline

Introduction

Public officials in federal and state governments have been concerned with protecting the country's critical infrastructure for many years, even prior to the terrorist attacks of September 11, 2001. In these early years, the government began to take a closer look at the safety of the nation's infrastructure after early acts of both foreign and domestic terrorism.

One early event that began to change the way officials thought of the need for critical infrastructure protection happened in September 1984. Members of the Rajneeshee cult sought to influence county elections in Oregon by contaminating restaurant salad bars with *Salmonella*. Their goal was to prevent residents from voting in the election so the group members could more easily take over political control of the county. The group also attempted to add *Salmonella* to the water supply. This was the first, and the largest, bioattack in the U.S. About 750 people became sick, but there were no fatalities.[1] Two members of the group were charged and convicted of attempted murder and were sentenced to 20 years in federal prison.[2]

Another critical event occurred a few years later, on December 21, 1988. Just 27 minutes after it took off from London, Pan Am Flight 103 exploded over Lockerbie, Scotland. All of the 270 people on board the flight were killed, including additional victims on the ground. An investigation

DOI: 10.4324/9781003434887-2

discovered that terrorists had placed a bomb on the flight that appeared to be a radio cassette recorder held inside a suitcase. Abdelbaset al-Megrahi, an intelligence officer from Libya, and Lamin Khalifah Fhimah were accused of murder after the bombing. Megrahi was convicted of murder and sentenced to life imprisonment, serving ten years in prison. Fhimah was acquitted of the charges against him. A third person, Abu Agila Mohammad Mas'ud Kheir Al-Marimi, from Libya, was taken into custody in 2022 and charged with the destruction of an airplane causing death, among other offenses.[3]

These events, and others including the assault on the World Trade Center in 1993 in New York City and the bombing of the federal building in Oklahoma City in 1995, increased fears of further attacks. Fears were also increasing as technology improved and the potential for computer hacking surged, highlighting the need for more cybersecurity efforts. In the midst of this, President Clinton signed an executive order that identified critical infrastructure sectors, which was the start of a new era in critical infrastructure protection. Since that time, policies for protecting the country's assets have expanded to meet ever-growing threats.

Today, critical infrastructure protection is a top priority for officials at all levels of the government, including federal, state, local, and tribal structures. It is also a priority for both public and private owners and operators of critical infrastructure. No matter who is involved, the goal is to protect our infrastructure from damage resulting from a man-made incident such as a terrorist attack but also from a natural event such as a hurricane or earthquake. In either case, the object is to return a community to a normal (or close to normal) state of affairs as quickly as possible. This chapter provides a history of critical infrastructure protection policies, providing a background to the current policies that exist in the U.S. today.

Career Focus: Elizabeth Sherwood-Randall, Homeland Security Advisor to President Biden

Elizabeth Sherwood-Randall became the Homeland Security Advisor to President Biden in 2021. A native of California, Sherwood-Randall earned a bachelor's degree from Harvard and a PhD in international relations from Oxford University where she was a Rhodes Scholar. Then-Senator Joe Biden chose her to serve as the chief foreign affairs and defense policy advisor, after which she became the co-founder and associate director of the Strengthening Democratic Institutions Project at the Belfer Center at the Harvard Kennedy School. When Clinton became president, Sherwood-Randall became the Deputy Assistant Secretary of Defense for Russia, Ukraine, and Eurasia, and then returned to Harvard and formed the Harvard-Stanford Preventive Defense Project. In 2009, she became the Special Assistant to President Obama and Senior Director for European Affairs, later becoming the White House Coordinator for Defense Policy. In July 2014, she became the Deputy Secretary of Energy.

When Biden was elected to the presidency, Sherwood-Randall was appointed to be the White House Homeland Security Advisor and the Deputy National Security Advisor. She has focused on strengthening critical infrastructure and increasing the resiliency of critical assets. Sherwood-Randall works closely with leaders of other countries to counter threats of terrorism and maintain global security. Randall is responsible for coordinating federal support to leaders on the state and local levels, nongovernmental organizations (NGOs), and those in the private sector. A primary responsibility is to ensure that the country is prepared for attacks and then be able to respond to those if they happen, whether it is a terror attack, natural disaster, or pandemic. Sherwood-Randall described her job in a tweet she sent on January 14, 2021: I will give everything I've got to lifting up our values and building a more secure U.S. and a safer world for our children.

Early Years of Critical Infrastructure Protection

Critical infrastructure protection is not new in the U.S. During the Cold War, U.S. officials recognized threats of possible attacks and created initiatives to prepare for these possibilities. Officials at the time identified critical infrastructure assets as power plants and grids, oil and gas pipelines, and other critical facilities that, if harmed, might affect the continuity of government and services. These plans were relatively minor and never serious enough to warrant much attention from the federal government. Most threats were seen as insignificant and were left to state and local law enforcement or individual companies.[4]

More "modern-day" critical infrastructure protection can be traced back to July 15, 1996, when President Clinton signed Executive Order 13010. In this document, the president revealed his plans to establish a new organization he called the President's Commission on Critical Infrastructure Protection. The commission was one of the first federal attempts to assess legitimate threats against the U.S.[5] The members of the commission would, according to Clinton, investigate the scope and nature of potential vulnerabilities and threats to the country's critical infrastructure, with particular attention given to the potential impact of cyber threats. The committee members were asked to recommend a comprehensive national plan or strategy for protecting critical infrastructure.[6]

Clinton also identified eight critical infrastructure sectors for the country in Executive Order 13010. These eight sectors were telecommunications, transportation, electric power, banking and finance, gas and oil storage and delivery, water supply, emergency services, and government operations. These sectors could be thought of as "categories" of assets or simply a way to organize the country's critical infrastructure protection plans.

One other change that Clinton made in this document was to expand the definition of what was considered to be infrastructure. Under the new approach, "infrastructure" was defined as a

> framework of interdependent networks and systems comprising identifiable industries, institutions (including people and procedures), and distribution capabilities that provide a reliable flow of products and services essential to the defense and economic security of the United States, the smooth functioning of government at all levels, and society as a whole.[7]

After holding many meetings and completing intensive research, the commission released its findings in its final report, which was released in 1997. In general, the report indicated that there was no immediate crisis that posed a significant threat to the nation's infrastructure, but there was a need for the government to take some action to protect its assets, especially with regard to cybercrime. Moreover, the report included recommendations geared toward securing the country's infrastructure. One key finding throughout the report was the need for more communication and increased information sharing between the private sector and the government.[8]

After the report was made public, Clinton sought a new plan that would be completed within the upcoming five years that would ensure that the country's critical infrastructure would be protected from any intentional disruption. The new plan was to include new plans for critical infrastructure protection, both physical and cyber. Moreover, according to Clinton, the plan would address the potential for attacks so that any interruption in service delivery would be brief, manageable, and geographically isolated. These plans were described in Presidential Decision Directive 63 (PDD-63), which was announced in May of 1998. PDD-63 was the first unclassified presidential national security directive that presented a new approach to protecting the country's critical infrastructure. It supported partnerships between the government and private ownership that had not been tried before.

In this new document, the term "critical infrastructure" was used to refer to the physical and cyber-based systems that people relied on for operating the economy and the government. PDD-63 defined the critical infrastructure as composed of five essential domains: banking and finance, energy, transportation, telecommunications, and government services. Further, the report identified specific critical infrastructure assets that required protection. These were referred to as "sectors." The sectors included information and communications; banking and finance; water supply; aviation, highways, mass transit, pipelines, rail, and waterborne commerce; emergency and law enforcement services; emergency, fire, and continuity of government services; public health services; electric power, oil and gas production, and storage.

Each sector was assigned a lead agency that would be responsible for securing and protecting the critical infrastructure in their particular sector. Each lead agency, in turn, was asked to appoint a sector liaison official. This person had the task of communicating with any appropriate private sector organizations and including them in any interactions about protecting assets. Members of the private sector were then asked to choose a person who would agree to serve as the sector coordinator. This person would work with the agency's sector liaison official to ensure cooperative efforts. Most agencies appointed their sector liaison official quickly, but it took longer for the agencies to appoint the sector coordinators. PDD-63 also set up a National Plan Coordination Staff to support the plan's development. The staff would be housed in the Critical Infrastructure Assurance Office (CIAO), which would be located in the Commerce Department. CIAO provided support to the sectors as they developed their plans but also helped the National Coordinators as they integrated the sector plans into a National Plan. Originally, CIAO was to exist for only three years to study the vulnerability of the country's critical infrastructures, but the work done by CIAO was so critical that it continued to operate past the three-year deadline.

In most cases, an individual from a relevant trade organization was chosen to serve as the sector coordinator. For example, the Environmental Protection Agency opted to have the Executive Director of the Association of Metropolitan Water Agencies serve as the sector coordinator for the water sector. Similarly, in the law enforcement sector (which is no longer a separate sector), the National Infrastructure Protection Center helped create an Emergency Law Enforcement Services Forum, consisting of senior state, local, and non-FBI law enforcement officials. In the case of banking and finance, the sector coordinator was chosen from a major banking/finance institution, who also served as the chairperson of the Financial Services Sector Coordinating Council. This was an agency that was created by industry officials to coordinate critical infrastructure protection activities with the federal government.

Personnel within the lead agencies were asked to develop plans to help ensure the protection of their assets. The plans were to include an assessment of any possible vulnerability that existed for each asset and plans to reduce the sector's vulnerability to an attack. They were also to develop response plans in the case of an attack, remediation plans, and reconstitution plans. The agencies were also asked to include needs for future research and development and possible opportunities for international cooperation. The plan was to be completed within 180 days, which were then to be fully implemented within two years and updated every two years after that.[9]

Another position created in the PDD-63 was the National Coordinator for Security, Infrastructure Protection, and Counterterrorism. This person was given the task of serving as the chair of the Critical Infrastructure Coordination Group. They reported through the assistant to the president for National Security Affairs. Also created in PDD-63 was a National Infrastructure Assurance Council, an advisory group that included private owners, representatives from state and local government, and representatives from relevant federal agencies. The council was to meet and provide the president with reports about the progress.[10]

A new office, the National Infrastructure Protection Center (NIPC), was created as part of PDD-63. This office, which was housed within the Department of Justice and the FBI, was given the job of defending the nation's public and private computer systems from possible cyberattacks and responding to illegal acts carried out by the use of computers and information technologies.[11]

In addition to setting up many new positions and agencies, PPD-63 recognized the need for a national capability to detect and respond to cyberattacks as they occurred. To do this, Clinton sought a Federal Intrusion Detection Network (FIDNET) that would work with the Federal Computer Intrusion Response Capability. Additionally, the Federal Bureau of Investigation was asked to expand its computer crime activities and form the NIPC. The NIPC would become the lead agency for federal threat assessment, vulnerability analysis, early warning capability, law enforcement investigations, and coordination of responses. Clinton asked that all agencies provide the NIPC with any information they had about cyberthreats or actual attacks. The NIPC would also work with the private sector through the Information Sharing and Analysis Center (ISAC) that would be operated by the private sector. Because the NIPC was housed in the Department of Justice (DOJ), the DOJ became the lead agency for protecting the country's critical infrastructure.

In December 1999, some of the sectors established an organization they called the Partnership for Critical Infrastructure Security. The agency's goal was to share information and strategies for infrastructure protection and to identify any possible interdependencies across sectors. This group was led by members of the private sector, but the Department of Homeland Security (and CIAO before that) acted as a liaison and provided administrative support. Sector liaisons from lead agencies were considered ex officio members. The partnership helped coordinate input from the organization to many of the national strategies and plans for infrastructure protection.

In January 2000, just prior to leaving the presidency, the Clinton administration released the National Plan for Information Systems Protection. In keeping with the focus of PDD-63, the new plan focused on protecting the cyber infrastructure, which was just emerging at the time.

Bush Administration

When George W. Bush was elected to the presidency, he chose to make some changes in the policies for critical infrastructure protection. The Bush administration shifted the focus of infrastructure protection, adding more of a focus on cybersecurity. The concern with asset protection was brought to the forefront after the terrorist attacks of September 11, 2001. At that time, while the administration still had a concentration on maintaining cybersecurity, its focus also now included the protection of physical threats, especially ones that might cause mass casualties. This was not always a popular approach, and there was some debate among officials about the direction that American policies for protecting the nations should take.

Pre-September 11, 2001

The Bush administration expanded the policies toward critical infrastructure protection that were first established by President Clinton. President Bush continued to stress that protecting the infrastructure was essential for citizens, the economy, the government, and national security. Bush also talked about the importance of protecting critical infrastructure not only to protect those structures but also to boost the national morale. Administration officials spoke more often about ensuring that any disruption in services be infrequent, of minimal duration, and manageable. The definition of critical infrastructure was expanded during this time to include those targets that, if attacked, would result in many casualties. The emphasis on protection also changed to include more collaboration between private sector owners/operators and the federal government. The Bush administration promised that they would work to help identify critical assets and create a plan for protecting those assets.

Fairly soon after entering office, Bush decided to consolidate the responsibilities of the many groups and agencies within the National Security Council (NSC) into 17 Policy Coordination Committees (PCCs). The responsibility of protecting critical infrastructure was given to the Counterterrorism and National Preparedness PCC. There was some debate surrounding a proposal to establish a federal chief information officer, who would oversee the security of all federal non-national security-related computer systems. At the same time, this person would also coordinate with the private sector on tasks relating to protecting privately owned computer systems. In the end, Bush chose not to create this position, instead relying on the deputy director of the Office of Management and Budget to do that job.

The president also turned to Congress for ideas about the best way to protect the country against acts of terrorism. The Hart-Rudman Commission, otherwise known as the U.S. Commission on National Security/21st Century, suggested the creation of a National Homeland Security Agency. Within this new agency would be a directorate that would oversee critical infrastructure protection. At first, Bush did not support this plan, but his plans changed after the terrorist attacks on the U.S. on September 11, 2001.

Post September 11, 2001

After the terrorist attacks of September 11, 2001, President Bush signed the USA PATRIOT Act (PL 107-56) and the Homeland Security Act of 2002 (PL 107-296). In the PATRIOT Act, the definition of critical infrastructure was updated to be

> Systems and assets, whether physical or virtual, so vital to the United States that the incapacity or destruction of such systems and assets would have a debilitating impact on security, national economic security, national public health or safety, or any combination of those matters.[12]

New terms were also mentioned for the first time, including "key assets" which were described as "individual targets whose destruction would not endanger vital systems, but could create local disaster or profoundly damage or Nation's morale or confidence."[13] An example of this would be the Statue of Liberty, which has both historic and symbolic meaning to people who live in the U.S.

The Infrastructure Protection Executive Notification Service, located within the DHS, was established as a way for officials in that agency to directly communicate with the Chief Executive Officers of major industrial firms. With the new system, the DHS was able to alert companies to any incidents affecting infrastructure or to make them aware of any pertinent threats. The department also oversaw the Critical Infrastructure Warning Network, which is a way to provide secure lines of communication between the DHS and other federal, state, and local agencies, the private sector, and international agencies.

Executive Order 13228

President Bush signed Executive Order 13228 on October 8, 2001, less than a month after the terrorist attacks on the U.S. Through this action, Bush created the Office of Homeland Security that would be overseen by the Assistant to the President for Homeland Security. The mission of the new agency was to "develop and coordinate the implementation of a comprehensive national strategy to secure the United States from terrorist threats and attacks."[14] The primary task of the new office was to coordinate efforts to protect the U.S. and its critical infrastructure from another attack and maintain efforts at recovery. When the DHS was subsequently established, many of the functions of the Office of Homeland Security were transferred there.

Additionally, the Homeland Security Council was added to the administration through the executive order. The council comprised the President; Vice-President; Secretaries of Treasury, Defense, Health and Human Services, and Transportation; the Attorney General; the Directors of FEMA, FBI, and CIA; and the Assistant to the President for Homeland Security, and later the Secretary of Homeland Security. Other officials from the White House and other officers would be invited to attend the meetings at night. The role of the newly developed council was to provide advice to the president about all aspects of protecting the country and its critical infrastructure.

A related document signed by President Bush in the days after 9/11 was Executive Order 13231, which he signed on October 16, 2001. In this executive order, Bush stated that it was U.S. policy "to protect against the disruption of the operation of information systems for critical infrastructure ... and to ensure that any disruptions that occur are infrequent, of minimal duration, and manageable, and cause the least damage possible."[15] When he signed the document, Bush established the president's Critical Infrastructure Protection Board. The board was made up of the CIAO and other federal officials and was tasked with recommending policies and coordinating programs "for protecting information systems for critical infrastructure." The board also was asked to write a National Plan for protecting infrastructure.

The board was chaired by a Special Advisor to the President for Cyberspace Security. The Special Advisor reported to both the Assistant to the President for National Security and the Assistant to the President for Homeland Security. Besides presiding over board meetings, the Special Advisor proposed policies and programs as needed if they would help ensure the nation's information infrastructure was protected.

The National Infrastructure Advisory Council was established in Executive Order 13231. The council was to advise the President on the security of information systems for critical infrastructure. Moreover, the council was to work toward increasing partnerships between public and private agencies, monitor the development of ISACs, and encourage the private sector owners to carry out vulnerability assessments of critical information and telecommunication systems.

National Strategy for Homeland Security

In July 2002, the Office of Homeland Security released a report entitled a "National Strategy for Homeland Security." The strategy described all government efforts geared toward protecting the nation against terrorist threats of all kinds. It identified actions for protecting the nation's critical infrastructure and key assets as one of six critical mission areas. This strategy added public health, the chemical industry, hazardous materials, postal and shipping, the defense industrial base, and agriculture and food to the list of sectors that have critical infrastructure in them. It also combined emergency fire service, emergency law enforcement, and emergency medicine as emergency services. It also eliminated those functions that belong primarily to the federal government (e.g., defense, intelligence, law enforcement). Some of the sectors were reassigned to different agencies. Many of the sectors were placed into the DHS (postal and shipping services and the defense industrial base).

The report also introduced a new type of assets, called key assets, which were identified as potential targets that, if destroyed, may not endanger vital systems across the nation but would nonetheless create a local disaster or affect the nation's morale. These would be things such as national monuments or historic attractions, dams, large commercial centers, or even sports stadiums.

The strategy reinforced the need to work closely with the private sector to assess vulnerabilities and develop a plan to deal with those vulnerabilities. In the strategy, the need to set priorities was stressed, explaining that not all assets are equally critical. The strategy was updated in October 2007, with few changes to the strategy for protecting critical infrastructure.

The "National Strategy for the Physical Protection of Critical Infrastructures and Key Assets" was published in February 2003, by the Office of Homeland Security. This document also helped to define what was meant by "key assets." This new strategy used a broader perspective on issues related to organizing the nation's efforts to protect its critical assets. It also identified the roles and responsibilities of agencies and people, any actions that needed to be taken, and guiding principles. In this document, the definition of what is critical infrastructure was expanded to include historical attractions, centers of government and commerce, and facilities that are associated with our national economy. This would include Wall Street in New York, chemical plants, and events where there would be large numbers of people in attendance.

The national Strategy also indicated that protecting critical infrastructure and key assets was a way to reduce the nation's vulnerability to possible terrorist acts. National goals and objectives were set to protect and secure specific vital infrastructures. The goals were intended to identify any assets that were deemed to be critical in preserving the health and safety of the nation, the government, economic and national security, and public confidence. In the event of an attack, the primary focus would be on protecting those assets and infrastructure that were identified as critical.

That same month, February 2003, the President's Critical Infrastructure Protection Board released a report, entitled "The National Strategy to Secure Cyberspace." This document focused on methods to protect information and data stored electronically or available on the Internet.

HSPD-7

On December 17, 2003, the Bush administration released Homeland Security Presidential Directive 7 (HSPD-7), called Critical Infrastructure Identification, Prioritization, and Protection. In this directive, the responsibilities of various agencies in protecting critical infrastructure were outlined. In addition, the role of sector-specific agencies (i.e., lead agencies; SSAs) was more clearly defined. The SSAs were asked to collaborate with officials from their particular sectors and identify, prioritize, and coordinate measures designed to protect the country's infrastructure. The directive also reiterated the need to create effective relationships between the DHS and agencies in other areas.

One change that was made was to appoint the DHS as the lead agency for the sector on chemical and hazardous materials. This person would report to the Secretary of Homeland Security annually regarding their relationships with the private sector. The directive also reinforced the need for all federal agencies to develop plans for protecting their critical infrastructure.

Through HSPD-7, President Bush asked for a comprehensive National Plan for Critical Infrastructure and Key Resources Protection to be completed by the end of 2004. This document was to include (a) a strategy to identify, prioritize, and coordinate the protection of critical infrastructure and key resources, including how the department will work with other stakeholders; (b) a summary of activities to be undertaken to carry out the strategy; (c) a summary of initiatives for sharing critical infrastructure information and threat warnings with other stakeholders; and (d) coordination with other federal emergency management activities.

In HSPD-7, the Secretary of Homeland Security was to serve as the principal federal official to lead critical infrastructure protection across the country. The responsibility for sectors was assigned to SSAs. The plans developed by the agencies, called the sector-specific plans (SSPs), were supposed to create a more coordinated approach to protect critical infrastructure and key resources.

The National Infrastructure Protection Plan (NIPP) was not completed by the December 2004 deadline. Instead, in February 2005, the DHS published an Interim NIPP. Then, in November 2005, the department released a "draft" NIPP. The final version of the NIPP was approved on June 30, 2006. The NIPP was then revised in early 2009 to reflect the evolution of the process, including concepts of all-hazards and resiliency. These changes did not represent major shifts in policy or programs.

The Bush administration devised a new Critical Infrastructure Protection Partnership Model. The new plan expanded the sector liaison and sector coordinator into Government Coordinating Councils and Sector Coordinating Councils for each sector. The goal was to increase representation of both owner/operators and government representation within the sectors. For example, the Water Sector Coordinating Council expanded to include two owner/operator representatives, along with one non-voting association staff member from each of the following participating organizations: the Association of Metropolitan Water Agencies, the American Water Works Association, the American Water Works Association Research Foundation, the National Association of Clean Water Agencies, the National Association of Water Companies, the National Rural Water Association, the Water Environment Federation, and the Water Environment Research Foundation. The Water Government Coordinating Council is chaired by the Environmental Protection Agency, the lead agency, and also includes the DHS, the Food and Drug Administration, the Department of Interior, and the Center for Disease Control. Government Coordinating Councils can also include state, local, and tribal government entities. The Sector Coordinating Councils were asked to establish their own organizational structures and leadership that would act independently from the federal government. Also, under this model, the Partnership for Critical Infrastructure Security was designated the Private Sector Cross-Sector Council. The Sector Coordinating Councils were to provide input into both the NIPP and the individual SSPs.

In March 2006, the Department of Homeland Security used its authority under the Homeland Security Act to form the Critical Infrastructure Partnership Advisory Council. However, the council was not required to meet the standards outlined in the Federal Advisory Committee Act that requires advisory committees to meet in public and to make written materials available to the public. This was done so that the members of the commission would feel free to share information that they would otherwise be hesitant to do. The DHS served as the head of the committee, and other members include owner/operators who are members of their respective Sector Coordinating Councils and federal, state, local, and tribal government representatives who belong to their Government Coordinating Councils.

The Obama Administration

President Barack Obama became president in 2009 and made a few changes to the infrastructure protection policies that were generated by Presidents Clinton and Bush. Instead, he continued to build on existing structures. Early in his presidency, in February 2009, President Obama asked for a review of the homeland security and counterterrorism structures that were located within the White House. Based on these reviews, the president merged the Homeland Security Council and the NSC into one agency, which he called the National Security Staff. The report also included a recommendation to appoint one person from the White House who would be responsible for overseeing federal policies regarding cybersecurity.

One of President Obama's first acts in office was to sign Executive Order 13527, "The Medical Countermeasures Following a Biological Attack." This was signed on December 30, 2009. Under this executive order, Obama outlined plans to provide medical services in the event of a biological attack. Working with state, local, territorial, and tribal governments, Obama sought to mitigate illness and death while also maintaining critical infrastructure. He ordered the U.S. Postal Service, the Secretaries of Health and Human Services, and the DHS to prepare for delivering medical countermeasures to communities to respond to a large-scale attack, with anthrax being the primary threat. These agencies were also to develop a plan to ensure that "mission-essential" functions of the government would continue.

SRNA

In 2010, the then-Secretary of Homeland Security wrote the Strategic National Risk Assessment (SNRA). This was a classified assessment that formed the basis of Presidential Policy Directive 8, which was announced by President Obama in 2011 (described in the next section). The goal of the SNRA was to help identify the types of incidents that posed the greatest threat to the security of the nation. The committee that assisted the investigation included officials from the Director of National Intelligence and the Attorney General. An unclassified version of the report was released to the public in December 2011. The report provided details on weaknesses in the nation's security and provided suggestions for how those could be addressed.

The committee drew from multiple sources, including historical records and experts from different disciplines. The members assessed the frequency and consequence of risks to answer the question, *with what frequency is it estimated that an event will occur, and what are the consequences of the incident(s) if it does occur*? The committee examined the threats and consequences associated with six categories of harm: loss of life, injuries and illnesses, direct economic costs, social displacement, psychological distress, and environmental impact.

The risks from possible threats and hazards that had the potential to have a significant impact on the nation's assets were discussed. The members identified risk factors and then identified core capabilities and capability targets that would be described in the National Preparedness Goal. The committee members relied on a new approach to asset protection that relied on collaborative thinking about strategic needs for prevention, protection, mitigation, response, and recovery requirements. It also promoted the necessity for all levels of the government to share a common understanding and awareness of threats and hazards so that they could prepare for, and respond to, events both independently and collaboratively. This was critical because, as noted by the committee, preparation and response are often more effective when multiple responders from local, state, and federal agencies are involved. It was also recognized that the whole community should be involved.

Possible events that could affect the nation's security were grouped into three categories: natural hazards; technological/accidental hazards; and adversarial, human-caused threats/hazards. The report also created six possible harms: loss of life, injuries and illnesses, direct economic costs, social displacement, psychological distress, and the environment. The SNRA Committee found that there were a wide range of threats and hazards that posed a significant risk to the nation, affirming the need for an all-hazards, capability-based approach to preparedness planning. Some of the key findings reported included:

1 Natural hazards, including hurricanes, earthquakes, tornados, wildfires, and floods, present a significant and varied risk across the country;
2 A virulent strain of pandemic influenza has the possibility of killing hundreds of thousands of Americans and affecting millions more, resulting in economic loss;
3 Technological and accidental hazards, such as dam failures or chemical substance spills or releases, could result in devastating fatalities and severe economic impacts. The likelihood of this happening may increase because of aging infrastructure;

4 Terrorist organizations or their affiliates may attempt to acquire, build, and use weapons of mass destruction . Conventional terrorist attacks, including those by "lone actors" employing explosives and armed attacks, present a continued risk;
5 Cyberattacks can have their own catastrophic consequences and can also cause other hazards, including power grid failures or financial system failures, which magnify the potential impact of cyber incidents.[16]

The SNRA Committee identified events that had the potential to pose the greatest risk to the security of the nation. The committee recognized that it was possible that many of the events they listed could potentially occur more than once every ten years, meaning that the nation's preparedness would probably be tested at some point in the next ten years. They also stressed that risks to critical infrastructure are always changing, and the nation must always be prepared for new hazards.

Executive Order 13563

On January 18, 2011, President Obama issued Executive Order 13563, entitled "Improving Regulation and Regulatory Review." In this document, Obama reaffirmed the mandates set forth in Executive Order 12866, known as "Regulatory Planning and Review." In the new executive order, Obama directed all federal agencies to develop a preliminary plan to review their regulations to determine whether any of the existing rules should be updated or altered in any way to make the agency's regulatory program more effective.

One example of this was the DHS Preliminary Plan, which became public on May 26, 2011. A primary focus of their plan was to include members of the public as part of the review process. They also sought to include members of the public in the development of the plan and then its implementation.[17]

PPD-8

President Obama signed Presidential Policy Directive-8, entitled "National Preparedness," on March 30, 2011. This document replaced Homeland Security Presidential Directive-8 (HSPD-8), which was signed by President George W. Bush in 2003. In the new document, Obama developed a way to further strengthen the nation's security and resilience by making the nation better prepared for events. He concentrated on an all-hazards approach to national security that included planning for possible terrorist acts, including cyberattacks, pandemics, technological events, and also natural disasters. Additionally, the president recognized that national preparedness and security must involve all people who have a personal stake in critical infrastructure or security, including those in government, the private sector, and individual citizens. The document requires that everyone be involved in the process of protecting assets instead of just government officials, which was primarily what was done in the past.

Obama asked the federal government to design a national preparedness plan that would both initiate new capabilities and improve existing capabilities that they deemed necessary to maintain the nation's preparedness for disasters. Five mission areas were identified in PPD-8. They are Prevent, Protect, Mitigate, Respond, and Recovery, as described below.

a Prevent: This was recognized as the most important of the mission areas. While people cannot prevent natural weather-related events, it is possible to prevent man-made events. This includes taking any actions necessary to avoid, prevent, or stop a threatened or actual act of terrorism or preventing imminent threats of any kind. Prevention-related activities may include increased inspections; more surveillance and security operations; efforts geared toward increased public health (i.e. immunizations); surveillance and testing of agricultural products; and law enforcement operations aimed at deterring or disrupting illegal activity.

b Protect: This involves taking actions necessary to secure the homeland against acts of terrorism and man-made or natural disasters. Key words in this area are "defense," "protection," "protect," "security," and any kind to include "cybersecurity." This refers to efforts to protect all citizens, residents, visitors, as well as physical assets against threats and hazards in a way that allows people to continue their way of life.

c Mitigate: This refers to actions geared toward reducing the loss of life and property that could occur after an event by lessening the impact of disasters. Key words are "risk reduction," "improve resilience," and "reduce future risk."

d Respond: In this category, the focus is on ensuring that people have the services needed after an event to save lives, protect property and the environment, and meet basic human needs. This includes responding quickly and ensuring that people have the services they need to survive. To do that, it is essential that policies are created that coordinate federal, state, and local activities.

e Recovery: This stage focuses on providing services needed to assist affected communities to return to a "normal" state as quickly as possible. Key words are "rebuilding," "restoring," "promoting," "interim," and "long term."[18] Efforts here focus on the timely restoration of services, strengthening and rebuilding of infrastructure, housing, and health facilities, as well as social, cultural, and historic elements of a community.[19]

In addition, there are six elements noted in PPD-8: National Preparedness Goal; National Preparedness System; National Preparedness Report; National Planning Frameworks; Federal Inter-Agency Operational Plans; and Build and Sustain Preparedness.

National Preparedness Goal: PPD-8 required the Secretary of Homeland Security to create a new National Preparedness Goal.[20] According to Obama, "The National Preparedness Goal shall be informed by the risk of specific threats and vulnerabilities—taking into account regional variations—and include concrete, measurable, and prioritized objects to mitigate that risk."[21]

The goal identifies and defines core capabilities that, according to the president, the country needs to achieve preparedness and, in the end, better national security. When met, the core capabilities will help the country to be prepared for all types of incidents that could pose a risk to the nation's security. These core capabilities are essential for officials to implement the five mission areas described earlier. The goal defines success as "a secure and resilient nation with the capabilities required across the whole community to prevent, protect against, mitigate, respond to, and recover from the threats and hazards that pose the greatest risk."[22]

A fundamental concept throughout the document is the emphasis on the whole community approach. This means that all interested groups and organizations need to work together in a variety of ways and make the best use of resources to be fully prepared for an event.[23] In December 2011, FEMA released a report entitled "A Whole Community Approach to Emergency Management: Principles, Themes and Pathways for Action." In it, officials describe the idea of a whole community approach. It means that all members of a community, including emergency management practitioners, community leaders, organizations, government officials, private business owners, and operators, and citizens can each help to assess the needs of their own community and decide the best method to organize and strengthen their assets, capacities, and interests. They will decide the best ways to prepare for potential threats and hazards.

The National Preparedness System refers to a document published in November 2011 that outlines an approach, resources, and tools needed to assist communities and the nation in meeting the National Preparedness Goal. It includes national planning frameworks that cover the five areas of prevention, protection, mitigation, response, and recovery, found above. The frameworks each use a common terminology and approach and are each built around the all-hazards approach to preparedness. In addition, the system is built on an "All-of-Nation" approach to preparedness. The system has six parts: (1) Identifying and Assessing Risk; (2) Estimating Capability Requirements; (3) Building and Sustaining Capabilities; (4) Planning to Deliver Capabilities; (5) Validating Capabilities; (6) Reviewing and Updating.[24]

National Preparedness Report: Under the PPD-8, the Secretary of Homeland Security must submit a National Preparedness Report to the president each year. This report is to include a summary of the progress that has been made toward achieving the National Preparedness Goal. The secretary is also required to identify any gaps in activities. The report could be used by the president when establishing the annual budget so that funds could be allocated to support existing activities or create new activities to fill the gaps.[25]

National Planning Frameworks: There are five frameworks that focus on the mission areas of PPD-8 (Prevention, Protection, Mitigation, Response, and Recovery). The frameworks demonstrate how different groups and agencies will cooperate to meet the needs of individuals, families, communities, and states in their efforts to prevent, protect, mitigate, respond to, and recover from any disaster or event.[26]

The frameworks were created in a way that they are scalable and could be adapted to each individual community or event. The frameworks establish only the overall theme or strategy (coordinating structure) for communities, which must then build and deliver the core capabilities identified in the National Preparedness Goal. It was stressed that there is a need for a common terminology that will be used across all of the frameworks as a way to ensure interoperability across all mission areas. The frameworks address the roles of individuals, nonprofit entities, government agencies, NGOs, the private sector, and communities in planning for response to events. Most importantly, the frameworks contain detailed information on the 31 core capabilities, which help to define desired outcomes, set capability targets, and specify appropriate resources.

Federal Inter-Agency Operational Plans: These plans describe the federal government's strategies to deliver the core capabilities outlined in the five frameworks described above. These plans help to define how federal policies and officials can provide support to state and local officials as they establish plans for responding to an event. The federal plans will also describe essential tasks and responsibilities and specific provisions for integrating resources and personnel with other governments.[27]

Build and Sustain Preparedness: This element stresses that the effort to build and maintain the country's preparedness is ongoing and will constantly build on existing activities.[28]

This element has four key sections:

a A comprehensive campaign, including public outreach and community-based and private sector programs;
b Federal preparedness efforts;
c Grants, technical assistance, and other federal preparedness support; and
d Research and development.

Executive Order 13636

Executive Order 13636, "Improving Critical Infrastructure: Cybersecurity," was issued on February 12, 2013. In it, Obama stressed that the nation's security depends on a reliable and functioning critical infrastructure and a secure cyber environment. He also stressed that the best way to achieve a safe environment is with better communication and cooperation with the owners and operators of critical infrastructure. Clearly, increased communication could lead to more efforts to collaborate on, develop, and implement risk-based approaches to cybersecurity. For these reasons, Obama asked the federal government to coordinate their activities with the owners and operators of critical infrastructure and improve communication between the groups. As another way to improve information sharing, Obama asked the Attorney General, the Secretary of Homeland Security, and the Director of National Intelligence to develop unclassified reports on any cyber threats that are reported and to share those reports with the targeted group. At the same time, Obama stressed the need to incorporate protections to protect individual privacy and civil liberties in each initiative.

In another part of the executive order, Obama sought to develop a technology-neutral Cybersecurity Framework to improve the nation's cybersecurity. This framework would establish standards, methodologies, procedures, and processes that the owners of critical infrastructure could voluntarily use to reduce their cybersecurity risks. The new plan, when done, would be a cost-efficient approach to helping owners and operators identify, assess, and manage cyber risks. The plan would help owners and operators identify potential vulnerabilities in a timely manner, and then provide innovative suggestions for addressing those risks. The process of creating the framework would be overseen by the Director of the National Institute of Standards and Technology (NIST), but other interested people would be allowed to participate. This would include Sector Coordinating Councils, owners and operators of critical assets, SSAs, regulatory agencies, universities, and other relevant groups. The framework would be the basis for the Voluntary Critical Infrastructure Cybersecurity Program, which was a way to encourage businesses to adopt strong cybersecurity practices to defend against cyberattacks.[29]

The framework was released in February 2014. Upon its release, all owners or operators of critical infrastructure were encouraged to use the framework to improve the security of their networks. Any agencies that had the responsibility of regulating the security of critical infrastructure were asked to review their policies to determine if they were sufficient and, if not, they were asked to consider adopting the recommended ones, or at least modifying what they had to align more with the standards found in the framework. The Secretary of Homeland Security was also asked to create incentives for participating in the voluntary program.

The Enhanced Cybersecurity Services program was expanded through Obama's executive order. This program allows classified information on cybersecurity threats and other technical information to be shared with infrastructure network service providers. Obama asked government agencies to expand the program to all critical infrastructure sectors so that more information about threats and other technical information would be shared with a bigger audience more quickly. For this to work, Obama asked to change the way security clearances to those employed by infrastructure owners and operators were granted, making the process quicker.

At the same time, the president wanted to ensure that the privacy and civil liberties of all individuals were protected. He asked that the Chief Privacy Officer and the Officer for Civil Rights and Civil Liberties of the DHS oversee the programs and recommend ways to ensure that citizens' rights were protected.

Executive Order 13691

In February 2015, Obama issued Executive Order 13691, called "Promoting Private Sector Cybersecurity Information Sharing." In this document he addressed the importance of sharing information pertaining to cybersecurity. Obama gave the Secretary of the DHS the responsibility of establishing Information Sharing and Analysis Organizations (ISAOs), which are very much like the ISACs in PDD-63. There would also be an ISAO that would work with all of the critical infrastructure stakeholders to develop voluntary standards and guidelines for establishing and operating

ISAOs. In the executive order, Obama also designated the National Cybersecurity and Communications Integration Center (NCCIC) as a critical infrastructure protection program, allowing it to receive and transmit cybersecurity information between the federal government and the ISAOs as protected critical infrastructure information.

PPD-21

In February 2013, President Obama announced Presidential Policy Directive 21 (PPD-21), which had the title "Critical Infrastructure Security and Resilience." This new plan superseded HSPD-7 from the Bush administration. PPD-21 reflected the increased interest in resilience and the all-hazard approach that has evolved in critical infrastructure protection policy over the past few years. The purpose of the directive was to establish a national policy to strengthen and maintain a secure critical infrastructure that is also resilient to attacks. This is important for the continuity of national essential functions. Both physical and cyber infrastructure were included.

In PPD-21, the need for increased and effective cooperation between public and private organizations was stressed. Companies were asked to cooperate not only with the government but sometimes with competing industries in efforts to increase the nation's security. As noted by Obama, critical infrastructure protection must be a shared responsibility between federal, state, local, tribal, and territorial entities, along with public and private owners and operators of the assets. He also noted the importance of working with international partners to strengthen infrastructure that was physically located outside the U.S.

The number of sectors and how they are organized were changed in PPD-21. In the 2006 NIPP, there were 17 critical infrastructure sectors established, as outlined in Homeland Security Presidential Directive/HSPD-7. However, since PPD-21 revoked HSPD-7, the 18 sectors were reorganized into 16 critical infrastructure sectors. Not only were the sectors reorganized, but the importance of how the sectors were linked to each other was stressed, along with the "cascading consequences" of infrastructure failures.

PPD-21 identifies the energy and communications sectors as uniquely critical and deserving of extra protection.[30] National Monuments and Icons was designated as a subsector of Government Facilities; Postal and Shipping was designated as a subsector of Transportation; Banking and Finance was renamed Financial Services; and Drinking Water and Water Treatment was renamed Water and Wastewater Systems. In March 2008, the DHS announced the creation of an additional sector, Critical Manufacturing. The sector encompasses groups from the primary metal, machinery, electrical equipment, and transportation equipment manufacturing industries. PPD-21 also gave the energy and communications sectors a higher profile, because of the administration's assessment of their importance to the operations of the other infrastructures.

PPD-21 also called for other federal departments and agencies to play a key role in critical infrastructure security and resilience activities through their appointment as SSA. An SSA is a federal department or agency that is responsible for, among other things, security and resilience programs and related activities of designated sectors. Each sector was assigned an SSA. For example, the DHS is the SSA for the Commercial Facilities and Dams sectors, and the Department of Energy and the Environmental Protection Agency are the SSAs for the energy and water sectors, respectively. The DHS also shares SSA responsibilities with the Department of Transportation for the transportation sector and the General Services Administration for the government facilities sector.[31]

Obama asked for an evaluation of the existing public–private partnership model to determine if it could be improved. To do this, he sought to collect baseline data and existing system requirements that would be the starting point for a more efficient exchange of information. After this was established, a new plan would be developed, called the Research and Development Plan for Critical Infrastructure, which would be a working document that would be updated every four years.

Throughout PPD-21, the president outlined the roles and responsibilities of different groups as the following:

1 The Secretary of Homeland Security "shall provide strategic guidance, promote a national unity of effort, and coordinate the overall Federal effort to promote the security and resilience of the nation's critical infrastructure." This person should evaluate national capabilities and challenges to protecting assets, analyze threats and vulnerabilities, and develop a national plan. The DHS was also asked to identify and prioritize infrastructure, maintain centers that provide situational awareness (i.e. emerging trends, imminent threats, or status of incidents), provide information and analysis on information, assess vulnerabilities, and coordinate federal government responses to significant incidents.
2 SSAs were asked to provide sector-specific information and expertise about their specific sector and then coordinate their activities and plans with the DHS and other agencies. They can also provide technical assistance if needed either to mitigate incidents or respond to incidents.
3 The Department of State was asked to work with representatives from other countries to strengthen the security and resilience of any critical infrastructure located outside of the U.S.
4 The Department of Justice (the FBI) was identified as the organization that would lead counterterrorism and counterintelligence investigations to disrupt and reduce foreign intelligence and actual or attempted attacks on the nation's infrastructure.
5 The Department of Interior should identify and coordinate security and resilience efforts for all monuments and icons.
6 The Department of Commerce was given the responsibility to engage the private sector, research, and academic organizations as a way to improve the security of cyber-based systems and help develop new ways to protect critical infrastructure.
7 The Intelligence Community should provide intelligence assessments regarding possible threats.
8 General Services Administration was given the task of providing contracts for critical infrastructure systems that include audit rights for the security and resilience of assets.
9 The Nuclear Regulatory Commission should oversee the protection of commercial nuclear power reactors, as well as the transportation of nuclear waste.
10 The Federal Communications Commission was asked to identify vulnerabilities in the Communications sector and work to address those.

In addition to all of this, there were three Strategic Imperatives outlined by Obama in the directive. These were:[32]

1 Refine and clarify functional relationships across the federal government as a way to advance critical infrastructure security and resilience. If needed, relationships among stakeholders should be defined or even redefined. The functions of federal agencies need to be clarified to reflect an increase in knowledge and changes in threats. To do this, Obama asked for two national centers operated by the DHS that would work to enhance critical infrastructure protection. One would focus on physical infrastructure and the other on cyber protection.
2 Enable efficient exchange of information between all levels of government and with all owners and operators of critical infrastructure. There is a need for more information sharing within the government and with the private sector.
3 Implement an analysis of incidents or threats to inform planning and operational decisions regarding critical infrastructure protection. This should include operational and strategic analysis.

2006 NIPP

In PPD-21, President Obama required that the NIPP be updated and revised. The NIPP had originally been published in 2006, but the administration believed it was time to update that plan. The updated plan was to include a focus on how the sectors rely on the energy and communications infrastructure and ways to mitigate the associated risks. Clearly there have been significant changes in the risk, policy, and operating environments surrounding the country's critical infrastructure since the NIPP was first published.

The 2006 version of the NIPP outlined an integrated national plan for managing risk for the country's critical infrastructure. The process included identifying assets and threats and then conducting threat assessments in which vulnerabilities were analyzed considering consequences and risk mitigation activities. Those activities would be prioritized based on cost-effectiveness. The 2006 NIPP also called for implementation plans for these risk reduction activities.

Each lead agency was asked to work in collaboration with other agencies in its sector to write an SSP. When the plans were completed, the DHS was to integrate the individual SSPs into a national plan. This could then be used to identify the assets that, if damaged, could pose a significant risk to the entire nation. Any risk reduction plans that required federal assistance would also be identified. The sector officials were asked to review the plans every three years and reissue revised plans if needed. This would help ensure that the plans would remain current and relevant to all security partners.

Only seven plans were made public, and the others were given the designation "For Official Use Only." The Government Accountability Office (GAO) reviewed nine of the plans and found that all complied with the NIPP process. However, some of the plans were more complete than others and provided more analysis than others. There were significant differences in the amount of detail provided and the general thoroughness of the reports. Moreover, while all of the plans provided details about the threat analyses conducted by the sector, eight of the plans described no incentives that the sector could use to encourage owners and operators to carry out voluntary risk assessments, as required by the NIPP. These incentives were needed since many of the companies in the sectors were privately owned; they were not regulated by the government. Instead, the government was forced to rely on voluntary compliance with the NIPP.

The GAO finished its report by making two key recommendations to the DHS. First, they recommended that the DHS provide better definitions of critical infrastructure information needs, and, second, that there be a better explanation of how this information could be used to attract more users.

2013 NIPP

After a brief revision in 2009, the NIPP was again revised in 2013 after Obama, in Presidential Policy Directive 21 (PPD-21), called for officials to update the document. He requested the update based on a belief that there had been significant changes in the critical infrastructure risk, policy, and operating environments, as well as our general knowledge about critical infrastructure protection. In essence, government officials and others were to complete a "gap analysis" to fix any gaps that may exist in asset protection. The 2013 National Plan builds upon previous NIPPs and emphasizes goals of critical infrastructure security and resilience. The ultimate goals were to (1) identify, deter, detect, disrupt, and prepare for threats and hazards to the nation's critical infrastructure; (2) reduce vulnerabilities of critical assets, systems, and networks; and (3) mitigate the potential consequences of incidents or adverse events that do occur to infrastructure.[33]

The revised NIPP was developed through a collaborative process that included stakeholders from the 16 critical infrastructure sectors, all 50 states, and from all levels of government and industry. The committee members worked to identify priorities and articulate goals that would help to mitigate risk to infrastructure and help be resilient in the case of an attack. As published, the following are the vision, mission, and goals:

Vision: A national in which physical and cyber infrastructure remain secure and resilient, with vulnerabilities reduced, consequences minimized, threats identified and disrupted, and response and recovery hastened.

Mission: Strengthen the security and resilience of the nation's critical infrastructure, by managing physical and cyber risks through the collaborative and integrated efforts of the critical infrastructure community.

Goals:

1. Assess and analyze threats to, vulnerabilities of, and consequences to critical infrastructure to inform risk management activities;
2. Secure critical infrastructure against human, physical, and cyber threats through sustainable efforts to reduce risk, while accounting for the costs and benefits of security investments;
3. Enhance critical infrastructure resilience by minimizing the adverse consequences of incidents through advance planning and mitigation efforts and employing effective responses to save lives and ensure the rapid recovery of essential services;
4. Share actionable and relevant info across the infrastructure community to build awareness and enable risk-informed decision making; and
5. Promote learning and adaptation during and after exercises and incidents.[34]

The 2013 NIPP, entitled "Partnering for Critical Infrastructure and Resilience," is largely the same as the two earlier versions but with more integration of resiliency and the all-hazard approach. However, the basic partnership model and the risk management framework were maintained. The revised NIPP stresses the importance of developing partnerships between national, regional, state, and local governments and owners and operators. It was made clear that coordination with infrastructure stakeholders is necessary to protect the public's safety and ensure national security.

The revised NIP made it clear that managing the risks from threats and hazards requires an integrated approach as a way to identify, deter, detect, disrupt, and prepare for threats and hazards to the nation's critical infrastructure; reduce vulnerabilities of critical assets, systems, and networks; and then mitigate the potential consequences to the critical infrastructure of incidents or adverse events that do occur.[35]

The new report recognized that the country's well-being relies on the security and resiliency of critical infrastructure, so the primary goal of any program must be the efforts to protect critical infrastructure. To do that, the NIPP establishes a procedure to define what assets are considered to be national critical infrastructure and how to protect them. International collaboration is also part of asset protection efforts. The new report also gives attention to cyber security.[36]

Cooperation with all partners was stressed, particularly with private sector owners and operators of critical infrastructure, alongside federal, state, local, tribal, and territorial governments, regional entities, NGOs, and academia.[37] The document stressed that these groups should work together to manage risks and achieve better security. Because everyone is involved in the process, many perspectives will be included, resulting in better information sharing.[38] To increase cooperation, many groups were included in the process. These include Sector Coordinating Councils, Government Coordinating Councils, and cross-sector councils.

Better communication was also needed by federal agencies to help to prevent the "silo effect" whereby an agency carries out a program but doesn't communicate that with other agencies. This can lead to wasted resources, but also inefficiencies and gaps in services.

The updated NIPP relied on a five-step risk management framework that was applicable to the general threat environment as well as to specific threats or incidents. The five steps could be applied to physical (tangible property), cyber (electronic communications and information), and human security (knowledge of people susceptible to attack). The five steps are:[39]

a Set Goals and Objectives: Define specific outcomes, conditions, end points, or performance targets that collectively constitute an effective risk management posture
b Identify Infrastructure and Assets: Build, manage, refine, and improve a comprehensive inventory of the assets, systems, and networks that make up the nation's critical infrastructure.
c Assess and Analyze Risks: Evaluate the risk, taking into consideration the potential direct and indirect consequences of all-hazards threats and known vulnerabilities. These risks can be compared to develop a more complete view of asset, system, and/or network risks and associated mission continuity, where applicable. It is also possible to establish priorities based on risk attached to an asset.
d Implement Protective Programs and Resilience Strategies (implement risk management activities): Select appropriate actions or programs to reduce or manage the risk identified and identify and provide the resources needed to address priorities.
e Measure Effectiveness: Use metrics and other evaluation procedures at the appropriate national, State, local, regional, and sector levels to measure progress and to assess the effectiveness of the Critical Infrastructure Protection programs. In this case, those involved can track their progress and use the data as a baseline for comparison and continuous improvement through program implementation.

Throughout his presidency, Obama assigned one month a year to be Critical Infrastructure Security and Resilience Month. In 2014, while signing a Presidential Proclamation to declare an event, he said, "The Security of our Nation is my top priority … we are working to protect our critical infrastructure from cyber threats, while promoting an open and reliable cyberspace … we are taking steps to reduce our vulnerabilities." He signed similar documents in 2010, 2012, and 2016.

Conclusion

In the early documents and plans for protecting the nation's critical infrastructure, it was clear that the emphasis on preparedness and protection was primarily on the federal government's response to disasters and emergency events such as earthquakes, floods, or hurricanes. There was only limited preparedness for potential acts of terrorism. However, this approach changed in the months after the terrorist attacks of September 2001. Then, the attention of the federal response shifted to preparing for possible terrorist attacks and mitigating potential effects so that service interruption was limited. Plans and documents such as the NIPP sought to make America's critical infrastructure and assets more secure and more resilient.[40] Obama continued to expand policies, organizations, and programs to protect critical infrastructure. He encouraged the DHS and SSAs to create collaborative partnerships to develop more effective plans and expanded the policies, organizations, and programs that govern the protection of the nation's critical infrastructure, thereby expanding the involvement of all interested parties in protecting assets. Obama also expanded protection efforts into cybersecurity policies, an emerging area of concern for critical infrastructure protection.

Key Terms

Executive Orders 13010, 13228, 13231, 13537, 13563, 13636, 13691
PPD-8, 21
Presidential Decision Directive 63
Sectors
Sector Liaison Official

- Sector Coordinator
- Critical Infrastructure Assurance Office
- National Coordinator for Security, Infrastructure Protection and Counterterrorism
- National Infrastructure Assurance Council
- Federal Computer Intrusion Detection Network
- National Infrastructure Protection Center
- Partnership for Critical Infrastructure Security
- National Plan for Information Systems Protection
- USA PATRIOT ACT
- Homeland Security Act of 2001
- Critical Infrastructure Warning Network
- Office of Homeland Security
- President's Critical Infrastructure Protection Board
- National Infrastructure Advisory Council
- National Strategy for Homeland Security
- National Strategy for the Physical Protection of Critical Infrastructures and Key Assets
- The National Strategy to Secure Cyberspace
- Homeland Security Presidential Directive 7
- National Plan for Critical Infrastructure and Key Resources
- National Infrastructure Protection Plan (NIPP) 2006, 2009, 2013
- Critical Infrastructure Protection Partnership Model
- Critical Infrastructure Partnership Advisory Council
- National Security Staff
- Strategic National Risk Assessment
- A Whole Community Approach to Emergency Management
- National Preparedness Report
- National Planning Framework
- Federal Inter-Agency Operational Plans
- Build and Sustain Preparedness
- NIST
- Enhanced Cybersecurity Services

Review Questions

1. What was Executive Order 13010, and why was it important?
2. What was President Clinton's approach to Critical Infrastructure Protection?
3. Explain the significance of PDD-63.
4. Compare and contrast the different approaches that Presidents Clinton, Bush, and Obama took toward protecting the nation's assets.
5. How did the concept of critical infrastructure change over time?
6. Describe the key findings of the SNRA Committee. Why were these important?
7. List the five mission areas of PPD-8.
8. What was the NIPP? How did it change over time?

Notes

1 McDade, Joseph E., and Franz, David (1998). "Bioterrorism as a Public Health Threat." *Emerging Infectious Diseases*, Vol. 4, Issue 3, pp. 493–494.
2 "1984 Ranjneeshee Bioterror Attack." *Bionity.* https://www.bionity.com/en/encyclopedia/1984_Rajneeshee_bioterror_attack.html
3 US Justice Department, "Pan Am Flight 103 Terrorist Suspect in Custody for 1988 Bombing over Lockerbie, Scotland," December 12, 2022. https://www.justice.gov/opa/pr/pan-am-flight-103-terrorist-suspect-custody-1988-bombing-over-lockerbie-scotland
4 Cordesman, Anthony H., and Cordesman, Justin G. (2002). *Cyber-Threats, Information Warfare, and Critical Infrastructure Protection: Defending the U.S. Homeland.* Westport, CT: Praeger.
5 Cordesman, Anthony H., and Cordesman, Justin G. (2002). *Cyber-Threats, Information Warfare, and Critical Infrastructure Protection: Defending the U.S. Homeland.* Westport, CT: Praeger.
6 Moteff, John D. (2015). "Critical Infrastructures: Background, Policy and Implementation." Congressional Research Service, 7-5700, June 10. www.crs.gov
7 Clinton, Bill (1996). "Executive Order 13010-Critical Infrastructure Protection," July 17.
8 Moteff, John D. (2015). "Critical Infrastructures: Background, Policy and Implementation." Congressional Research Service, 7-5700, June 10. www.crs.gov

9 Moteff, John D. (2015). “Critical Infrastructures: Background, Policy and Implementation.” Congressional Research Service, 7-5700, June 10. www.crs.gov
10 Moteff, John D. (2015). “Critical Infrastructures: Background, Policy and Implementation.” Congressional Research Service, 7-5700, June 10. www.crs.gov
11 Cordesman, Anthony H., and Cordesman, Justin G. (2002). *Cyber-Threats, Information Warfare, and Critical Infrastructure Protection: Defending the U.S. Homeland.* Westport, CT: Praeger.
12 107th U.S. Congress, *Uniting and Strengthening America by Providing Appropriate Tools Required to Intercept and Obstruct Terrorism Act of 2001* (Public Law 107-56), Section 1016(e), 2001.
13 Office of Homeland Security, “National Strategy for Homeland Security,” 2002, p. 31.
14 Bush, George W. (2001). “Executive Order 13228 - Establishing the Office of Homeland Security and the Homeland Security Council,” October 8. Online by Peters, Gerhard, and Woolley, John T. *The American Presidency Project.* http://www.presidency.ucsb.edu/ws/?pid=61509
15 Bush, George W. (2001). “Executive Order 13231 - Critical Infrastructure Protection in the Information Age,” October 16. Online by Peters, Gerhard and Woolley, John T. *The American Presidency Project.* http://www.presidency.ucsb.edu/ws/?pid=61512
16 DHS, FEMA, “CIKR Awareness AWR-213, Participant Guide,” September 2010, 1:18-19; DHS, “The Strategic National Risk Assessment in Support of PPD-8: A Comprehensive Risk-Based Approach toward a Secure and Resilient Nation,” December 2011.
17 US Department of Homeland Security, “DHS Implementation of Executive Order 13563,” July 27, 2015. http://www.dhs.gov/dhs-implementation-executive-order-13563
18 DHS, FEMA, “Critical Asset Risk Management, Participant Guide,” September 2014, pp. 1–17.
19 US Department of Homeland Security, “National Response Plan Brochure,” 2005; see also PPD-8 (2011).
20 US DHS FEMA, “Learn about Presidential Policy Directive,” 2012, p. 8; also DHS, FEMA, “Critical Asset Risk Management, Participant Guide,” September 2014, 1:12.
21 Obama, 2011, PPD-8.
22 US DHS, “National Preparedness Goal,” 2011, p. 1.
23 DHS, FEMA, “Critical Asset Risk Management, Participant Guide,” September 2014, 1:13.
24 DHS, FEMA, “Critical Asset Risk Management, Participant Guide,” September 2014, p. 1–13.
25 DHS, FEMA, “Critical Asset Risk Management, Participant Guide,” September 2014, p. 1–13.
26 DHS, FEMA, “Critical Asset Risk Management, Participant Guide,” September 2014, p. 1–13.
27 DHS, FEMA, “Critical Asset Risk Management, Participant Guide,” September 2014, p. 1-13-14.
28 DHS, FEMA, “Critical Asset Risk Management, Participant Guide,” September 2014, p. 1–14.
29 Source: EO 13636 (February 2013).
30 DHS, FEMA, “Critical Asset Risk Management, Participant Guide,” September 2014, p. 1–7.
31 US Department of Homeland Security, and US Department of Justice, Global Justice Information Sharing Initiative, “Critical Infrastructure and Key Resources, Protection Capabilities for Fusion Centers,” December 2008. https://it.ojp.gov/documents/d/CIKR%20protection%20capabilities%20for%20fusion%20centers%20s.pdf
32 The White House: Office of the Press Secretary, “Presidential Policy Directive—Critical Infrastructure Security and Resilience,” 2013.
33 DHS, FEMA, “Critical Asset Risk Management, Participant Guide,” September 2014, p. 1–9.
34 US DHS, “National Infrastructure Protection Plan,” 2013, p. 5; DHS, FEMA, “Critical Asset Risk Management, Participant Guide,” September 2014, p. 1–11.
35 US DHS, “National Infrastructure Protection Plan,” 2013.
36 DHS, “NIPP,” 2013.
37 DHS, FEMA, “Critical Asset Risk Management, Participant Guide,” September 2014, p. 1–8.
38 US DHS, “National Infrastructure Protection Plan,” 2013, p. 1.
39 DHS, FEMA, “CIKR Awareness AWR-213, Participant Guide,” September 2010, p. 3–8.
40 DHS, FEMA, “CIKR Awareness AWR-213, Participant Guide,” September 2010, p. 1–8.

3 Critical Infrastructure Protection Today

Chapter Outline

Introduction

Chapter 2 provided an early history of federal attempts to protect assets, and this chapter provides a description of some of the more current policies for protecting the nation's infrastructure that have been supported by recent presidents. These policies are in a continual state of flux as the government continues to improve critical infrastructure protection in response to ever-changing threats. This happened to President Biden who responded to the devastation caused by Hurricane Ian, a category 5 hurricane that hit Florida in late September 2022. Like other severe storms, Ian left a path of destruction that was estimated to cause $112 billion in damages and left 150 people dead.[1] President Biden asked the Federal Emergency Management Agency (FEMA) to provide flood relief funds to those affected,[2] but he also asked the U.S. Small Business Administration to provide funds to anyone who owned small businesses that were impacted by the hurricane.[3]

Protecting the nation's infrastructure and the safety of the citizens has become a top priority of recent presidents Trump and Biden. Their policy approaches to critical infrastructure protection are diverse and continue to adapt as new threats emerge. In this chapter, these different approaches are described.

Career Focus: Pete Buttigieg

Peter "Mayor Pete" Buttigieg is a Democratic politician, former mayor of South Bend, Indiana, and current Secretary of Transportation for President Joe Biden. Buttigieg was born in South Bend, Indiana, but left to earn a BA from Harvard University in history and literature and to study at Oxford University where he was a Rhodes scholar. There he earned a degree

DOI: 10.4324/9781003434887-3

in Philosophy, Politics, and Economics. After college, Buttigieg became a consultant for McKinsey & Company, a company that handled issues pertaining to economic development and energy development. Buttigieg's political career began when he worked for Democrat John Kerry, a presidential candidate in 2004, and for Jill Long Thompson, a candidate for governor in 2008. Buttigieg himself ran a campaign to be the state treasurer for Indiana in 2010 but lost to the Republican candidate. At the age of 24, he chose to run for mayor of South Bend, winning with 74% of the vote. This made him the youngest mayor of the city of that size. In 2014, Buttigieg took a break from being mayor to serve in Afghanistan as a member of the U.S. Navy Reserve, which he joined in 2009. Buttigieg was awarded a Joint Service Commendation Medal for his service. After seven months overseas, Buttigieg returned to Indiana in 2015 to run again for mayor of Indiana. This time he received 80% of the vote. It was during this campaign that he came out publicly as gay.

Buttigieg served as the president of the Indiana Urban Mayors Caucus and was on the Board of Directors of the Indiana Association of Cities and Towns. He was a candidate to become the chair of the Democratic National Committee in 2017 but declined the position before a decision was made.

Mayor Pete announced he was running for the presidency of the U.S. in 2020. As a candidate, he promised to fund the development of renewable energy, to pass new gun-control legislation, and to make healthcare more widely available. Buttigieg withdrew from the presidential race to support Joe Biden's campaign. When Biden became president, he chose Buttigieg to be his Secretary of Transportation. Buttigieg was sworn in as the 19th Secretary of Transportation in 2021, being the first openly gay person to serve in the presidential cabinet and the youngest to be the Secretary of Transportation. In this position, Buttigieg oversees the transportation system throughout the nation (roadways, rails, air travel). As Secretary, Buttigieg worked to protect the supply chains during the pandemic to ensure products were available. He supports legislation passed by Congress to hold airlines responsible for flight delays and cancellations. In 2023, Buttigieg was criticized for not visiting East Palestine, Ohio, the site of a train derailment and chemical spill for three weeks after the accident.

Buttigieg has been married since 2018 to his partner Chasten, and they have two children.

Trump

During his campaign for the presidency, Republican Donald Trump promised to spend more funds on protecting the country's infrastructure and protecting Americans. Once he was elected to the office, there were multiple threats made by adversaries to harm the country's critical infrastructure, particularly to the computer networks in different federal agencies. A malware attack on the power grid in Ukraine made it clear to many officials in the U.S. that this or similar malware could be used against the U.S. Trump found himself under public pressure to take action to protect the nation's critical infrastructure. Throughout his time in office, President Trump signed many executive orders to protect critical infrastructure and citizens of the U.S.

2017

Not long after taking the oath of office, President Trump signed Executive Order (EO) 13800 on May 11, 2017, called "Strengthening the Cybersecurity of Federal Networks and Critical Infrastructure." In this document, Trump stressed the need to reduce the nation's cybersecurity risks and a possible attack on computer networks. He underscored the idea that a strong, safe, and resilient cyber infrastructure can foster economic growth and advance national security. The president began by emphasizing the need to strengthen the cybersecurity of federal computer systems and related critical infrastructure.[4] When signing this EO, Trump recognized that it is "the policy of the United States to manage cybersecurity risk as an executive branch enterprise." He indicated that the "executive branch has for too long accepted antiquated and difficult-to-defend IT" and that "(k)nown vulnerabilities include using operating systems or hardware beyond the vendor's support lifecycle, declining to implement a vendor's security patch, or failing to execute security-specific configuration guidance." He also stressed the importance of protecting individuals and businesses from cyberattacks.[5]

Trump's EO outlined a plan to modernize information technology infrastructure. To do this, the head of each federal agency was asked to address their existing practices for managing cyber risks in their agencies and standardize them with the National Institute of Standards and Technology (NIST) Framework for Improving Critical Infrastructure Cybersecurity. They were asked to submit information about their agency's cyber risk management practices to the Secretary of Homeland Security and the Director of the Office of Management and Budget within 90 days. The resulting report was intended to outline the risk mitigation that had been made by each agency head and identify any gaps. Trump named the Department of Homeland Security (DHS) to be the coordinator for the nation's cybersecurity measures.

Trump's second step to modernize technology was to order an assessment made by the Secretary and Director of the DHS. These officials were asked to jointly assess each agency's risk management report to determine whether their plans to mitigate any possible risk were appropriate and sufficient to prevent a cybersecurity risk. Upon completing this task, these two officials were asked to submit a comprehensive plan to the president that would describe a blueprint for protecting the executive branch from possible harm. The report was also to include ideas for an ongoing, continuing process for reassessing federal cybersecurity efforts.

Third, Trump also wanted to assess the potential scope and duration of a power outage that could result from a cyberattack. He asked the Secretaries of Energy and Homeland Security to cooperate with state, local, tribal, and territorial governments to assess the readiness of the U.S. to manage such an incident.

A fourth critical part of Trump's EO was to increase cooperation with foreign allies to strengthen cybersecurity. Trump noted that the U.S. is "highly connected" and therefore dependent on a globally secure and resilient Internet. For this reason, Trump explained that the U.S. must work with foreign allies and government officials to maintain an open and reliable Internet that is secure for all users. Trump ordered that the Secretaries of State, Treasury, Defense, Commerce, and Homeland Security, in coordination with the Attorney General and the Director of the Federal Bureau of Investigation, to submit reports on their priorities for an international cybersecurity plan, including items regarding the investigation of cybercrime, sharing information on cyber threats and response, or capacity building. He asked that officials work with counterparts in other countries in diplomatic, military, financial, intelligence, information sharing, and law enforcement capacities.

The fifth key part of Trump's EO was to establish a more robust workforce across the nation that is skilled in cybersecurity. Trump ordered the Secretaries of Commerce and Homeland Security to "assess the scope and sufficiency efforts to educate and train the American cybersecurity workforce of the future, including cybersecurity-related education curricula, training, and apprenticeship programs, from primary through higher education." The officials were tasked with making recommendations to meet this goal. The final report from the Secretary of Commerce and the DHS was titled "Supporting the Growth and Sustainment of the Nation's Cybersecurity Workforce."[6]

President Trump signed an additional EO on May 11, 2017. EO 13799, called Establishment of the Presidential Advisory Commission on Election Integrity, created an agency that was tasked with ensuring that the American electoral system was safe and secure. In doing so, the commission members were asked to review the possibility of fraudulent voter registration and voting fraud. This EO was later rescinded by President Trump on January 3, 2018, in EO 13820, titled EO on the Termination of Presidential Advisory Commission on Election Integrity. In this document, Trump revoked the earlier EO and terminated the commission. The commission had found no evidence of voter fraud in the presidential campaign.

2018

On May 15, 2018, Trump announced EO 13833 that was designed to advance America's cybersecurity workforce. The document was entitled "Enhancing the Effectiveness of Agency Chief Information Officers." In it, Trump sought to increase the development of cybersecurity staff in the executive branch as a way to establish a more robust federal cybersecurity workforce. He also established the President's Cup Cybersecurity Competition as a way to help train those individuals who sought to pursue a career in this field. The challenge would help to identify, challenge, and reward the federal government's best cybersecurity personnel (see Chapter 4).

About two weeks later, on May 30, 2018, the Office of Management and Budget (OMB) and the DHS published the Federal Cybersecurity Risk Determination Report and Action Plan to the President of the U.S. (also known as the Risk Report) that was mandated by EO 13800 (above). The two agencies analyzed the ability of 96 civilian agencies to identify, detect, respond to, and recover from cyber incidents. The results demonstrated that 74% of civilian agencies had cybersecurity programs that were either "At Risk" or "High Risk" of an attack. Additionally, the analysis found that most agencies who were victims of an attack were not capable of discovering how the cyberattackers attempted to gain access to their online data, that they did not collect adequate information about threats, nor did they fully understand the nature of the potential for threats against them. In all, the report concluded that there were major gaps in the cybersecurity policies of most agencies.

This report outlined recommendations to address existing gaps in cybersecurity, or an action plan intended to address the identified security weaknesses. The report identified four core actions that could be taken to address cybersecurity concerns. They included:

1 Increase cybersecurity threat awareness among federal agencies by implementing the Cyber Threat Framework to prioritize efforts to combat cybercrime and manage potential risks to the agency's cybersecurity;
2 Standardize IT and cybersecurity capabilities across agencies as a way to manage costs and improve asset protection;
3 Consolidate agency Security Operation Centers to improve the ability to detect incidents and the ability to respond; and
4 Stress accountability for cybersecurity among the agencies through improved governance, annual risk assessments, and increased communication with leadership in the agencies.

To support these ideas, President Trump allocated $1.5 billion for cybersecurity within the DHS.[7]

In November 2018, Trump signed the Cybersecurity and Infrastructure Security Agency (CISA) Act of 2018 that was passed by Congress. One of the changes made in this bill was to change the role of the DHS National Protection and Programs Directorate. Many of the responsibilities of the National Protection and Programs Directorate were transferred to a new agency called the CISA. This agency became the primary federal organization to confront the cyber threats facing the U.S. (see Chapter 4).

Another Congressional act that President Trump signed was the Geospatial Data Act, which pertained to the role that geospatial data and technology play in American society. Geospatial data pertains to information that is collected on objects, events, locations, or patterns of behavior. It can be used by local government officials to understand zoning and traffic patterns. Weather forecasters also rely on it to predict storms, and retailers use this data to have a better understanding of customer purchasing patterns. It is also used by the military for intelligence purposes. In sum, geospatial data is used daily for business, health, transportation, and safety of citizens.

The bill gave support to an existing federal agency, the Federal Geographic Data Committee (FGDC), which helps to develop and implement policies related to geospatial data. It is chaired by the Secretary of the Department of the Interior. The National Geospatial Advisory Committee is also located within the Department of the Interior. This agency ensures that non-federal organizations are included when policies are being created, and their perspectives are relayed to the FGDC.

The bill also advanced the sharing of geospatial data between the federal, state, local, and tribal governments and the private sector. This data, and any technology or policies that are needed to support or promote the sharing of information, is referred to as the National Spatial Data Infrastructure. The Congress and the president recognized the need to support the development of new technology and the application of geospatial capabilities as critical infrastructure, which needs to be protected.

2019

On May 15, 2019, Trump signed EO 13873, entitled "Securing the Information and Communications Technology and Services Supply Chain." Many believed that the document could potentially prevent Chinese telecommunications companies from selling equipment in the U.S. that might compromise computer systems in the country. When he signed the order, Trump said, "Foreign adversaries are increasingly creating and exploiting vulnerabilities in information and communications technology and services." He continued by saying that these adversaries were using this technology "in order to commit malicious cyber-enabled actions, including economic and industrial espionage against the United States and its people."[8]

Specifically, the EO that Trump signed made it a crime for a company to purchase or use any technology made by entities that are controlled by a "foreign adversary" and could be used to create an "undue risk of sabotage" of communications systems located in the U.S. In short, U.S.-based companies would be banned from using equipment from manufacturers who are identified as posing a risk to national security. Trump asked the Commerce Department to compile a list of countries that would be considered to be foreign adversaries. One Chinese company in particular, Huawei, was singled out by the Commerce Department when it added the company to the list of entities controlled by foreign adversaries. It was alleged that Huawei, a company that is financially supported by China, is used by the Chinese government for espionage and theft of property and therefore a threat to national security.[9]

2020

Trump began 2020 by signing EO 13905, "Strengthening National Resilience through Responsible Use of Positioning, Navigation and Timing Service," in February. This was intended to protect any critical infrastructure that relies on GPS. In the EO, Trump asked all federal agencies to reduce the possible disruption of critical infrastructure assets that rely on GPS services such as transportation, electricity, or communications. Trump also ordered the DHS to establish a plan to assess how vulnerable existing critical infrastructure systems were to disruption and then create plans on how these systems could be made more secure. The Office of Science and Technology was also asked to create a plan to create a GPS system that would not depend on global navigation satellite systems and would be more difficult to hack.[10]

Trump followed that action with another EO 13920, "Securing the United States Bulk-Power System," signed May 1, 2020. In this, President Trump prohibited any person from acquiring, importing, transferring, or installing any bulk-power electric system equipment that was designed or manufactured by a person with whom a foreign country has an interest or who may be under the direction of a foreign adversary. Trump directed the Secretary of Energy to work with the Departments of Commerce, Defense, Homeland Security, and Interior, to ensure that the acquisition of bulk-power systems is in line with national security demands.

While discussing the EO, Trump made it clear that foreign adversaries were increasingly attacking the bulk-power system in the U.S. through malicious acts on cyber systems. The power systems support national defense, citizen healthcare, and the economy and the way of life in the U.S. Therefore, a successful attack on the power system may present a significant risk to the economy, health, and safety of the U.S. The use of bulk-power equipment that was designed and manufactured or supplied by people who are in or under the direction of foreign adversaries has the potential for harm as they can use this to attack our systems and therefore may pose a risk to the nation's security.

A third EO passed in 2020 was 13913, "Establishing the Committee for the Assessment of Foreign Participation in the United States Telecommunications Services Sector," signed by President Trump on April 4. This document created the Committee for the Assessment of Foreign Participation in the U.S. telecommunications services sector. The task of the committee was to assist the Federal Communications Commission in reviewing the national security concerns that are raised by foreign governments who participated in the U.S. telecommunications services sector. The committee members were tasked with reviewing applications and licenses to discover potential risks to national security posed by the applicants and then respond to any risks that are found. In general, the committee would help to address concerns that may be raised by foreign participation in the U.S. telecommunications services sector. The committee members include the secretaries of Defense, Homeland Security, the Attorney General, and other agency heads as the president deemed appropriate.

During the COVID pandemic that started in early 2020, Trump issued an EO on April 28, 2020 (EO 13917) that established meat plants as "critical infrastructure" so that they would remain open. The title of this document was rather lengthy: "Delegating Authority Under the Defense Production Act with Respect to Food Supply Chain Resources During the National Emergency Caused by the Outbreak of COVID-19." When he signed this bill, he relied on the Defense Production Act of 1950, which was created during the Korean War to force companies to manufacture products that support national defense.

In June 2020, Trump announced that he opposed an undersea Internet cable involving Google and Facebook from Hong Kong to the U.S. over national security concerns. Trump claimed that a proposed underwater cable network would permit Chinese intelligence to have direct access to U.S. agencies and the personal information of users.[11]

In August 2020, Trump signed EO 13942, "Addressing the Threat posed by TikTok, and Taking Additional Steps to Address the National Emergency with Respect to the Information and Communications Technology and Services Supply Chain," in response to concerns that the application called TikTok, which is owned by the Chinese government, was being used by the Chinese government to collect data on American users. Trump also described TikTok as a threat to national security, foreign policy, and the economy of the U.S. Trump prohibited any transaction by any person in the U.S. to the company that owns TikTok or has any interest in the company, to the extent that the law allowed it.

Biden

Upon becoming president, Democrat Joseph R. Biden faced many incidents of cyberattacks, including a ransomware attack on Colonial Pipeline and JBS Foods. He also faced cyberattacks on

Solar Winds and Microsoft Exchange. These events all increased the public's attention to the potential for serious cyberattacks on federal agencies. The Biden administration made cybersecurity and cyber resiliency a top concern of the DHS, focusing particularly on the threat of ransomware and expanding the qualified cybersecurity workforce. Biden said, "It is the policy of my Administration that the prevention, detection, assessment, and remediation of cyber incidents is a top priority and essential to national and economic activity."[12]

Biden began his administration's battle with cybercrime on May 12, 2021, by signing EO 14028, "Improving the Nation's Cybersecurity." Biden noted that much of the country's cybersecurity defenses are outdated and weak, leaving agencies vulnerable to cybercrimes. He sought to strengthen the country's cybersecurity while protecting the federal agency networks and critical infrastructure. There were multiple points made by President Biden in this document.

First, Biden stated that the country must remove barriers to sharing information about threats between government agencies and the private owners and operators of critical infrastructure. The EO sought to guarantee that service providers share information with government agencies, including information pertaining to network breaches.

Second, Biden stressed the need to encourage federal government agencies to apply stronger cybersecurity standards to their computer systems. This included secure cloud services, multifactor authentication processes, and the use of encryption when transferring data.

Third, Biden sought to update the security of supply chains. For this reason, the EO mandated baseline security standards be developed for any software that was purchased by the federal government such as increased transparency in the software programs. All software programs that were deemed to be developed securely would have an "energy star" label attached to them.

Fourth, Biden acted to establish a Cyber Safety Review board that will be led by officials from both the government officials and the private sector. The board was asked to meet after a cyber event occurs to investigate the events surrounding the attack and then make recommendations to prevent similar events in the future. The board would also improve technology to improve the detection of cybersecurity and malicious activity on federal networks.

Fifth, Biden described the need to create a "Standardized Playbook" for incident response so that all federal departments and agencies take uniform steps to identify threats and have a coordinated response.

Finally, Biden promised to improve investigative techniques and remediation and recovery in the case of an attack. To help with this task, federal departments and agencies were required to keep an event log as a way to identify patterns of intrusions. Mitigation techniques would then be developed as a way to reduce the impact of an event, according to Biden.

In May 2021, there was a ransomware attack on a major petroleum pipeline in the U.S. (Colonial Pipeline). In response, in mid-July, the Transportation Security Administration, an agency located within the DHS, announced that they were issuing a Security Directive for the owners of critical pipeline owners. The Directive required pipeline owners and operators to report their cybersecurity practices. It also mandated that any owners and operators of pipelines in which hazardous liquids and natural gas are transported must employ additional protections such as mitigation measures to protect against cyberattacks. They must also develop and implement a cybersecurity contingency and recovery plan to ensure their networks are safer. In addition, owners and operators must conduct an annual cybersecurity architecture design review to identify gaps in their security plans.[13]

Biden continued his efforts to protect cybersecurity on June 16, 2021, when he told reporters that he gave Russian leader Vladimir Putin a list of 16 critical infrastructure entities that were "off limits" to a cyberattack. These 16 entities included energy, water, healthcare, emergency, chemical, nuclear, communications, government, defense, food commercial facilities, IT, transportation, dams, manufacturing, and financial services. He did not make any threats of a military response.[14]

At the end of July 2021, President Biden signed a National Security Memorandum on "Improving Cybersecurity for Critical Infrastructure Control Systems." He noted that most critical infrastructure in the nation is owned privately so it is their responsibility to protect it. And, because it is often run by individual owners, the cybersecurity policies are inconsistent and often sector specific. However, Biden also noted that the threat to infrastructure is the most significant threat to the nation's security and that the threats are increasing in number. He said, "The degradation, destruction, or malfunction of systems that control this infrastructure could cause significant harm to the national and economic security of the United States."[15] For this reason, Biden stressed the need for new approaches to cybersecurity.

When he signed this document, Biden required the CISA in the DHS and the NIST from the Commerce Department to come up with performance goals pertaining to cybersecurity for critical infrastructure. In doing so, Biden was confident that the private owners of Critical Infrastructure such as power companies, water companies, and transportation companies would then be influenced to strengthen their cybersecurity measures as well.

Biden instructed immediate actions by federal agencies and private organizations. One suggestion was to require monitoring of networks to detect potential malicious acts before they cause much damage. If something was detected, this information would be shared between the government and owners. The initiative was first attempted in the Electricity subsector. Biden then established the need for some baseline or standard set of goals for cybersecurity that would be consistent among all CI sectors, noting that there was still a need for specific standards based on the individual or specific needs of a sector. Biden repeated the requirement described in EO 13636, originally issued by President Obama, that the Secretaries of Homeland Security and Commerce, along with other agencies, should create a common baseline set of security goals that all critical infrastructure owners and operators should follow to protect critical infrastructure and national security. The Secretary of Homeland Security would then issue a final report that describes the goals for both individual sectors and all sectors.

Biden additionally formally established the President's Industrial Control System Cybersecurity (ICS) Initiative in this document to increase cybersecurity found within the infrastructure. The ICS was described as a voluntary program between the federal government and the private owners of critical infrastructure to jointly create and use technology that delivers improved threat visibility, as well as indicators, warnings, and detections of ongoing events.[16] It will also help to coordinate responses to events.[17]

On October 29, 2021, the Biden administration published "A Proclamation on Critical Infrastructure Security and Resilience Month," 2021, announcing that November 2021 would be Critical Infrastructure Security and Resiliency Month. In describing this action, Biden said,

> We must do everything we can to safeguard and strengthen the systems that protect us; provide energy to power our homes, schools, hospitals, businesses, and vehicles; maintain our ability to connect; and ensure that we have reliable access to safe drinking water.

I am committed to protecting our critical infrastructure; most of our critical infrastructure depends on cooperation between federal, state, tribal, and local governments and private partners. He said,

> At home, my Administration is committed to making a once-in-a-generation investment to prioritize security, sustainable, and resilient infrastructure. … The threats against our critical infrastructure are increasingly complex and nuanced, and we all must be prepared to better protect ourselves from malicious actors threatening our cyber and physical security. That means staying vigilant, investing in new security measures, being prepared to respond to threats, and

> collaborating more with our partners. During Critical Infrastructure Security and Resiliency Month, we reaffirm our commitment to protecting our infrastructure today and securing it for tomorrow.[18]

2022

On March 21, 2022, Biden made an announcement about the nation's cybersecurity. He reminded people,

> From day one, my Administration has worked to strengthen our national cyber defenses, mandating extensive cybersecurity measures for the Federal Government and those critical infrastructure sectors where we have authority to do so, and creating innovative public-private partnerships and initiatives to enhance cybersecurity across all our critical infrastructure.

He then warned the public about the potential for malicious cyberattacks from Russia. He said, "My administration will continue to use every tool to deter disrupt, and if necessary, respond to cyberattacks against critical infrastructure."[19] Table 3.1 provides a statement made by the Secretary of Homeland Security that provides more detail on Biden's approach to cybercrime.

Biden recognized that the federal government needed the help of owners and operators of private critical infrastructure to defend against these and other cyber threats. He called on private

Table 3.1 Statement from Secretary Mayorkas on President Biden's National Cybersecurity Strategy, March 2, 2023

U.S. Department of Homeland Security
Office of Public Affairs

DHS will continue to bolster partnerships and strengthen the nation's cyber resilience

WASHINGTON—Today, Secretary of Homeland Security Alejandro N. Mayorkas released the following statement on President Biden's National Cybersecurity Strategy:

This National Cybersecurity Strategy establishes a clear vision for a secure cyberspace. The Department of Homeland Security continuously evolves to counter emerging threats and protect Americans in our modern world. We will implement the President's vision outlined in this Strategy, working with partners across sectors and around the globe to provide cybersecurity tools and resources, protect critical infrastructure, respond to and recover from cyber incidents, and pave the way for a more secure future.

The Department of Homeland Security (DHS) and its components play a leading role in strengthening cybersecurity resilience across the nation and investigating malicious cyber activity.

Today's National Cybersecurity Strategy emphasizes the department's critical role in the nation's cybersecurity by:

- Empowering the Cybersecurity and Infrastructure Security Agency (CISA) to bolster its work to defend Federal Civilian Executive Branch systems and modernize our networks;
- Calling for Congress to pass legislation to codify DHS's Cyber Safety Review Board (CSRB), a public–private partnership through which cyber leaders from the federal government and private sector conduct fact-finding and issue recommendations in the wake of significant cyber incidents;
- Committing to further disrupting ransomware actors, including through investigations by the U.S. Secret Service and U.S. Immigration and Customs Enforcement, and CISA's work as co-chair of the Joint Ransomware Task Force to increase the resilience of our critical infrastructure from these attacks;
- Underscoring the value of operational public–private sector collaboration and the successes of targeted and calibrated regulatory approaches like those that the Transportation Security Administration and U.S. Coast Guard are planning to leverage to protect the transportation sector; and
- Identifying CISA to lead the update to the National Cyber Incident Response Plan, which will provide clarity across the federal incident response landscape.

DATE: 3/2/2023

owners to accelerate their efforts to expand their cybersecurity and "lock their digital doors." He encouraged private owners to

> harden our cyber defenses immediately by implementing the best practices we have developed together over the last year. You have the power, the capacity, and the responsibility to strengthen the cybersecurity and resilience of the critical services and technologies on which Americans rely … your vigilance and urgency today can prevent or mitigate attacks tomorrow.[20]

Conclusion

Recent presidents Trump and Biden both put protecting critical infrastructure at the top of their administrations' agenda. Despite being from different political parties, they both sought ways to protect critical infrastructure, with attention being put on cybercrime. Both presidents sought to increase the security of computer networks, both publicly and privately owned. They also sought to increase communication between all agencies so that any threat to a network could be thwarted and any damages mitigated.

Key Terms

Executive Orders 13800, 13799, 13873, 13833, 13905, 13920, 13913, 13917, 13942, 14028
Presidential Advisory Commission on Election Integrity
Cybersecurity and Infrastructure Security Agency Act of 2018
National Spatial Data Infrastructure
Committee for the Assessment of Foreign Participation in U.S. Telecommunications
National Security Memorandum on Improving Cybersecurity for Critical Infrastructure Control Systems
President's Industrial Control System Cybersecurity Initiative
Proclamation on Critical Infrastructure Security and Resilience Month

Review Questions

1. Describe the five elements of Executive Order 13800.
2. What are the four recommendations made to address the gaps in cybersecurity that were identified in the 2018 Risk Plan?
3. What were the points made in Executive Order 14028?
4. Compare and contrast the approaches toward protecting the nation's critical infrastructure supported by Presidents Trump and Biden.
5. If you were president, what policies would you support to protect critical infrastructure?

Notes

1 Bucci, Lisa, Alaka, Lauren, Hagen, Andrew, Delgado, Sandy, and Beven, Jack (2023). "Hurricane Ian," April 3. https://www.nhc.noaa.gov/data/tcr/AL092022_Ian.pdf
2 Kustura, Katie (2023). "Which Florida County had the Most Flood Insurance Claims with FEMA after Hurricane Ian?" *Daytona Beach News Journal*, May 9. https://www.news-journalonline.com/story/news/2023/05/09/fema-flood-insurance-payouts-to-fl-counties-post-ian-exceed-3-6-billion/70196244007/
3 Dorsainvil, Aja (2023). "SBA Offering Hurricane Ian Disaster Loan Assistance for Small Businesses," May 9. https://www.wptv.com/weather/hurricane/sba-offering-hurricane-ian-disaster-loan-assistance-for-small-businesses

4. "President Donald J. Trump Is Unleashing America's 5G Potential – The White House." archives.gov
5. Trump, Donald J. (2017). "Presidential Executive Order on Strengthening the Cybersecurity of Federal Networks and Critical Infrastructure," May 11. https://trumpwhitehouse.archives.gov/presidential-actions/presidential-executive-order-strengthening-cybersecurity-federal-networks-critical-infrastructure/
6. https://www.nist.gov/document/eowfreporttopotuspdf https://www.nist.gov/itl/applied-cybersecurity/nice/resources/executive-order-13800/findings-and-recommendations
7. Chalfant, Morgan (2017). "Trump Pressed to Secure US Critical Infrastructure." *The Hill*, March 29. https://thehill.com/policy/cybersecurity/326218-trump-pressed-to-secure-us-critical-infrastructure/
8. Trump, Donald J. (2019). "Executive Order on Securing the Information and Communications Technology and Services supply Chain." *The White House*, May 15. https://trumpwhitehouse.archives.gov/presidential-actions/executive-order-securing-information-communications-technology-services-supply-chain/
9. Trump, Donald J. (2019). "Executive Order on Securing the Information and Communications Technology and Services Supply Chain." *The White House*, May 15. https://trumpwhitehouse.archives.gov/presidential-actions/executive-order-securing-information-communications-technology-services-supply-chain/
10. Chalfant, Morgan, and Miller, Maggie (2020). "Trump Signs Executive Order to Guard Critical Infrastructure that Relies on GPS," February 12. https://thehill.com/homenews/administration/482738-trump-signs-executive-order-to-guard-critical-infrastructure-that/
11. Hendel, John, and Woodruff Swan, Betsy (2020). "Justice Department Opposes Google, Facebook Cable Link to Hong Kong." *Politico*, June 17. https://www.politico.com/news/2020/06/17/justice-department-hong-kong-google-facebook-cable-326688
12. Biden, Joseph (2021). "Executive Order on Improving the Nation's Cybersecurity." *The White House*, May 12. https://www.whitehouse.gov/briefing-room/presidential-actions/2021/05/12/executive-order-on-improving-the-nations-cybersecurity/
13. "Fact Sheet: Biden Administration Announces Further Actions to Protect U.S> Critical Infrastructure," July 29, 2021.
14. Phillips, Morgan (2021). "Biden Gave Putin List of 16 Critical Infrastructure Entities 'off limits' to Cyberattacks." *FoxBusiness*, June 16.
15. Biden, Joseph (2021). "National Security Memorandum on Improving Cybersecurity for Critical Infrastructure Control Systems," July 28. https://www.whitehouse.gov/briefing-room/statements-releases/2021/07/28/national-security-memorandum-on-improving-cybersecurity-for-critical-infrastructure-control-systems/
16. "Fact Sheet: Biden Administration Announces Further Actions to Protect U.S. Critical Infrastructure," July 29, 2021.
17. Jercich, Kat (2021). "Biden Calls for Improved Critical Infrastructure Cybersecurity," July 29. Global edition. See also July 28, 2021: "Fact Sheet: Biden Administration Announces Further Actions to Protect US CI." https://www.whitehouse.gov/briefing-room/statements-releases/2021/07/28/fact-sheet-biden-administration-announces-further-actions-to-protect-u-s-critical-infrastructure/
18. Biden, Joseph R. (2021). "A Proclamation on Critical Infrastructure Security and Resiliency Month, 2021," October 29. https://www.whitehouse.gov/briefing-room/presidential-actions/2021/10/29/a-proclamation-on-critical-infrastructure-security-and-resilience-month-2021/
19. Biden, Joseph (2022). "Statement by President Biden on Our Nation's Cybersecurity," March 21. https://www.whitehouse.gov/briefing-room/statements-releases/2022/03/21/statement-by-president-biden-on-our-nations-cybersecurity/
20. "Statement by President Biden on our Nation's Cybersecurity," March 21, 2022.

4 The Department of Homeland Security

Chapter Outline

Introduction

The 117th Boston Marathon was held on April 15, 2013. On that day, thousands of people lined the streets of the city as participants or spectators. At about 2:49 PM, two bombs detonated, 14 seconds

DOI: 10.4324/9781003434887-4

apart near the finish line. Three people were killed, including an eight-year-old boy, and 281 people were injured, with 14 people requiring amputations. Multiple agencies responded including local law enforcement, state police, and federal police such as the FBI; Alcohol, Tobacco and Firearms;, the U.S. Marshals; and Immigration and Customs Enforcement (ICE).

The two men responsible for planting the bombs were brothers Dzhokhar (19 years old) and Tamerlan (26 years old) Tsarnaev who built pressure-cooker bombs and left them in backpacks on the sidewalk. In the ensuing chase, the brothers killed a police officer at MIT, kidnapped another man, and had a shootout with officers in which two were injured. Tamerlan was killed in the shootout, and Dzhokhar was later found in a boat that was being stored in a backyard. Dzhokhar was convicted of 30 criminal charges, including the use of a weapon of mass destruction (WMD) and malicious destruction of property that resulted in death. He was sentenced to death, but his sentence was overturned by the First Circuit in 2020 because of errors by the judge. The U.S. Supreme Court agreed to hear the case and, in March 2022, decided the First Circuit improperly ruled and reinstated the death penalty. He is currently being housed in a maximum-security federal institution in Colorado.[1]

This was the first time multiple improvised explosive devices were used by domestic terrorists in the U.S. Members of the Office of Bombing Prevention (part of the Cybersecurity and Infrastructure Security Agency) agree that the threat of similar bombings in the U.S. remains high. Threats of bombings continue to increase and threaten the lives of possible victims.[2] The Department of Homeland Security (DHS), along with its many offices and agencies, is responsible for investigating these threats and stopping them from causing damage and deaths. More information about an employee of the Office of Bombing Prevention is in Table 4.1.

Table 4.1 Interview with Member of CISA's Office of Bombing Prevention

Employee Spotlight: 28 Years Later, CISA OBP's Leo Bradley Recounts Firsthand Experience of the Oklahoma City (OKC) Bombing Attack

On April 19, 1995, the deadliest act of domestic terrorism in U.S. history occurred as a Vehicle-Borne Improvised Explosive Device detonated in front of the Alfred P. Murrah Federal Building in downtown Oklahoma City. The blast killed 168 people and left hundreds more injured. Twenty-eight years later, CISA OBP continues to work relentlessly to prevent and mitigate future threats. OBP Senior Policy Advisor Leo Bradley sat down to discuss his firsthand experience responding to this incident and how the bombing prevention community has adapted in the years since.

Q: What was your specific role at the time of the bombing?

A (Leo): I was the commander of the 61st Ordnance Detachment (Explosive Ordnance Disposal) based at Fort Sill, Oklahoma. We were responsible for responding to military ordnance for the Army base and calls for assistance from local authorities in the State of Oklahoma and northern Texas. Our unit served as the informal liaison for Fort Sill until appropriate chains of command were established during this event.

Q: In terms of historical incidents and based on your professional experience, what was different about this attack at the time?

A (Leo): Most improvised explosive devices (IEDs) at the time were small pipe bombs used for criminal activity, or IEDs used to booby trap marijuana fields. The sheer size of this bomb was shocking, as was the surprise of the target location. Typically, high-value targets in New York City or Washington, D.C., were more probable, so a federal building in the middle of the country was unprecedented.

Q: What impact did this attack have on you personally and within the local community?

A (Leo): The people of Oklahoma City were amazing. I remember that as the emergency response went on, the first responders would run out of things as simple as flashlight batteries. A radio station would announce this request and within minutes, countless individuals would flood that location with batteries. Additionally, local shop keepers opened their doors and told the first responders to take whatever they needed free of charge. Restaurants brought food to designated spots. Local construction companies called in their workers and drove bulldozers to wherever they were needed. Also, however, are vivid memories of responders removing bodies from the rubble the day after the bombing. It was heartbreaking. We

(*Continued*)

Table 4.1 (Continued)

Employee Spotlight: 28 Years Later, CISA OBP's Leo Bradley Recounts Firsthand Experience of the Oklahoma City (OKC) Bombing Attack
surveyed the destruction, and none of us had ever seen an entire building destroyed and damage for blocks around the site. Looking back, I realize it influenced my career, as I spent the next 30 years warning people about the dangers of bombings and building capability to prevent it from happening again.
Q: How has the bombing prevention community adapted since the attack?
A (Leo): The OKC bombing generated significant force protection regulations for federal buildings. Many of the regulations regarding parking and planters preventing cars from getting too close to federal buildings date back to this event. Considerations to remove childcare facilities on-site for federal offices became part of the conversation as well. Additionally, we have stronger regulatory procedures over chemical explosive precursors. Response capabilities are better integrated and exercised routinely. Emphasis on bombing prevention methods to both the law enforcement and first responder community, as well as the public, have been prioritized on a broader scale to better secure facilities. Overall, this was a watershed moment that spurred America to realize we were not immune to any attack and should remain diligent through a whole-of-nation approach to mitigating potential bombing incidents.

The DHS was created by President George W. Bush in the days after the terrorist attacks of September 11, 2001. The new agency was intended to be the foremost organization in the U.S. responsible for protecting American citizens from significant events, including both natural disasters and man-made disasters. The DHS has also been given the task of securing U.S. borders against terrorists, transnational criminal groups, or any other person or group attempting to cause harm.[3] In sum, the DHS is responsible for

> counterterrorism, cybersecurity, aviation security, border security, port security, maritime security, administration and enforcement of our immigration laws, protection of our national leaders, protection of critical infrastructure, cybersecurity, detection and protection against chemical, biological and nuclear threats to the homeland, and response to disasters.[4]

The DHS is the lead agency responsible for identifying national critical infrastructure assets and securing them from possible attack or damage. They do this by relying on a risk-informed, all-hazards approach that also ensures the privacy and civil liberties of all citizens. By working with private owners and operators of critical infrastructure, other federal agencies, and international partners, the DHS helps the government and private agencies prepare for, prevent, protect against, mitigate, respond to, investigate, and recover from potential threats. A primary function here is to ensure that the country is prepared for, and resilient to, acts of terrorism, cyberattacks, pandemics, and catastrophic natural disasters.

This chapter provides background into the DHS and describes the many roles it plays in protecting the nation's critical infrastructure.

History

Just 11 days after the terrorist attacks of September 11, 2001, President Bush formed the Office of Homeland Security, which would be located within the executive branch, naming former Pennsylvania Governor Tom Ridge to be the director. In November 2002, Congress passed the Homeland Security Act of 2002 (PL 107-296) that elevated the Office of Homeland Security into a cabinet-level department, now called the DHS. The intent of the new office was to coordinate all efforts to protect the homeland. Parts or all of 22 different existing federal agencies were combined to form the new department, which was the most comprehensive reorganization of the federal government in many years. The new agency began operating on March 1, 2003. Congress gave the new department the responsibility to prevent terrorist attacks, reduce the vulnerability of the nation to such attacks, and respond rapidly in the aftermath of an attack.

When it came to protecting the nation's critical infrastructure, Congress assigned the DHS the responsibility to reduce the vulnerability of critical infrastructure to attack and other hazards.[5] One of the primary tasks assigned to the new agency was to develop a comprehensive national plan for securing the country's assets and for recommending "measures necessary to protect the nation's critical infrastructure and key resources in coordination with other agencies of the Federal Government and in cooperation with state and local government agencies and authorities, the private sector, and other entities."[6] This entailed increased cooperation with private industry and officials from state, local, tribal, and territorial governments. This way, consistent policies could be created to keep the nation's critical infrastructure and information systems safe. The DHS continues to analyze and reduce cyber threats and potential vulnerabilities in computer systems.

Another primary goal of the DHS is to ensure that services are provided to citizens in the event of a disaster. If an event occurs, the DHS will oversee the government's response to ensure a coordinated and comprehensive response to provide assistance to those in need. They have made cooperative agreements with other federal, state, local, and private sector agencies to ensure that recovery efforts will be quick and effective. The response can be increased sharing of critical information, providing financial assistance, providing training for personnel and law enforcement partners, and assisting with any rebuilding and recovery efforts that are needed.[7]

Over the years, the DHS has continued to evolve as it seeks to protect the country from terrorist acts and, at the same time, protect the country's critical infrastructure from harm. It recently improved the nation's attention to protecting the country's cybersecurity in light of multiple attacks on the computer networks owned by the federal government but also by private owners and operators. The department's current priorities and goals to advance its mission are listed in Table 4.2.

Table 4.2 DHS Priorities, 2023

1	**Support and champion** our workforce and advance a culture of excellence
2	**Recruit, hire, and retain** a world-class, diverse workforce to create an inclusive, representative, and trusted department
3	**Advance cohesion** across the department to improve mission execution and drive **greater** efficiency
4	**Innovate and transform** our delivery of services to advance mission execution, improve the customer experience, and increase access to services
5	**Enhance openness and transparency** to build greater trust with the **American** people and ensure the protection of the privacy, civil rights, civil liberties, and human rights of the communities we serve
6	**Transform the department's infrastructure** to ensure it is a more productive and **flexible** workplace responsive to our workforce's and the public's needs
7	**Combat** all forms of terrorism and targeted violence
8	**Increase cybersecurity** of our nation's networks and critical infrastructure, including election infrastructure
9	**Secure our borders** and modernize ports of entry
10	**Build** a fair, orderly, and humane immigration system
11	**Ready the nation** to respond to and recover from disasters and combat the climate crisis
12	**Combat** human trafficking, labor exploitation, and child exploitation

Source: Department of Homeland Security, "Priorities." https://www.dhs.gov/priorities

Leadership

The head of the DHS is the Secretary of Homeland Security. This person coordinates the activities of the many offices that comprise the department and its employees. Former Governor Tom Ridge of Pennsylvania served as the first Secretary of Homeland Security, from January 24, 2003, to his resignation on February 1, 2005.

Ridge was succeeded as Secretary of the DHS by Federal Judge Michael Chertoff, who became the Secretary on February 15, 2005, after being approved by the Senate by a vote of 98–0. He remained in that office until January 21, 2009. On July 13, 2005, after completing a review of the agency, Chertoff announced a significant reorganization of the offices or Directorates.[8] He wanted to restructure the Information Analysis and Infrastructure Protection (IA/IP) Directorate and rename it the Directorate of Preparedness. The new directorate would include portions of the Office of State and Local Government Coordination and Preparedness, including the grant functions and some of the preparedness functions. Moreover, Chertoff's reorganization involved the creation of a new position, the Assistant Secretary for Infrastructure Protection. The mission of the new directorate became managing grants and overseeing other national preparedness efforts, such as training for first responders, increased citizen awareness, protecting public health, and ensuring the safety of infrastructure and cybersecurity.

The third DHS Secretary was Janet Napolitano, who held the post from January 20, 2009, until she left on September 6, 2013. During her tenure, Napolitano oversaw the completion of a Quadrennial Homeland Security Review of the department. To do this, agency personnel worked closely with the White House staff and members of other federal, state, local, and tribal agencies to develop goals for homeland security, along with a unified, strategic framework to help agencies meet those goals.[9]

The fourth Secretary of the DHS was Jeh Johnson, who became the leader of the department on December 23, 2013. Before becoming the head of the DHS, Johnson held the positions of General Counsel for the Department of Defense, General Counsel of the Department of the Air Force, and Assistant U.S. Attorney for the Southern District of New York. While at the Defense Department, Johnson helped to develop counterterrorism policies and assisted in reforms to the military commissions system at Guantanamo Bay. Johnson received his legal training at Columbia Law School, graduating in 1982.

The DHS continued to evolve under Johnson. In 2010, the Federal Protective Service moved from ICE to the National Protection and Programs Directorate. In 2013, the tasks performed by the Office of Risk Management and Analysis were transferred to the Office of Policy. Additionally, the U.S.-VISIT program was incorporated into the Office of Biometric Identify Management. And then, in 2014, some responsibilities of the Office of Infrastructure Protection were handed over to the Office of Cyber and Infrastructure Analysis.

Alejandro Mayorkas became the fifth head of the DHS in February 2021, after being nominated for the position by President Biden. Mayorkas was born in Cuba and eventually became a lawyer. He was a deputy U.S. Attorney in California where he prosecuted white-collar crimes, becoming the Deputy Secretary of the DHS from 2013 to 2016. He was also the Director of U.S. Citizenship and Immigration Services from 2009 to 2013. He has helped to develop the Deferred Action for Childhood Arrivals (DACA) program, dealt with officials from foreign governments to establish cybersecurity agreements; helped create policies against human trafficking (the Blue Campaign); and created the Fraud Detection and National Security Directorate about immigration.

Since becoming the DHS Secretary, Mayorkas has made cybercrime a top priority for the DHS. He has intensified the fight against ransomware and other forms of malware that have been increasingly used against both federal and private networks. To help increase cybersecurity, Mayorkas created an internal task force that included representatives from CISA, the U.S. Secret Service, the

U.S. Coast Guard, and legal experts to identify gaps in existing security. He also sought to build a more robust workforce who could identify cybersecurity threats before they cause damage.[10]

After attacks were made on a water treatment facility in Florida and the Colonial Pipeline, Mayorkas created the "industrial control systems" as a way to mobilize action to improve the resilience of industrial control systems. He also recognized the need to improve the cyber resilience of the nation's transportation systems (rail, aviation, pipelines, and marine transport). And he focused on the security of the country's election system, and the need to protect the nation's electoral system.

Mayorkas has outlined a series of "90-day sprints" that are geared toward bringing attention to a problem and launching new initiatives. One of his sprints involves elevating the federal fight against ransomware, a malicious code that infects computers and prevents them from operating until a ransom has been paid, usually in cybercurrency. Another sprint revolves around creating a more robust workforce in cybersecurity. A third sprint concerns the need to support the cyber resilience of transportation systems such as rail, pipelines, and aviation. A fourth sprint has to do with building the resilience of the country's elections system.

DHS Budget

The Fiscal Year (FY) 2021 Enacted Total Budget for the DHS was $87,816,178.[11] The FY 2023 Budget, listed in Table 4.3 and described fully in Table 4.4, invests in immigration processing, IT modernization, cybersecurity, climate investment, and research infrastructure, as well as hiring and retaining the DHS workforce.[12] Table 4.5 shows how the funds were allocated for each department within the DHS. Finally, a statement made by the Secretary of the DHS, Alejandro Mayorkas, is presented in Table 4.6.

Table 4.3 President Biden's Budget Allocation for the DHS

FY 2021 Enacted	*FY 2022 President's Budget*	*FY 2023 President's Budget*	*FY 2022 to FY 2023 Total Changes*	*FY 2023 FY 2022%*
$87,816,178	$90,982,133	$97,290,726	$6,308,593	6.9%

Table 4.4 Details of President Biden's FY 2023 Budget

- $19.7B for the Disaster Relief Fund to allow for response and recovery to major disasters.
- $309M for border security technology and assets.
- $176M for trade and travel enforcement.
- $319M for hiring 300 additional Border Patrol Agents with improvements in basic training; hiring 33 Intelligence Specialists; two new stations, Air and Marine Operations facilities, the build-out of a mission support facility in Indianapolis, and facilities to house Office of Professional Responsibility agents.
- $230M for noncitizen processing and care requirements, medical claims costs, and establishment of the Office of the Chief Medical Officer. Includes funds necessary to update the Unified Immigration Portal core capabilities and service architecture, two body-worn cameras, hiring of 300 Border Patrol Processing Coordinators, transportation of detainees at the Southwest border, and funding for the Digital Immigrant Processing and rescue beacons.
- $375M for U.S. Citizenship and Immigration Services (USCIS) for asylum adjudications.
- Requests $389M for continued support to reduce the application and petition backlog, and support the International and Refugee Affairs Division.
- Includes $527M for continued expansion of U.S. Immigration and Customs Enforcement's (ICE) Alternatives to Detention monitoring programs.
- Provides ICE $1.4B for detention beds.
- Includes $25M for ICE's Emergency Family Shelter beds.
- $2.3B for ICE Homeland Security Investigations including $15M for the Department's Center for Countering Human Trafficking and $11M for the Victim Assistance Program.

(*Continued*)

Table 4.4 (Continued)

- $421M for ICE's transportation and removal program.
- $140M to support the construction of a Joint Processing Center along the Southwest border.
- $174M to continue the work established through the American Rescue Plan Act of 2021.
- $407M for the National Cybersecurity Protection System.
- $425M for the Continuous Diagnostics and Mitigation program to fortify the security of Federal Government networks and systems.
- $55M to combat the threat of climate change.
- $37M for FEMA- and CISA-identified research and development requirements for communities and infrastructure.
- $871M to ensure TSA employees are paid at a level that is no less than their counterparts on the General Schedule pay scale.
- $243M to enable TSA to recruit and retain a workforce to meet the increasing demands of passenger travel.
- $108M for the Checkpoint Property Screening System program.
- $19M for In-Person Screening Algorithm Development to address capability gaps to detect new and evolving threats to civil aviation.
- $20M in transfers and increased funding for REAL-ID.
- $5.1B in net discretionary funding for FEMA, including money for risk reduction to mitigate the effects of climate change, community-level climate resilience projects, and flood hazard mapping programs.
- $3.2B in FEMA grants supporting state, local, tribal, and territorial partnerships to improve the nation's disaster resilience and implement preparedness strategies. This includes $80M to establish a critical infrastructure cybersecurity grant program to support risk reduction strategies to protect critical infrastructure from cyberattacks.
- $16M for the readiness of FEMA's Incident Management Workforce.
- $4.3M to establish a non-Stafford Act Incident Management Assistance Team.
- $2.7B in net discretionary funding for the U.S. Secret Service and $42M to hire 200 additional USSS staff.
- $34M to prepare for the 2024 Presidential Campaign.
- $23M for armoring 25 Fully Armored Vehicles and 10 Camp David limousines.
- $11.5B in net discretionary funding for the U.S. Coast Guard.

Table 4.5 Biden's DHS Budget by Organization

FEMA: 30%
Customs and Border Patrol (CBP) 18%
U.S. Coast Guard: 14%
Transportation Security Administration 11%
U.S. Immigration and Customs Enforcement: 9%
U.S. Citizenship and Immigration Services: 5%
U.S. Secret Service: 3%
Cybersecurity and Infrastructure Security Agency: 3%
Other: 7%

Table 4.6 Statement by Secretary Mayorkas on the President's Fiscal Year 2024 Budget

U.S. DEPARTMENT OF HOMELAND SECURITY
Office of Public Affairs

The FY 2024 Budget invests in border security, cybersecurity, maritime security, natural disaster recovery and resilience, and other critical DHS missions.

The Biden-Harris Administration today released the President's Budget for Fiscal Year 2024, which provides $60.4 billion in discretionary funding for the Department of Homeland Security (DHS), and an additional $20.1 billion for the Disaster Relief Fund to address major disasters.

(*Continued*)

Table 4.6 (Continued)

"The threats facing the homeland are more diverse and complex than they were twenty years ago, when the Department of Homeland Security was first created. The President's Budget seeks to protect the security of the American people amidst a very dynamic threat environment," said **Secretary of Homeland Security Alejandro N. Mayorkas.** "This budget invests in programs that protect us against the threat of terrorism here and from abroad, strengthen the security of our borders, ensure the swift response to and recovery from natural disasters, and so much more. It equips our department to address the threats of today and prepare for the threats of tomorrow."

At the Department of Homeland Security, the Budget will:

Invest in Cybersecurity and Infrastructure Security Protection. The President's Budget supports the Cybersecurity and Infrastructure Security Agency's (CISA) mission to protect our cyberspace and serve as the national coordinator for critical infrastructure security and resilience. Building upon investments made in FY 2023, an additional $149 million results in a total of $3.1 billion dedicated to programs reinforcing cybersecurity and infrastructure security. Of note, $98 million is provided to implement the Cyber Incident Reporting for Critical Infrastructure Act and $425 million for the new Cyber Analytics Data System which is a robust and scalable analytic environment capable of providing advanced analytic capabilities to CISA's cyber operators.

Invest in Climate and Natural Disaster Resilience. The budget provides for the allocation of resources to the Federal Emergency Management Agency (FEMA) to strengthen the ability to prepare for and respond to disasters of various magnitudes. The budget includes $4.0 billion for FEMA climate resilience programs, over a $150 million increase from the FY 2023 enacted budget. As part of the Administration's efforts to address climate change, the department will invest in programs to reduce pollution and protect public health. The DHS budget includes $123 million to support integrated investments into Zero-Emission Vehicles and charging infrastructure.

Modernize Coast Guard Operational Capability and Presence in the Arctic Region. As a maritime nation, the U.S. depends on a strong and agile Coast Guard to enhance the nation's maritime safety, security, and economic prosperity. The President's Budget provides $12.1 billion in net discretionary funding to sustain readiness, resilience, and capability while building the Coast Guard of the future. The Budget continues efforts for the Coast Guard's highest acquisition priorities and includes $579 million for the Offshore Patrol Cutter program and $55 million to advance the Great Lakes Icebreaker acquisition—an asset ensuring America's continued economic prosperity on our domestic waterways. The budget also includes $170 million for a third new Polar Security Cutter and $150 million to purchase a commercial polar icebreaker for Coast Guard operations. These assets will also increase access to the Arctic region and strengthen the U.S.'s presence in polar waters, a critical national security priority.

Modernize Transportation Security Administration (TSA) Pay and Workforce Policies. The FY 2024 President's budget honors previous promises to compensate the TSA workforce at rates comparable to other federal workers. Enhancements to TSA pay support the President's and Secretary Mayorkas's commitment to fostering diversity, equity, and inclusion in the workforce. The budget includes $1.4 billion to ensure TSA employees are paid at a level that is comparable to counterparts on the General Schedule pay scale.

Help Secure the Border and Build a Safe, Orderly, and Humane Immigration System. The President's Budget provides the resources to continue securing our border and support a fair, orderly, and humane immigration system. The Budget provides $865 million for U.S. Citizenship and Immigration Services to process increasing asylum caseloads, address the backlog of applications for immigration benefits, support the Citizenship and Integration Grant Program, and improve refugee processing. It invests $535 million in U.S. Customs and Border Protection for border technology, to include $305 million for Non-Intrusive Inspection Systems, with a primary focus on fentanyl detection at ports of entry. It also provides $113M to hire additional personnel, for an increase of 350 Border Patrol Agents and 150 Customs and Border Protection Officers, as well as an additional 460 processing assistants at CBP and ICE and 39 positions to strengthen the transportation and removal program. The Budget also provides $40 million to combat human smuggling as well as illicit drug operations such as the production and distribution of fentanyl through the Repository for Analytics in a Virtualized Environment (RAVEN), to help special investigative units disrupt and dismantle Transnational Criminal Organizations and their networks.

(*Continued*)

Support 2024 Presidential Campaign Security. The Budget provides a total of $3.0 billion in net discretionary funding to support the protection and investigation missions of the U.S. Secret Service. To prepare for the upcoming 2024 Presidential Campaign, the budget includes $191 million to ensure that the 2024 Presidential Campaign is adequately resourced for the protection of major candidates, nominees, their spouses, and nominating conventions.

Less than two years after 9/11, 22 agencies and 180,000 public servants came together with a mission to safeguard the American people and our way of life. 20 years later, the Department of Homeland Security has grown in size, capability, and complexity. As new threats emerge with increasing speed, our work to reinforce our homeland security has never been more important to our national security.

Source: Department of Homeland Security, March 9, 2023. https://www.dhs.gov/news/2023/03/09/statement-secretary-mayorkas-presidents-fiscal-year-2024-budget

Organization of DHS

There are many agencies within the DHS that help protect the country's critical infrastructure. Each one plays a specific role in identifying assets, recognizing threats, and planning for recovery in the case of an event. Many agencies fall within the category of Operational and Support Components, whereas others have more administrative tasks.

Operational and Support Components of DHS

The Operational and Support Components of the DHS are the offices that work most closely with agencies and individuals to protect the country's critical infrastructure. Probably the most recognized agency in this realm is the Cybersecurity and Infrastructure Security Agency, otherwise known as CISA. Other critical and well-known federal agencies, including Federal Emergency Management Agency (FEMA), TSA, and others, also play a vital role in protecting the nation from harm. These agencies are all described below.

Cybersecurity and Infrastructure Security Agency

A new agency was created in 2018 by President Trump to strengthen the nation's cybersecurity and be able to defend against cyber threats is the CISA, established when he signed the CISA Act of 2018. The new agency was formed to be the nation's top cyber defense agency and coordinator for critical infrastructure security. The agency provides help to other federal, state, local, and tribal agencies, businesses, and organizations to build a more secure and resilient cyber infrastructure by managing risk and increasing their resilience to attacks. They work with all of the infrastructure sectors and subsectors to oversee national efforts to protect the security, resilience, and reliability of computer networks across the U.S. The mission of CISA is to reduce or minimize the "risk of systemic cybersecurity and communications challenges in our role as the nation's flagship cyber defense, incident response, and operational integration center."[13] The agency works with federal state, and local governments and agencies, private sector owners and operators, law enforcement, intelligence units, and international partners. In 2020, the agency was given a budget of $3.16 billion.

CISA Divisions

There are five divisions in CISA, including Cybersecurity, Infrastructure Security, Emergency Communications, the Integrated Operations Division (IOD), and the Stakeholder Engagement

Division. The National Risk Management Center is also included within these Divisions. Each of these is described below.

The CISA Cybersecurity Division has the mission to "lead the nation's efforts in protecting the federal government networks and to collaborate with the private sector to increase the security of critical network."[14] There are multiple offices to help do this. One is Threat Hunting, otherwise known as the Hunt and Incident Response Team (HIRT). This agency provides assistance to identify and stop threats and attacks on computer systems in critical infrastructure sectors, or in federal, state, local, tribal, and territorial governments, and to develop mitigation plans for removing malware. The employees of HIRT provide expertise to respond to incidents. Their goal is to handle each emergency in a way to reduce risk, limit damages, and reduce recovery costs.[15] Other agencies found within the Cybersecurity Division include Capacity Building (designed to help federal agencies build their capacity for increased cybersecurity), Mission Engineering (an agency that collects data on cyber threats and attacks to support planning and mitigation), Joint Cyber Defense Collaboration (supports the creation of collaborations and public-private partnerships to protect critical infrastructure), Vulnerability Management (designed to identify and reduce the impact of vulnerabilities in computer systems), and the Office of Technical Director (helps to coordinate activities across CISA).

The second division in CISA is the Infrastructure Security Division. This agency cooperates with other government agencies and private sector agencies to conduct vulnerability assessments so that owners and operators have a better understanding of risks and mitigation techniques. This way, appropriate actions can be taken to reduce harm to their assets. The Infrastructure Security Division includes seven offices, as shown in Table 4.7.

The third Division in CISA is the Emergency Communications Division. This division was created in 2007 after it became clear that there were communications gaps during the terrorist attacks of September 11, 2001. This agency promotes better communication between and among emergency responders and agencies to maintain American safety, security, and resiliency. CISA provides training, tools, and guidance to federal, state, local, tribal, territorial, and private sector agencies as they develop plans for communications. CISA helps to ensure that these governments and agencies have plans, resources, and training that are needed to support increased communication. They also ensure that agencies have the technology they may need to maintain communications after an event.

Table 4.7 Offices Located Within the Infrastructure Security Division

1. **Security Programs**: provide public and private agencies with experts and guidance to help them build plans to keep their assets safe in an ever-evolving threat environment.
2. **CISA Exercises**: provide assistance to agencies as they conduct security exercises and training for both cyber and physical threats.
3. **Bombing Prevention**: seeks to protect critical infrastructure by building capabilities within the general public and private sector to prevent, protect against, respond to, and mitigate bombing incidents.
4. **Infrastructure Assessments and Analysis**: helps to gather and manage vital information to protect the nation's critical infrastructure by making data available to agencies.
5. **Chemical Security**: is responsible for implementing regulatory programs to protect critical infrastructure and facilities that handle hazardous chemicals.
6. **School Safety**: strengthens the safety and security of schools by developing products and resources that are designed for the K-12 community.
7. **Strategy, Performance, and Resources**: provides support for the Division to meet any statutory requirements and objectives.

The fourth division in CISA is the IOD. This agency prepares and plans operations carried out by the CISA and geared to protecting critical infrastructure. They provide operational awareness and intelligence analysis and help to develop operational plans to mitigate damage that could result from an attack.

An essential part of the IOD division is CISA CENTRAL. This agency shares technical information about cyber threats and related topics with public and private sectors and international partners as a way to reduce the risks of malicious attacks on the nation's networks. They assist with incident response, intrusions, incidents, mitigation, and recovery. Intrusions into computer systems are tracked by the National Cybersecurity Protection System. They also operate a 24/7 Response Center that provides situational awareness, analysis, and response to events. The Advanced Malware Analysis Center is available to carry out analysis of cyberattacks and malicious code. Experts will analyze samples provided by the agency and provide recommendations for removing the malware and/or possible recovery actions.[16]

The fifth key agency in CISA is Stakeholder Engagement. The Stakeholder Engagement Division helps to create strategic partnerships with stakeholders on the federal, state, local, tribal, and territorial levels, and international groups. It supports conversations and activities to develop strong partnerships, both public and private, so that all agencies are informed about threats. This way, the nation can build a strong and secure computer network and reduce the national risk to infrastructure.

This agency oversees the Critical Infrastructure Cyber Community Voluntary Program (C3VP). This uses the Cybersecurity Framework from the National Institute of Standards and Technology to form a way to base cyber information for the 16 sectors of critical infrastructure that want to improve their cybersecurity. These include technical assistance, implementation guidance on the Cybersecurity Framework, and other guidance on conducting risk assessments. The program also attempts to increase information sharing and develop workforce development programs.

The National Risk Management Center, created in 2018, has expert employees who work alongside public and private owners of critical infrastructure to identify and analyze critical risks and oversee resiliency actions. CISA experts carry out risk analysis that can be used to establish a more resilient critical infrastructure. Some of the initiatives that the Center works on include 5G, election security, the effects of electromagnetic pulses, national critical functions, and pipeline security.[17]

CISA Roles

IMPROVING CYBERSECURITY

As the government's primary agency in charge of reducing the nation's cyber risk, CISA plays a key role in the implementation of President Biden's Executive Order 14028, *Improving the Nation's Cybersecurity*. The elements of the executive order and CISA's implementation strategy are listed in Table 4.8.

INCREASED COMMUNICATION/PARTNERSHIPS

An important role for CISA, as defined in Biden's executive order, is to create collaborative partnerships with both public and private sector partners, federal departments and agencies, state and local governments, international governments, and representatives from government, industry, and academia, all as a way to promote increased communication and share information on threats and technology to improve resilience. Information can be shared and distributed on actionable reactions with partners who may share similar risks.

Table 4.8 Implementation of Executive Order 14028

EO Goal	*CISA Actions*
Remove barriers to sharing threat information between government and the private sector	CISA will work with OMB to ensure critical data related to cyber threat will be shared between public and private agencies. CISA will work to ensure that cyber incident reports are shared quickly among federal agencies to allow for a fast response.
Modernize and implement stronger cybersecurity standards in federal government agencies	CISA will consider efforts to develop a federal cloud security program for agencies that store information in the cloud. They will also create cybersecurity and incident response protocols for cloud technology. CISA will work with the General Services Administration and OMB to modernize the Federal Risk and Authorization Management Program that helps offices meet established standards for cybersecurity. CISA will help to establish stronger standards for cybersecurity for all steps of an agency's planning process. This may involve the use of multifactor authentication and encryption policies.
Improve software supply chain security	CISA will help to establish criteria for security measures for all software used by the Federal Government. They will also help to establish requirements for software and products that are eligible for federal purchase.
Establish a cyber safety review board	CISA will support the creation of the Cyber Incident Review Board that will review cybersecurity incidents on federal agencies and make recommendations for improving actions for incident response.
Create a standardized playbook for response to cybersecurity vulnerabilities and incidents	CISA will create a standardized program for agency responses to cyber threats and incidents. This plan will improve and standardize responses to an attack.
Improve detection of cybersecurity incidents that occur in federal networks	CISA will support practices that help detect malicious activity on federal networks quickly and identify the offender. This information will be shared with other vetted agencies or private organizations.
Improve capabilities for investigation and remediation	CISA will create ways to get a more complete analysis of incidents so that effective mitigation techniques can be established and future attacks deterred.

To make this process easier, CISA has developed multiple ways for agencies to share sometimes sensitive or propriety information with each other or with the government and ensure that it will not be exposed through their Protected Critical Infrastructure Information (PCII) program.[18] The agencies accepted into this program are vetted and must agree to privacy standards. This means that information shared with the government or with other agencies related to a threat is protected from public disclosure. All information that is submitted to the PCII Program office is considered to be protected.[19]

CISA has other ways to share sensitive information. One of those is a program called Traffic Light Protocol. Here, CISA labels information with different colors to indicate if it can be shared, how, and with whom.[20] Another program to share materials is the Cyber Information Sharing and Collaboration Program (CISCP). Here, members can share intelligence about cyber threats, incidents, and vulnerability information in real time and then use that information to prevent, mitigate, or recover from cyber incidents. Knowledge shared via CISCP can allow a participant to secure their network prior to an attack. This service is free of charge to members.[21]

ISACS, or Information Sharing and Analysis Centers, are an additional way to share information in a confidential way. ISACS are nonprofit groups that were created by the owners and operators of critical infrastructure as a way to share data between government and industry groups. These are typically affiliated with sectors.[22] The multi-state ISAC (MS-ISAC) provides incident response for state, local, and tribal governments and shares real-time information on threats and attacks. Similarly, ISAOs, or Information Sharing and Analysis Organizations, were established as a way to collect, analyze, and then circulate information on cyber threats. ISAOs are also sector-affiliated. These were noted in Executive Order 13691 as an effective way to promote more effective sharing of information related to cybersecurity between members of the private sector and the government.[23]

Automated indicator sharing provides for an exchange of machine-readable cyber threat indicators and defensive measures to help protect participants and limit the prevalence and harm from a cyberattack. This program does not cost the participants anything. As soon as a company notices a possible threat, it is shared with others.[24]

The Homeland Security Information Network (HSIN) is one more way for agencies to share unclassified, but sensitive information. Federal, state, local, tribal, and territorial governments, and private sector partners, often rely on HSIN to manage their operations, analyze data, send notices, or other information they need to perform their duties.[25]

EDUCATION AND TRAINING

Since 2009, CISA has been the country's hub for information on cyber and communications information and technical expertise. They collect and analyze risk information and then use that to improve understanding of current risks and mitigation strategies and to predict future risks. Using this data, CISA can then support activities to protect and strengthen critical infrastructure. They provide training on cybersecurity incident response and tools at no cost, publish technical guidance, and provide incident response to aid in minimizing impacts of incidents.

CISA regularly conducts exercises to train members of the government and private industry to increase their security. These exercises allow stakeholders to consider methods to improve existing cybersecurity procedures. One example of this is the Federal Virtual Training Environment. This is a series of online training courses that are available for free. They also offer Certification Prep Courses, courses on defense skills, and incident response. Courses are also available in Continuous Diagnostics and Mitigation. Tabletop training programs are also available.

One way information is disseminated is through a publication called Insights (available on their website) through which they provide background information on threats and mitigation actions that agencies can apply.[26] A second publication is Cyber Essentials, which is intended to assist small businesses and local government agencies and help them develop cybersecurity practices. The Cyber Essentials are often the starting point to cyber readiness for small companies.[27]

Educating the public is accomplished by CISA in multiple ways. One is the National Cyber Security Awareness Month, which occurs each October. This becomes an opportunity to educate public and private sector partners about cybersecurity through events and initiatives. The ultimate goal of Awareness Month is to raise the public's understanding of cybersecurity and to increase the resiliency of the nation in the event of a cyber incident. CISA's goal is to help people protect themselves as technology and private information is more available online.

The Stop.Think.Connect campaign is another way to help the public learn about the importance of cybersecurity. This is a national public awareness campaign aimed at increasing the understanding of cyber threats and empowering the American public to be safer and more secure online. This campaign provides resources that are tailored to multiple demographics, including small businesses, students, educators, parents, and others and resources including blogs, newsletters, and tip cards to give citizens the information they need to be safer when using the Internet.

CISA has created a series of video games for both adults and kids that teach about cybercrime. The games present simulated cybersecurity threats, and the players must devise defenses and responses. In one game called Defend the Crown, the player owns a castle, but cyber ninjas are attempting to steal valuable items and secrets. To win, players must develop a basic understanding of cyberattacks and defense strategies.[28]

Professors in Practice is one more way that CISA helps to educate the public. CISA, along with the National Defense University and the College of Information and Cyberspace, hosts a program called *Professors in Practice*. Each week in August, a professor will discuss a topic. These sessions are open to all who wish to attend. In each session, attendants will be able to engage with the professor as they learn important aspects and grasp necessary actions that should be taken for strong, sustainable, and resilient cybersecurity. The program showcases cybersecurity professors who are currently working in cybersecurity. Their backgrounds and experiences provide unique, expert perspectives on the knowledge, skills, and essential management strategies needed by both public and private sector professionals to establish sound security and resilience for their organizations.

CISA educates the public about how to stay safe while shopping online, especially during the holiday shopping season. CISA provided Americans tips for safe shopping, including the use of strong passwords, updating software, using caution before opening suspicious links, and turning on multifactor authentication. They referred to these tips as "cyber hygiene" and stressed that they would drastically improve safe online shopping.

Each year, CISA holds its Annual National Cybersecurity Summit that has over 15,000 attendees who are stakeholders located throughout the world. This is held through the four weeks during the month of October. The Summit provides a forum for collaboration about how we can protect our physical and cyber infrastructure. Each day features presentations from targeted leaders across government, academia, and industry. There is no cost and the summit is open to all.[29]

The CISA Cybersecurity Awareness Program is a national program designed to increase the public's awareness of cyber threats so they can make changes needed to make them safer online. It stresses the need to be safe at home, at work, and in their community.[30]

Career Focus: Shields Up! Program

It is critical that every organization is prepared to identify and respond to a potential cyber incident. Through the Shields Up program, CISA assists organizations as they prepare for, respond to, and mitigate the impact of a cyberattack. CISA encourages stakeholders to voluntarily share information about threats to help prevent harm to critical infrastructure.

Organizations: One part of the Shields Up program provides guidance for all organizations. CISA recommends that all agencies increase protections surrounding their critical assets. These recommendations to reduce the likelihood of a cyberattack include the use of multifactor authentication, updated software, the implementation of strong control for the use of the cloud, and the use of vulnerability scanning to reduce possible exposure to threats. Conduct training exercises or tabletop exercises so that all employees understand their roles if there is an attack. They also suggest that IT or cybersecurity personnel are monitoring online behavior to identify potential threats; and confirm that all networks are protected by antivirus and antimalware software. If an attack occurs, ensure that the company or organization has a crisis-response team to investigate the incident.

Businesses/CEOs: Corporate owners play an important role in ensuring that their organization has strong security to protect their computer systems. To help them, CISA has created a series of recommendations for business leaders. One is to empower the company's Chief Information Security Officers by including them in the planning process to reduce risks. Each leader must help all employees understand that security must be a top priority. CISA recommends that companies lower the threshold for reporting cyber incidents to officials. All malicious activity should be reported to officials so that issues can be quickly identified. Top business leaders should participate in a simulation of an incident and the plans for a response to determine their effectiveness. The plans should focus on continuity of services to people and businesses.

CISA provides recommendations for any organization that has been the victim of a ransomware attack. They recommend isolating the impacted system or powering it down to avoid the spread of the malware. Agencies should document the events of the attack, including what happened, and conduct an investigation to understand the event.

Source: Shields Up! CISA, https://www.cisa.gov/shields-up

VULNERABILITY ASSESSMENTS

CISA carries out cost-free vulnerability assessments and CDM upon request to agencies, federal organizations, and private organizations to help evaluate operational practices and resilience to manage risk. This identifies risks to operations, assets, and individuals and allows network administrators to understand the risk level of their systems. CISA also offers the Enhanced Cybersecurity Services program that helps to detect intrusions and prevention services. They provide ideas for how an agency can improve its cybersecurity and thereby protect the nation's CI.[31] They also offer cyber hygiene services to reduce cybersecurity risks.[32]

CISA also has Cybersecurity Incident and Vulnerability Response Playbooks that are available to the public. These playbooks that provide the steps that should be used when planning for and conducting training activities for cyberattacks. The Incident Response Playbook involves planning for an event that involves malicious cyber activity that causes a major incident. The second playbook, The Vulnerability Response Playbook, is centered on a network vulnerability that is identified and used by cyber-offenders who gain unauthorized entry into computer networks to cause damage. Participants must devise ways to plan to mitigate the consequences of these attacks as a way to train for a real-life attack.

DEVELOPING A CYBERSECURITY WORKFORCE

It is necessary that the U.S. train new experts in cybersecurity for both private sector and federal organizations. There is a critical need to have skilled and qualified professionals who can help to detect malware, detect unauthorized use of networks, and mitigate the effects of an attack. CISA is assisting in developing this workforce in many ways and increasing the size and capability of the workforce.[33] Four of these methods are below.

Cyber Career Pathways Tool: This tool assists interested individuals in identifying a career path in a cyber-related field. It provides information on the skills that are needed to have a position in this field. This was created in conjunction with the Federal Cyber Career Pathways Working Group.

Educational Partnerships: CISA partners with middle schools and high schools, not-for-profit agencies, state school boards, and universities to include cybersecurity information into existing curricula. CISA and the National Security Agency (NSA) sponsor the National Centers of Academic Excellence program in which they identify colleges and universities as the top schools in Cyber Defense. CISA also co-sponsors the CyberCorps Scholarship for Service program. Students who receive the scholarships must, upon graduation, work in federal, state, local, or tribal governments for time.

National Initiative for Cybersecurity Careers and Studies: CISA has developed a website for the public to offer education to increase skills in cybersecurity. They provide more detailed information in courses on Critical Information Infrastructure Protection that help businesses and individuals review the security of the data in computer networks within the infrastructure sectors that are critical to national security, including banking, securities, and commodities markets, industrial supply chain, electrical/smart grid, energy production, transportation systems, communications, water supply, and health care.[34]

The President's Cup Cybersecurity Competition: CISA first created this in 2019 after Biden issued Executive Order 13870. The competition involves fictional storylines where participants must solve problems using their cyber skills. The competition is hosted by CISA as a way to identify, recognize, and train the best cyber talent across the federal workforce.

ELECTION SECURITY

The electoral process across the country is vital to the democratic system of government, and it requires a trusted and secure computer network. In January 2017, the national elections infrastructure was identified to be critical infrastructure and became a subsector under the Government Facilities Sector. Both the DHS and the CISA are dedicated to ensuring secure computer systems on the local, state, and federal levels to prevent risks to the electoral infrastructure. They provide services that are geared to reducing both cyber and physical risks to elections. This includes the safety of voter registration databases, voter registration, the systems used to carry out the voting process (i.e. counting, auditing, and displaying of results), voting systems, storage systems for election infrastructure, and polling places (including early voting locations). CISA officials will certify voting equipment as needed and provide additional funding for updating equipment. In doing so, CISA will help to secure the election infrastructure. They will also provide various education programs and materials to reduce the possible risks to the process. If any threats are detected that information will be shared with election officials so they can respond quickly and take any needed action. To do this, CISA works in conjunction with many different agencies including the National Association of Secretaries of State, the National Association of State Election Directors, the International Association of Government Officials, the National Association of Election Officials, and the Elections Assistance Commission.

CISA GLOBAL

Today's globally interconnected world presents a wide array of serious risks and threats to cyber infrastructure. It is critical that the U.S. government remain engaged to protect national security interests and economic interests. CISA works with foreign partners to build the nation's cyber capacity and strengthen our ability to globally defend against cyber incidents, enhance the security and resilience of critical infrastructure, identify and address the most significant risks to the national critical functions, and provide seamless and secure emergency communications. Sharing threat information, mitigation advice, and best practices with international partners helps to create an open, reliable, and secure interconnected global network.

Future of CISA

The CISA Strategic Plan for 2023–2035 outlines their activities for the upcoming years. The agency will continue to lead efforts to reduce risk and build resilience to threats and attacks. This plan involves four goals.[35] They are:

Lead the national effort to protect cyberspace and other U.S. critical infrastructure. CISA will continue to collaborate with shareholders and international partners to implement plans to reduce risk, mitigate risks as they emerge, and respond if a major incident occurs.

Reduce risks to, and strengthen the resilience of, America's critical Infrastructure to increase the safety and security of the country. Planning for recovery from major events is also essential. This includes identifying and establishing relationships with owners and operators to increase our understanding of vulnerabilities.

Strengthen efforts for collaboration and sharing of information with public and private agencies. This means actively communicating with stakeholders of critical infrastructure to better understand security risks and needs, and to provide CISA's products, services and resources when needed.

Build a culture of excellence within CISA based on core values and principles of teamwork, innovation, inclusion, transparency and trust.

Career Focus: Jen Easterly: Head of CISA

The current leader of CISA is Jen Easterly. Easterly was nominated to the position by President Biden in April 2021 and subsequently confirmed unanimously by the Senate on July 12, 2021.

Easterly graduated from the U.S. Military Academy at West Point and a master's degree in Philosophy, Politics, and Economics from the University of Oxford, where she was a Rhodes Scholar. Prior to serving in CISA, Easterly served in the U.S. Army where she had responsibilities in intelligence and cyber operations. She served in Haiti, the Balkans, Iraq, and Afghanistan. She helped to design the U.S. Cyber Command. She is a two-time recipient of the Bronze Star. Easterly retired after over 20 years of service.

Easterly oversaw Firm Resilience at Morgan Stanley where she was responsible for the cybersecurity of the networks. She oversaw the preparedness for and response to business-disrupting operational incidents and possible risks to the firm.

Easterly served as Special Assistant to President Obama and was the Senior Director for Counterterrorism in the White House. She also served as the Deputy for Counterterrorism at the NSA.

She is the recipient of numerous honors and awards, including the 2022 National Defense University Admiral Grace Hopper Award; the 2021 Cybersecurity Ventures Cybersecurity Person of the Year Award; the 2020 Bradley W. Snyder Changing the Narrative Award; and the 2018 James W. Foley Legacy Foundation American Hostage Freedom Award.

A member of the Council on Foreign Relations and a French-American Foundation Young Leader, Jen is the past recipient of the Aspen Finance Leaders Fellowship, the National Security Institute Visiting Fellowship, the New America Foundation Senior International Security Fellowship, the Council on Foreign Relations International Affairs Fellowship, and the Director, NSA Fellowship.

As Director of CISA, Jen leads the CISA's efforts to understand, manage, and reduce risk to the cyber and physical infrastructure Americans rely on every day.

U.S. CITIZENSHIP AND IMMIGRATION SERVICES

Citizenship and Immigration Services oversees the process of immigration and naturalization services for those seeking to enter the U.S. This agency collects and maintains a great deal of personal information about applicants, so they are in need of a secure and trusted cyberspace.[36] During the COVID-19 pandemic, this agency passed a temporary final rule that would ensure that the food chain was still open and functioning, thereby reducing the impact on public health. Under the rule, businesses could hire workers if they were "essential" to the food supply chain. The employer also had to prove that there was not a sufficient number of workers in the U.S. who were qualified and available to fill the position. The new policy impacted only aliens who were already in the U.S.[37]

U.S. COAST GUARD

The U.S. Coast Guard (USCG) is the federal agency responsible for securing and safeguarding the country's maritime territories. They must often rely on a stable and trustworthy cyberspace to carry out their operations at sea, in the air, on land, and in space. The Coast Guard was given the control to prevent or reduce cyber threats and thereby protect maritime interests. It adheres to best practices to identify cyber-related vulnerabilities and put risk management strategies into place as needed. They also play a role in coordinating cyber incident responses when needed and help protect the country against potentially dangerous cargo.

U.S. CUSTOMS AND BORDER PATROL

The U.S. Customs and Border Patrol (CBP) agency oversees securing American borders as a way to protect the country against illegal entry of unauthorized personnel and possible terrorist threats. At the same time, they support legitimate travel, economic tasks, and immigration. They also seek to disrupt transnational criminal groups who may seek to enter the country for illegal reasons, as well as the possible spread of diseases and agricultural diseases.

FEDERAL EMERGENCY MANAGEMENT AGENCY

The FEMA is the nation's emergency management and preparedness agency that oversees the federal response to disasters, both natural and man-made.[38] They coordinate with local, state, federal, tribal, private sector and nonprofit partners after a disaster occurs and ensure that all survivors have food, water, and other necessary services. FEMA's mission is to help people before, during, and after disasters. To that end, FEMA's mission "is to support our citizens and first responders to ensure that as a nation we work together to build, sustain and improve our capacity to prepare for, protect against, respond to, recover from, and mitigate hazards."[39]

Unlike most agencies that focus on disaster response, FEMA has existed for many years. Congress passed legislation in 1803 for federal disaster relief after a fire in Portsmouth. The legislation provided relief to merchants allowing for suspension of bond payments for a short period of time. In 1979, President Carter formally created FEMA by signing Executive Order 12127, which reorganized five different agencies that had a role in disaster response into a new agency.[40] In 1988, Congress passed the Disaster Relief and Emergency Assistance Amendments, which helped define more clearly the direction for emergency management.[41]

During the late 1980s and early 1990s, the public's opinions of FEMA were fairly negative after its slow response to Hurricane Hugo in 1989 and Hurricane Andrew in 1992. Victims complained that FEMA's response was disorganized.[42] In 1993, President Clinton elevated it to a Cabinet-level agency and hired a new director, James Lee Witt. Witt changed the agency's response to an "all-hazards, all phases" approach so that FEMA would be prepared to respond to all types of hazards, and in all phases of a crisis.[43]

With the new changes, FEMA was more effective in responding to disasters. In 1994, FEMA oversaw relief efforts after a major earthquake in Los Angeles, and in 1995, FEMA assisted the victims of the Oklahoma City bombings. After the terrorist attacks of September 11, 2001, FEMA expanded its response to terrorist acts. President Bush made the agency more focused on security, terrorism preparedness, and mitigation and when the Homeland Security Act of 2002 signed by President Bush, FEMA became part of the DHS.[44]

FEMA's slow response to help the victims of Hurricane Katrina led Congress to pass the "Post-Katrina Emergency Management Reform Act (PKEMRA) of 2006," more commonly referred to as the Post-Katrina Act.[45] Under this law, FEMA remained a part of the DHS but would have an elevated status and more autonomy. It also assigned the task of advising the President, the Homeland Security Council, and the Secretary of Homeland Security regarding emergency management to the FEMA administrator.

In 2012, FEMA was tested after Hurricane Sandy hit the Northeast U.S.[46] Hundreds of homes were destroyed, and many people were killed. There was general agreement that officials from the DHS and FEMA established strong partnerships with state and local governments and were more prepared for the disaster.[47] However, some residents were forced to wait for weeks for federal assistance.

Today, FEMA continues to prepare for, and respond to, disasters and terrorist attacks. They cooperate with local officials to prepare for events so that when they do occur, communities will be more resilient.[48] Personnel within the agency help to "identify actions that should be taken before, during, and after an event that are unique to each hazard."[49] FEMA has offices in ten regional offices around the country and a headquarters in Washington DC.

The 2022–2026 Strategic Plan outlined goals and objectives for the future success of the agency. Three goals of the agency are (1) to instill equity as a foundation of emergency management; (2) to lead communities in climate resilience; and (3) to promote and sustain a nation that is prepared for possible disasters.

FEMA's approach to preparing the nation is Ready.gov, an educational program for the public to help people be prepared for all types of emergencies. The program helps families be aware that different emergencies can occur and that they must be prepared for them. FEMA urges families to develop an emergency plan and have an emergency supply kit in their homes.

In addition to helping prepare families for a disaster, FEMA assists victims in the wake of an event. Through their disaster response and recovery assistance programs, FEMA helps to provide victims with immediate needs like food, water, and housing. They provide generators to supply power, heating or cooling to office buildings, businesses, or residential facilities. In many situations, FEMA operates disaster recovery centers in those areas that have been affected by an event to provide information, legal advice, and financial assistance to those who need help.[50]

An essential service that FEMA ensures is communication during and after an event. Public safety agencies like law enforcement, fire services, and medical services must have secure, open, and reliable communications systems. FEMA will provide communications support and ensure that emergency communications are available through their Mobile Emergency Response Support, or MERS. This is a moveable communications platform that can be moved to a disaster area when needed. They also provide radio and communication equipment as needed. FEMA also has Mobile Communications Office Vehicles to help FEMA with their response and recovery operations. The vehicles have workstations and satellite communications.

FEMA NATIONAL ADVISORY COUNCIL

An organization within FEMA is the National Advisory Council. Established in June 2007 by the Post-Katrina Emergency Management Reform Act, the Council is a way to develop better coordination of federal policies for preparedness, protection, response, recovery, and mitigation for all events. The Council comprises 40 members who have been appointed by the administrator and include state, local, and tribal governments; nonprofit; and private sector. These members advise the FEMA administrator on all aspects of emergency management.[51]

STRATEGIC FORESIGHT INITIATIVE

Strategic Foresight Initiative (SFI) is a joint effort within the emergency management community that is facilitated by FEMA. This organization assists members of the emergency management community in understanding changes to the field and the impact of those changes on future patterns of emergency management. Members include emergency managers, stakeholders, and experts at the federal, state, and local levels. SFI helps members consider what the world may look like in the future. This includes the changing role of the individual, such as the increased availability of information and methods to make their views public and the ever-changing role of technology. Climate change is also an area of concern, along with the impact on the availability of water, aging critical infrastructure, and changes in populations (mass migrations). The trust (or lack of trust) by the people in government is considered, along with an increased threat of terrorist attacks. Participants in SFI include first responders, emergency managers and professionals at the federal, state, and local levels (both public and private), subject matter experts on relevant topics, academics, and other stakeholders.[52]

SFI has focused some of its work on critical infrastructure. This is important because infrastructure will change moving into the future, which will probably have major impacts on emergency managers. Some trends include the availability of funding for private sector improvements to critical infrastructure; the possible move to centralized ownership of critical infrastructure; and the increased reliance on computers and technology in critical infrastructure. The age of existing infrastructure is a concern to many as the country's infrastructure is growing. The age of the infrastructure is an indicator of structure failure. For example, bridges, which are designed to past 50 years, have an average age of 43 years across the U.S.[53]

U.S. IMMIGRATION AND CUSTOMS ENFORCEMENT

Immigration and Customs Enforcement works to protect the U.S. and its citizens from cross-border crime and oversees the implementation of immigration practices in the nation. One agency within ICE is Homeland Security Investigations (HSI). This agency assists with cyber-related investigations on cybercrime. The Cyber Crimes Center, or C3, provides support and training for cybercrime to governments and international law enforcement groups. The organization also has a computer forensics lab that assists with the recovery of digital evidence.

U.S. SECRET SERVICE (USSS)

The U.S. Secret Service (USSS) has the mission to protect the nation's elected officials, foreign heads of state, and diplomats. They also protect national events. They also safeguard the U.S. financial infrastructure.

The USSS has an agency called the Electronic Crimes Task Forces, which helps to identify and locate international cyber criminals that may have committed cyber-enabled crimes related to the financial infrastructure such as intrusions, bank fraud, illegal financing operations, money laundering, or data breaches. The Secret Service's Cyber Intelligence Section has investigated and gathered evidence on the cybercrimes of multiple transnational cyber criminals who had illegal access to millions of credit card numbers and stole approximately $600 million from financial and retail institutions. The Secret Service also runs the National Computer Forensic Institute, which has law enforcement officers, prosecutors, and judges who have specialized training in cybercrime and related issues. The Cyber Fraud Task Forces are also found in USSS. Here, law enforcement, prosecutors, private industry, and other experts provide expertise to pursue threats.

TRANSPORTATION SECURITY ADMINISTRATION

The Transportation Security Administration (TSA) is charged with overseeing the security of the nation's transportation sectors to keep them safe from terrorist attacks while at the same time ensuring freedom of travel for people and commerce. This includes aviation, highways, railroads, pipelines, and mass transit. The TSA works with CISA and private partners to increase the cyber resilience of networks in all areas. They routinely assess cybersecurity policies and increase education and training, development of guidance and best practices, and other practices.

Aviation officials focus on the safety of airline travel and increasing security measures. They screen airline passengers and their baggage to ensure that no dangerous material is brought onboard an aircraft. TSA also regulates the installation and maintenance of equipment to detect explosives. TSA agents provide security for airport perimeters. They also oversee the Air Marshals program[54] As well as a Pre-Check system for trusted travelers who have been pre-approved to board a plane without going through screening process. The TSA has also created a "Secure Flight" program in which, TSA agents use technology to identify evolving threats pertaining to air travel. TSA agents attempt to screen 100% of cargo that comes into the U.S. and must screen all cargo to keep the nation safe and secure the global supply chain. In the Trusted Traveler program travelers can use expedited lanes when crossing international borders and at the airport. In the Visa Waiver Program, citizens from participating countries can enter the U.S. for either business or tourism and remain for up to 90 days without a visa.

DHS Administrative Offices

Management Directorate

The Management Directorate oversees the employees in the organization, making sure that each employee plays a specific role or has a specific task. The directorate also ensures that there are effective communication procedures between each group and with other non-governmental offices. The directorate also has authority over budget issues and the dispersing and expenditure of agency funds for protecting critical infrastructure. Clearly this is important to protecting critical infrastructure because of the high cost of protecting assets. Many grant programs that help fund protection plans originate or are managed by this office. This directorate includes human resources and personnel, property, and equipment. They also oversee biometric identification services. In general, this directorate is responsible for securing the federal infrastructure that exists throughout the nation.

The Federal Protective Service is part of the Management Directorate. It is a federal law enforcement agency that is responsible for protecting federal facilities and anyone in them (employees and visitors). On accession, the agency will provide protection for special events where large numbers of people may gather. In addition to this, the agency personnel conducts security assessments of buildings to ensure they are safe for those inside and also provide K–9 explosive detection if needed.[55]

The Office of Biometric Identity Management provides services for biometric identification to federal, state, and local government officials so that they can identify individuals and determine if they are a risk. These services include collecting biometric data, analysis, and storage.[56]

Science and Technology Directorate

The Science and Technology Directorate provides support for research and development regarding critical infrastructure protection. The agency carries out research on topics such as explosive detection, blast protection, and safe cargo containers. They monitor threats and develop ways to prevent those threats. The directorate also provides federal, state, and local officials with any necessary technology they need to protect their citizens.

The Science and Technology Directorate operates five laboratories in five different states (AL, FL, MD, NJ, NY). It helps to identify threats and detect and develop an appropriate response.

Another component of the National Laboratories is the Chemical Security Analysis Center, created in 2006 by the Presidential Directive. This agency identifies and assesses chemical threats that face the U.S. and appropriate responses. They provide a response team, called CSAT (Chemical Security Analysis Technical Assistance), that provides help in response to a chemical threat or chemical hazard. For example, they can give advice to a chemical facility. In a program called Jack Rabbit III, the directorate is conducting laboratory and field experiments to learn more about building security, safety, and resilience in the chemical supply chain. The National Biodefense Analysis and Countermeasures Center studies and prepares for biological threats to the country. They carry out intelligence assessments and help create preparedness plans, response plans, and forensic analyses.

The National Urban Security Technology Laboratory is part of the National Laboratories. Those who work in this lab develop tools and technologies that are used by first responders in the case of an event. Their mission is to test and evaluate, research, and develop tools needed for events. Similarly, the Transportation Security Laboratory has the mission to "enhance homeland security by performing research, development and validation of solutions to detect and mitigate the threat of improvised explosive devices."[57] In short, TSL helps those in aviation settings to detect explosives by use of specialized labs and with their expertise in chemistry, engineering, math, and other fields.

Finally, the Plum Island Animal Disease Center is part of the National Laboratories. The experts in this facility study animal diseases and viruses and how to contain any diseases that occur. They help the country prepare for emergencies with vaccines, diagnostics, and bioforensics.[58]

Countering Weapons of Mass Destruction Office

The Countering Weapons of Mass Destruction Office, part of the DHS, is focused on the prevention of attacks against the U.S. that result from a WMD. The mission of the office is to coordinate with other agencies, both domestic and international, to prevent attacks. They identify emerging or current WMD threats and strengthen the ability to detect a possible attack and disrupt it.

Office of Intelligence Analysis

The Office of Intelligence Analysis (OIA) equips security partners with the intelligence and information they need to ensure the safety of the country. The office disseminates intelligence to others in state and local government, and the private sector to identify, to assist them in mitigating and responding to threats. OIA has oversight of the intelligence functions of the DHS and other federal agencies.[59]

Found with OIA are fusion centers. These agencies help to gather information and then share information with other government agencies and private sector partners.[60] The Joint Terrorism Task Force (JTTF) is led by the FBI. They coordinate resources to investigate cases of suspected terrorism. There are 104 JTTFs located in different parts of the country.

Conclusion

The DHS is just one agency that is involved in critical infrastructure protection. Its goal is to make the U.S. able to prepare for or respond to an event of national significance. There are many other organizations that share this goal. There is no doubt that these groups will continue to adapt as they respond to new threats or adapt to changes in the security realm and respond to the security needs of the nation.

Key Terms

Office of Homeland Security
Homeland Security Act
Department of Homeland Security
CISA
Cybersecurity Division of CISA
CISA Central
PCII
ISACs/ISAOs
Federal Emergency Management Agency
STOP.THINK. CONNECT
Shields Up
Post-Katrina Emergency Management Reform Act (PKEMRA) of 2006
FEMA National Advisory Council
National Advisory Council (FEMA)
National Infrastructure Advisory Council (NIAC)
Transportation Security Administration

Review Questions

1. Provide a description of the history and development of the Department of Homeland Security. Who is the current Secretary?
2. What is CISA, and what are the responsibilities of this organization?
3. How does CISA share sensitive information with other agencies? Why is this important?
4. What is the role of FEMA in infrastructure protection?
5. What are some of the other offices found within the DHS that play a role in infrastructure protection?

Notes

1. Boston Marathon Bombing, "History Channel," April 13, 2013. https://www.history.com/topics/21st-century/boston-marathon-bombings
2. CISA, Office of Bombing Prevention. "Bombing Prevention Community Looks Back on the Boston Marathon Attack 10 Years Later." *Bulletin*, Issue 9. https://content.govdelivery.com/accounts/USDHSCISA/bulletins/34e61ea
3. DHS Budget
4. "DHS Website: Secretary of Homeland Security." https://www.dhs.gov/topics/secretary-homeland-security
5. US Department of Homeland Security, "Prevent Terrorism and Enhance Security," July 20, 2015. http://www.dhs.gov/prevent-terrorism-and-enhance-security
6. DHS, FEMA, "CIKR Awareness AWR-213, Participant Guide," September 2010, 2:7.
7. US Department of Homeland Security, "Building a Resilient Nation," July 17, 2015. http://www.dhs.gov/building-resilient-nation
8. US Department of Homeland Security, "Creation of the Department of Homeland Security," July 13, 2015. http://www.dhs.gov/creation-department-homeland-security
9. US Department of Homeland Security, "Creation of the Department of Homeland Security," July 13, 2015. http://www.dhs.gov/creation-department-homeland-security
10. "Fact Sheet with Summary of DHS Cybersecurity Workforce Spring Activities."
11. Source: "Budget in Brief Fiscal Year 2023." www.dhs.gov/sites/default/files/2022-03/22
12. "DHS Budget 2024." https://www.dhs.gov/dhs-budget
13. CISA, "Cybersecurity Incident Response." https://www.cisa.gov/topics/cybersecurity-best-practices/organizations-and-cyber-safety/cybersecurity-incident-response

14 CISA, "Cybersecurity Division, Mission." https://www.cisa.gov/about/divisions-offices/cybersecurity-division
15 "Detection and Prevention." https://www.cisa.gov/detection-and-prevention
16 "Detection and Prevention." https://www.cisa.gov/detection-and-prevention
17 CISA, "National Risk Management Center." https://www.cisa.gov/about/divisions-offices/national-risk-management-center
18 https://www.cisa.gov/pcii-program; see also Information Sharing and Awareness. https://www.cisa.gov/information-sharing-and-awareness
19 CII Act of 2002, 6 U.S.C. § 131, and www.dhs.gov/pcii-program
20 CISA, "Traffic Light Protocol." cisa.gov/TLP
21 "Information Sharing and Awareness." https://www.cisa.gov/information-sharing-and-awareness
22 "Information Sharing and Awareness." https://www.cisa.gov/information-sharing-and-awareness
23 "Information Sharing and Awareness." https://www.cisa.gov/information-sharing-and-awareness
24 "Information Sharing and Awareness." https://www.cisa.gov/information-sharing-and-awareness
25 "Information Sharing and Awareness." https://www.cisa.gov/information-sharing-and-awareness
26 https://www.cisa.gov/insights
27 https://www.cisa.gov/publication/cisa-cyber-essentials
28 https://www.cisa.gov/cybergames
29 https://www.cisa.gov/cybersummit2021
30 "CISA Cybersecurity Awareness Program." https://www.cisa.gov/resources-tools/programs/cisa-cybersecurity-awareness-program#:~:text=The%20CISA%20Cybersecurity%20Awareness%20Program,have%20a%20part%20to%20play
31 https://www.cisa.gov/cyber-assessments; Cyber Resource Hub. https://www.cisa.gov/cyber-resource-hub; cisa.gov/aes
32 CyberHygiene Services. https://www.cisa.gov/cyber-hygiene-services
33 https://www.cisa.gov/cybersecurity-education-career-development
34 health https://niccs.cisa.gov/education-training/catalog/national-defense-university/critical-information-infrastructure; or www.niccs.us-cert.gov
35 CISA, "2023–2025 Strategic Plan," September 2022. https://www.cisa.gov/sites/default/files/2023-01/StrategicPlan_20220912-V2_508c.pdf
36 https://www.uscis.gov/
37 "DHS Offers Flexibilities to Increase Food Securities, Stabilize U.S. Supply Chain During Covid-19," December 5, 2020. https://www.uscis.gov/news/news-releases/dhs-offers-flexibilities-to-increase-food-security-stabilize-us-supply-chain-during-covid-19
38 Federal Emergency Management Agency Website, "FEMA: 35 Years of Commitment." http://www.fema.gov/fema-35-years-commitment
39 FEMA, "About the Agency." http://www.fema.gov/about-agency
40 Carter, Jimmy (1979). "Executive Order 12127—Federal Emergency Management Agency," March 31. Online by Gerhard Peters and John T. Woolley, *The American Presidency Project*. http://www.presidency.ucsb.edu/ws/?pid=32127; Taylor, Andrew (1992). "Andrew Is Brutal Blow for Agency." *CQ Weekly*: 270, September 12. http://library.cqpress.com/cqweekly/WR102408159
41 https://www.fema.gov/about
42 Prah, P.M. (2005). "Disaster Preparedness." *CQ Researcher*, Vol. 15, pp. 981–1004. http://library.cqpress.com/cqresesarcher/document.php?id=cqresrre2005111800&type=hitli; Taylor, Andrew (1992). "Andrew Is Brutal Blow for Agency." *CQ Weekly*, 2703, September 12. http://library.cqpress.com/cqweekly/WR102408159; Roberts, Patrick S. (2006). "FEMA After Katrina." *Policy Review*, June/July, pp. 15–33.
43 Roberts, Patrick S. (2006). "FEMA after Katrina." *Policy Review*, June/July, pp. 15–33.
44 Adams, Rebecca (2005). "FEMA Failure a Perfect Storm of Bureaucracy." *CQ Weekly*, September 12, pp. 2378–2379. http://library.cqpress.com/cqweekly/weeklyreport109-000001853459; Roberts, Patrick S. (2006). "FEMA after Katrina." *Policy Review*, June/July, pp. 15–33.
45 Library of Congress, Congressional Research Service, "Federal Emergency Management Policy Changes after Hurricane Katrina: A Summary of Statutory Provisions," March 6, 2007. http://www.fas.org/sgp/crs/homesec/RL33729.pdf
46 National Oceanic and Atmospheric Administration, "Billion-Dollar Weather/Climate Disasters." National Climatic Data Center, 2013.
47 Naylor, Brian (2012). "Lessons from Katrina Boost FEMA's Sandy Response." *NPR*, November 3. http://www.npr.org/2012/11/03/164224394/lessons-from-katrina-boost-femas-sandy-response
48 Naylor, Brian (2010). "Has FEMA Recovered from Hurricane Katrina?" *NPR Morning Edition*, August 27. http://www.npr.org/templates/story/story.php?storyId=129466751

49 FEMA, "Plan, Prepare and Mitigate." http://www.fema.gov/plan-prepare-mitigate
50 Blair, Kimberly (2014). "FEMA Disaster Recovery Centers Opening This Weekend." *Pensacola News Journal*, May 9. http://www.pnj.com/story/news/2014/05/09/femas-disaster-recovery-center-opens-saturday/8911747/
51 DHS, FEMA, "National Security Council." https://www.fema.gov/about/offices/national-advisory-council
52 Source: FEMA-OPPA-SFI@fema.gov https://www.fema.gov/pdf/about/programs/oppa/findings_051111.pdf
53 https://www.fema.gov/pdf/about/programs/oppa/critical_infrastructure_paper.pdf
54 Elias, Bart, Peterman, David Randall, and Frittelli, John. *Transportation Security: Issues for the 114th Congress.* CRA Report RL33512.
55 Department of Homeland Security, "The Federal Protective Service." https://www.dhs.gov/topic/federal-protective-service
56 Department of Homeland Security, "Office of Biometric Identity Management." https://www.dhs.gov/obim
57 Transportation Security Laboratory, "Science and Technology." https:///www.dhs.gov/science-and-technology/transportation-security-laboratory
58 Plum Island Animal Disease Center, "Science and Technology." https:///www.dhs.gov/science-and-technology/plum-island-animal-disease-center
59 https://www.dhs.gov/intelligence-enterprise
60 https://www.dhs.gov/national-network-fusion-centers-fact-sheet

5 Other Federal Risk Management Agencies

Chapter Outline

Introduction

On February 3, 2023, community members in East Palestine, Ohio, were evacuated when, at 8:55 PM, a Norfolk Southern freight train derailed as it made its way through the town. The train, carrying hazardous materials that can cause cancer (vinyl chloride, benzene, and butyl acrylate), caught fire, releasing dangerous chemicals into the air. Some of the train cars burned for over two days, and others were set on fire as part of a "controlled burn."[1]

Multiple agencies from Ohio and neighboring states responded to help control the damage, and the mayor of the city declared a state of emergency. Thousands of small fish in nearby streams died. Residents experienced health problems such as coughing, bloody noses, nausea, headaches, and burning sensations in their lungs. Government officials reported that there were no dangerous chemicals existing in either the air or water. Contaminated soil was taken to Michigan for disposal, and contaminated water was sent to Texas. Property values plummeted so much that the members of the U.S. Senate Commerce Committee established a fund to pay homeowners for the decline of their property values.[2]

Many agencies responded to this incident. The Environmental Protection Agency (EPA) was concerned about protecting the air, soil, and water, from possible contamination and ensuring the safety of the residents. Officials from the EPA were on the site within hours of the accident to monitor the air quality. The National Transportation and Safety Board, which is responsible for investigating accidents in the nation's transportation system, was also on site to investigate the

DOI: 10.4324/9781003434887-5

derailment. The U.S. Secretary of the Department of Transportation, Pete Buttigieg, visited the site but was criticized for not visiting until three weeks after the disaster. President Biden also did not visit the community.[3]

When an event like this occurs, there are many federal agencies in addition to the Department of Homeland Security that play a role in critical infrastructure protection. This chapter will explore those other agencies and the federal responsibilities for critical infrastructure protection and risk assessment outside of the DHS.

Department of State

Created in 1789, the U.S. Department of State is the diplomatic wing of the federal government and oversees U.S. relations with other countries. The mission statement for the Department of State indicates that the department "is to shape and sustain a peaceful, prosperous, just, and democratic world and foster conditions for stability and progress for the benefit of the American people and people everywhere."[4] Personnel at the Department of State seek to advance U.S. national security through diplomacy with foreign countries, promote the country's economic interests, provide services as needed to protect citizens at home and abroad, and reaffirm the U.S.'s role throughout the world.[5]

The Department of State's attention to protecting the nation's critical infrastructure is in part centered on efforts to reduce terrorism. Personnel work closely with representatives from foreign governments to detect, prevent, and respond to terrorist events should they occur. Personnel attempts to reduce the availability of funds available for terrorist groups, increase law enforcement efforts to detect activities by terrorist groups, and increase information sharing between interested parties.

In 1984, the State Department established a program called Rewards for Justice through which they award up to $10 million for any information that helps them counter possible terrorist acts and at the same time and prevent harm to the nation's critical infrastructure. The department provides monetary rewards for information in four categories: terrorism, foreign election interference, North Korea, or malicious cyber activities. This means that if an individual provides officials with information that aids the department in identifying anyone who carries out malicious cyber activities against critical infrastructure in the U.S. (and violating the Computer Fraud and Abuse Act or CFAA) while acting in conjunction with a foreign government, they could be eligible for a reward. Some malicious cyber acts may violate CFAA. Examples of these include using a ransomware attack on an agency with an extortion threat and accessing a protected computer (i.e. a computer system owned by the federal government) without authorization to obtain information. To ensure the safety of any potential informant, the State Department has set up a Dark Web tip-reporting channel. The reward may be issued via cryptocurrency. Since the program began, it has paid over $200 million for information on cyber offenders who threatened U.S. national security.[6] More information on this program is presented in Table 5.1.

A more recently created State Department agency that focuses on preventing terrorism is the Bureau of Counterterrorism, which was created in 2012 when Congress passed HR 2333 (PL 103-236). This organization coordinates national efforts to block terrorist acts and works closely with international actors to coordinate approaches and information regarding terrorism. The agency also advises the secretary of events and trends.[7]

Another aspect of the State Department's efforts to protect critical infrastructure is their efforts to reduce cybercrime. One method used by the department is to partner with other countries and support international efforts to secure cyberspace for an open, secure, and reliable communication infrastructure. This is essential because computer networks play a critical role in the economy and

Table 5.1 Reward

Rewards for Justice is offering a reward of up to $10 million for information on Maalim Ayman, leader of Jaysh Ayman, an al-Shabaab unit that has conducted terrorist attacks in Kenya and Somalia. On January 5, 2020, al-Shabaab terrorists attacked Kenyan and U.S. personnel at Manda Bay Airfield, Kenya, killing two U.S. contractor pilots and a U.S. Army specialist acting as an air traffic controller. A third U.S. contractor and two other U.S. service members were injured in the attack.
Maalim Ayman was responsible for preparing for the January 2020 attack. In November 2020, the Department of State designated Ayman as a Specially Designated Global Terrorist under Executive Order (EO) 13224, as amended.
The Manda Bay facility is a Kenya Defense Forces military base utilized by U.S. armed forces to provide training and counterterrorism support to East African partners, respond to crises, and protect U.S. interests in the region.

international security of the U.S. The focus of this effort is the Bureau of Cyberspace and Digital Policy, which leads the department's programs for cyberspace and digital diplomacy. The mission of this agency is to coordinate the department's work in these areas, encourage responsible behavior in cyberspace, and advance policies that protect the integrity and security of the Internet.

The State Department has been involved in other efforts to protect the nation's infrastructure, including the nation's supply chain. President Biden called for a Supply-Chain Summit in October 2021, to address how to tackle challenges to the supply chain that resulted from the COVID-19 pandemic. Representatives from 14 countries and the European Union (EU) attended. This was followed by a Supply Chain Ministerial in July 2022, hosted by the Secretary of State Blinken. The goal was to build a reliable supply chain and reduce disruptions in the future.[8]

Recently, the Department of State recognized that the COVID virus was a threat to the nation's infrastructure and that there was a need to strengthen the global biosecurity infrastructure. To that end, the department published the U.S. Global COVID-19 Response and Recovery Framework, which was intended to help secure the global healthcare network. They encouraged every country to build back better to prevent, detect, and respond to the next infectious disease outbreak.

When it comes to the energy infrastructure, the Department of State works for net-zero emissions by 2050 to reduce the nation's carbon footprint. They work to advance clean energy and develop resilient and transparent energy systems to meet the climate change threat. The department seeks to develop and execute international energy policy through the Bureau of Energy Resources.

Department of Justice

The Department of Justice plays an essential role in protecting the nation's critical infrastructure. The primary role of the Justice Department is to keep the country and the citizens safe from all threats, foreign and domestic, and to protect the civil rights of all residents.[9] PDD-63 assigned the Department of Justice to be the lead sector liaison for emergency law enforcement services. In addition, Executive Order 13231 assigned the attorney general to serve on the President's Critical Infrastructure Protection Board and to be co-chair of the Incident Response Coordination and Physical Security committees. Consequently, there are many organizations within the Department of Justice that play a role in critical infrastructure protection responsibilities.

When it comes to cybercrimes, the Justice Department seeks to protect Americans from malicious offenders who rely on technology to victimize others in sometimes violent and coercive ways. Offenders sometimes commit cyberstalking, distribute intimate images or private information against a person's will, or victimize individuals through offenses such as sextortion, doxing, or swatting. Members of the Justice Department investigate and prosecute these acts through the Criminal Division's Computer Crime and Intellectual Property Section, as well as cyber-specialized prosecutors. If pertinent, the Civil Rights Division may prosecute offenses if they are found to be motivated by bias. Sex trafficking is conducted through online activity and is targeted by the Justice Department.

If needed, the Justice Department will support proposed legislation in Congress to protect citizens from cybercrimes. An example of this is the Stopping Harmful Image Exploitation and Limiting Distribution Act of 2021. The purpose of this law was to prohibit the distribution of intimate images if it was not consensual.

The Justice Department plays a critical role in protecting the safety of the Internet for all users, whether they are businesses or private individuals. Department personnel investigate cyber threats and attacks on the U.S. that cause victims to lose billions of dollars in damages each year. They are a critical actor in identifying cyber offenders, investigating their offenses, and holding them responsible for wrongdoing. The department then seeks to disrupt offenders' ability to fully carry out attacks and profit from their actions. In doing so, they attempt to deter others from similar actions. If possible, department personnel will recover property lost as a result of a crime. Recently, the department was able to recover $2.3 million that was paid as a ransom to hackers that attacked the Colonial Pipeline.

One agency found in the Justice Department that protects cyber infrastructure is the Computer Crime and Intellectual Property Section (CCIPS), which is found within the Department's Criminal Division. The CCIPS has the primary goal of detecting and deterring both computer crimes and intellectual property crimes. They do this through collecting evidence, investigating, and prosecuting cyberattacks on our nation's critical assets and by providing legal advice to investigators and prosecutors who are investigating and prosecuting offenses in the U.S. and around the world. This section also addresses policy and legislation issues such as information sharing among the military, the intelligence community, law enforcement, civilian agencies, and the private sector, as well as government network intrusion detection and strategic planning. CCIPS regularly coordinates its activities with the Department of Defense, the Critical Infrastructure Assurance Office, the National Security Council, and interagency groups that work on issues related to critical infrastructure protection.[10]

Within the CCIPS is the Cybersecurity Unit, which was created in December 2014. The personnel in this unit provide expert legal advice on criminal electronic surveillance statutes for law enforcement agencies that conduct cyberinvestigations. They also work with private sector officials and members of Congress as needed. The agency helps law enforcement obtain the resources they need to solve cybercrimes and prosecute them. Another key aspect of the cybersecurity unit is the prevention of cybercrimes and increasing the public's understanding of the issue.[11]

In 2018, the attorney general created a Cyber-Digital Task Force to carry out a review of the department's response to cyber threats and to identify how federal law enforcement could more effectively accomplish its mission of investigating and deterring cybercrimes. The resulting report focuses on multiple areas of concern, including threats to the electoral process. They reinforced the importance of being prepared for cyber events, including the importance of building relationships, sharing information, and engaging organizations and sectors that face significant risks.[12]

Another agency developed to battle cybercrime is the Ransomware and Digital Extortion Task Force, which has the goal of bringing to bear "the full authorities and resources of the department in confronting the many dimensions and root causes of this threat."[13] The task force members recognize that many cybercrimes and ransomware offenses are transnational, but they also recognize that it is necessary to enhance internal investigations and prosecutions of offenders and criminal organizations. They rely on input from the U.S. Attorneys' Offices, the Criminal Division's CCIPS and Money Laundering and Asset Recovery Section (MLARS), the National Security Division, and the Federal Bureau of Investigation (FBI), among other components within the Justice Department, to assist in identifying offenders who engage in these crimes. They also help to create ways to disrupt and dismantle the groups and the networks they rely on to carry out their attacks.

In April 2021, the Justice Department conducted a review of its response to cyber incidents. The investigation relied on the input from many organizations within the Justice Department such as the Criminal Division's CCIPS, the Civil Rights Division, the Child Exploitation and Obscenity Section, the MLARS, the FBI, among others. The Committee made multiple recommendations to expand the department's response to cyber offenses and disrupt attacks, including the creation of a National Cryptocurrency Enforcement Team (NCET) to concentrate on the illegal use of emerging cryptocurrency. The agency, created in October 2021, immediately hired prosecutors with expertise in related areas to enforce laws on money laundering, computer crimes, and other offenses.

The new agency, housed within the CRM, was set up to oversee investigations and prosecutions of offenders who violate cryptocurrency regulations. The NCET personnel work alongside members of the MLARS, CCIPS, and other department sections that have legal experts in these areas. The NCET team also attempts to trace and recover assets that were taken as a result of fraud and extortion, such as cryptocurrency ransom payments. The NCET's responsibilities also include assisting in investigations and prosecutions involving digital assets and strengthening relationships with federal, state, local, and international law enforcement agencies, along with private industries.

Other recommendations made by the Review Committee to protect the safety of computer networks include:

- Establishing a Civil Cyber-Fraud Initiative to be housed within the Department's Civil Division. This group would determine if civil actions would be necessary against any government employees or contractors who do not meet cybersecurity standards;
- Initiating a Cyber-Fellowship within the Justice Department that will support new prosecutors and attorneys who focus on offenses related to cybercrime and threats to national security;
- Generating new policies for increasing cybersecurity surrounding email, including encryption and anti-phishing training for department employees.[14]

A few months later, in July 2021, the Department of Justice, along with the Department of Homeland Security, set up a website they called the Ransomware Resource (StopRansomware.gov). This is a website for dissemination of cybersecurity resources consolidated from multiple federal agencies that are available to all businesses and individuals that will help combat threats of ransomware. The site is the first joint website to assist organizations detect, prevent, or mitigate the risk of cyberattacks. The material includes guidance, alerts about new attacks, and resources, which is now available in one place, which lessens the chances of missing information. The attorney general recognized that stopping these attacks must be a joint effort that includes business leaders and owners who play a role in recognizing these offenses and taking action to stop these crimes from occurring.

The department also recognizes the need to develop cooperative work environments with other government agencies, other nations, state, local, tribal, and territorial governments, and private sector owners and operators. As a way to support this, the department appointed a prosecutor within the Justice Department to be the Cyber Operations International Liaison. This new prosecutor will work with other experts within the department and with European allies to support actions against cyber offenders such as arrests, extraditions, and asset seizures.

To protect the Department of Justice itself from possible cyberattacks, the department has increased efforts to prevent any compromise to its networks. They relied on EO 14028, signed by President Biden to improve the nation's cybersecurity, which required additional measures for all federal departments to protect cybersecurity. The Justice Department agreed to institute multifactor authentication efforts, data encryption, and improvements to email and data-transfer methods.

Federal Bureau of Investigation

The FBI, located within the Department of Justice, is the nation's lead federal law enforcement agency whose mission is to investigate possible cyber-based crimes and intrusions to protect critical infrastructure. They investigate cybercrimes, identify offenders, and disrupt criminal groups who intend to harm America's infrastructure. They also assist in educating the public about ways to protect themselves and their businesses from attacks. The FBI's Criminal Investigative Division investigates and prosecutes Internet crimes such as cyber fraud, identity theft, child sexual exploitation, or cyber-terrorism. The FBI has located experts in cybercrimes in each of the field offices across the country and has created "cyber action teams" that travel around the world to identify and prosecute cybercrimes. The FBI works alongside other state and federal agencies to achieve their goals.

The FBI seeks to share its knowledge of cyber concerns with industry through a variety of outreach initiatives and information sharing programs. They publish multiple reports to provide information about cyber threats or possible actions to take to mitigate the effects of an attack. The FBI has developed ways to share cyber-related information with state and local governments that, despite being sensitive or classified, could also prevent cybercrimes from occurring. One of those methods is InfraGard, which started in Cleveland in 1996 and has become one of the largest and unique public–private partnerships (see Chapter 8). This is a cooperative, outreach effort between officials in the FBI and private sector businesses, academic institutions, state and local law enforcement agencies, security personnel, IT professionals, healthcare and emergency workers, members of the military, and other participants who use the program to communicate and work together to increase the security of the nation's assets. The goal of InfraGard is simply to increase the dialogue and the flow of timely communication and information between these multiple groups about threats and risks so that the owners and operators of infrastructure assets can better protect their resources. Their stated mission is "to enhance our nation's collective ability to address and mitigate threats to U.S. critical infrastructure by fostering collaboration, education, and information sharing through a robust private sector/government partnership."[15]

InfraGard maintains an ongoing, two-way discussion with its members to provide information to help protect their infrastructure, including threat advisories, alerts, and warnings as needed. They also help to educate its members by providing analytical reports, vulnerability assessments, or intelligence bulletins. They also provide training exercises on infrastructure and protection measures. This interaction also gives FBI personnel information as they engage with owners and operators who have expertise. Allows peers to interact and share information on threats or other matters.

The exchange of sensitive information is possible because each InfraGard member has been vetted and identified as a person who might need to know the information for security reasons. The agencies and the FBI have created a relationship based on trust and credibility. This process also allows the FBI to work with the members when they are investigating incidents or gathering intelligence about a threat or an event. InfraGard holds meetings within each chapter but also maintains a public Web site alongside a secure private Web site, which has critical infrastructure-related information for the members of InfraGard.

InfraGard currently has over 26,000 members in over 77 local Infraguard Member Alliances (IMAs) around the U.S. All 56 FBI field offices support at least one InfraGard Chapter. Each chapter has a Special Agent Coordinator associated with it who in turn works closely with the Cyber Division in the FBI. IMAs often host events. InfraGard maintains partnerships with the FBI's Directorate of Intelligence, Counterterrorism Division, Counterintelligence Division, Criminal Investigative Division, and Weapons of Mass Destruction Directorate. They also share information and have created partnerships with the DHS and other federal agencies.

In addition, the FBI's National White Collar Crime Center, created originally in 1978, provides assistance and training for law enforcement groups and regulatory agencies that carry out investigations into white-collar crimes and related economic crimes. The goal is to combat economic and cybercrimes. The training is provided on topics such as encryption, virtual currency, or digital forensics. They also provide training at no cost to other federal agencies and to state, local, tribal, and territorial law enforcement.[16]

An organization within the FBI is the Internet Crime Complaint Center (IC3), which processes complaints regarding Internet-related criminal acts. They have created a convenient way for the public to report offenses and encourage people to file complaints because that helps the FBI understand trends in cybercrimes, emerging threats, and new scams and then make those public. Upon receiving a complaint, agents will determine if there was an actual compromise and then determine if the information should be (or can be) shared with the public. If needed, they will refer the complaint to federal, state, local, and international law enforcement agencies. They will also issue an alert to affected entities, if pertinent.[17] They educate the public about safe behavior on the Internet. In 2021, IC3 reported that they received 847,376 complaints, a 7% increase from the previous year. The potential losses from those offenses could be more than $6.9 billion. These complaints included ransomware, schemes surrounding business e-mails, and illegal use of cryptocurrency.[18]

The FBI created the Virtual Assets Unit in early 2022, a new partnership between the FBI's Criminal Investigative and Cyber Divisions. These agencies will work together, to investigate crimes that rely on virtual money or cryptocurrency, such as drug trafficking, human trafficking, or tax evasion. This organization has been tasked with providing analysis, support, and training to law enforcement and intelligence agencies regarding cryptocurrency. Alongside agency staff who are experts in the field, agency personnel will track illicit funds and investigate offenses, all to identify and disrupt criminal networks. In general, the agency will provide a focused approach to identifying crypto crimes that can pose a threat to national security and harm the nation's economy.[19]

The National Cyber Investigative Joint Task Force (NCIJTF) was created in 2008. Noting that many "traditional" crimes such as terrorism, financial fraud, and identity theft are now being carried out online by tech-savvy offenders, the FBI brought together over 30 agencies to create a task force to share information and coordinate investigations. The members include law enforcement, international partners, and private agencies. They seek to identify offenders, pursue them, arrest them, and break up their organizations. At the same time, the NCIJTF works to protect the privacy rights of Americans.[20]

The FBI will, when needed, work with sector-specific agencies in an "action campaign" to assist particular stakeholders if an emerging cyber threat is identified and needs quick attention. Other times, the FBI supports Cyber Summits that are typically one-day events that bring corporate attorneys and employees together to discuss ways to share information.

The Domestic Security Alliance Council was founded in 2006 to increase collaboration with private owners to provide a forum to exchange information and increase communication between public and private groups regarding criminal activity that affects interstate commerce to protect employees, assets, infrastructure, and information.

Career Focus: Christopher Wray, Head of the FBI

The current Director of the FBI, Christopher Wray, was sworn in on August 2, 2017. Born in New York City, he graduated from Yale University and then Yale Law School. After serving as a law clerk for the U.S. Court of Appeals for the Fourth Circuit, he practiced law at an international law firm where he investigated white-collar crime.

Before this, he was an assistant U.S. attorney for the Northern District of Georgia. There, he successfully prosecuted federal offenses including gun trafficking, financial fraud, and drug trafficking. In 2001, Wray was appointed to serve as the associate deputy attorney general and then became the principal associate deputy attorney general. In 2003, President George Bush nominated Wray to be the assistant attorney general for the Criminal Division in the Department of Justice. In this position, he oversaw national and international investigations and prosecutions. He participated in the Counterintelligence and Export Control section of the Criminal Division. Wray was part of the task force that investigated the misdeeds of officials at Enron. He was also part of the investigation into cybercrime offenses being carried out by foreign governments.

As the FBI Director, Wray is responsible for managing the organization's day-to-day tasks. The agency investigates federal offenses such as domestic and international terrorism, cybercrimes, white-collar crimes, and kidnappings. Whoever serves in this position must ensure that cases are handled correctly. The director reports to the U.S. Attorney General and the President of the U.S.

Source: FBI History. Directors, Then and Now. https://www.fbi.gov/history/directors

Department of Commerce

The U.S. Commerce Department plays a role in protecting the nation's critical infrastructure, and the communications networks in particular as they relate to commerce, trade, and the economy. In PPD-63, President Clinton wrote that "the Department of Commerce, the General Services Administration, and the Department of Defense shall assist federal agencies in the implementation of best practices for information assurance within their individual agencies." He also wrote that "the Department of Commerce and the Department of Defense shall work together, in coordination with the private sector, to offer their expertise to private owners and operators of critical infrastructure to develop security-related best practice standards." From this, it can be seen that the Department of Commerce is given the task of increasing the public's awareness of cybersecurity, while also protecting privacy. This is a way to maintain the public's safety while also supporting economic security and national security.

One source of the department's infrastructure protection responsibilities is the Telecommunications Authorization Act of 1992, which created Commerce's National Telecommunications and Information Administration (NTIA). This agency serves as the main adviser to the president on telecommunications and other information policies. NTIA oversees grant programs that increase the use, availability, and reliability of broadband and other technologies. The NTIA leads outreach and aids the agencies in the communications sector. It was hoped that the NTIA would raise industry awareness about the true nature of threats and vulnerabilities to those in the sector, and then increase efforts for better information sharing.[21]

After the terrorist attacks of 9/11, investigations pointed to a breakdown of communications between agencies. In response, Congress in 2012 passed legislation called the First Responder Network Authority (FirstNet) that established a wireless network solely for first responders and public safety officials. FirstNet officials consult with local, state/territory, tribal, and federal public safety agencies across the country and other stakeholders, providing secure and safe communications network. FirstNet Authority is an independent agency that is housed within the Department of Commerce and NTIA.[22]

Another principal agency within the Department of Commerce is the National Institute of Standards and Technology (NIST). This agency is a non-regulatory agency that supports innovation and industrial competitiveness. This agency's role in protecting cyberspace became more apparent after President Obama, on February 12, 2013, issued Executive Order 13636. In this statement, Obama stated that the "cyber threat to critical infrastructure continues to grow and represents one of the most serious national security challenges we must confront." He then asked the Commerce Department to develop a Cybersecurity Framework to reduce potential cyber risks to the nation's critical infrastructure, thus allowing the economy to thrive. Today, NIST develops standards, guidelines and best practices for the federal government and the public to maintain a secure cyber world. At the same time, they protect privacy of Internet users.

The Department of Commerce and the European Commission worked jointly to create the EU–U.S. Privacy Shield Framework. By this policy, companies must transfer data between countries in transatlantic commerce in such a way that the data will be protected.

Department of Transportation

According to PDD-63, the Department of Transportation (DOT) serves as the lead sector liaison for aviation, highways, mass transit, pipelines, rail, and waterborne commerce. They work to protect the nation's transportation infrastructure from physical and cyberattacks. In President Obama's Executive Order 13231, the Secretary of Transportation was appointed to serve on the President's Critical Infrastructure Protection Board and to serve as the co-chair of the Infrastructure Independencies Committee. They often collaborate with the DHS, TSA, CISA, and private partners as needed to improve methods to protect critical infrastructure.

The ransomware attack on the Colonial Pipeline in May 2021 demonstrated the risks that exist to pipeline operations, the economy, and Americans. To tackle these concerns, the DOT created the Pipeline and Hazardous Materials Safety Administration (PHMSA). This organization is tasked with protecting the environment and citizens and the safe operations of pipelines, underground storage tanks, and the shipping of hazardous materials. To this end, PHMSA assists in developing and enforcing regulations pertaining to reliable and safe operations of pipeline systems and shipments of materials by land, sea, and air. Much of this infrastructure is held and operated by private owners and operators. PHMSA collaborates with owners and operators and publishes guidance documents to help owners and operators understand the rules and regulations that affect pipeline operations.[23]

The DOT has authority over cybersecurity as it pertains to different transportation modes. They must maintain a secure computer network to deter cybersecurity risks within the transportation system. To do this, the Office of Intelligence, Security, and Emergency Response helps protect the transportation system from those who may want to attack these critical systems while at the same time ensuring ease of mobility to allow for personal and business travel. This is accomplished by developing techniques for increasing intelligence, security, preparedness, and response.[24]

The department reaches out to the public to help them learn more about protecting transportation infrastructure. The DOT Cybersecurity Symposium of 2021 held sessions each Wednesday morning throughout the month of October (before CISA's Cyber Summit), that were open to the public. Different topics were discussed by experts to help the public learn more about the activities of the DOT surrounding cybersecurity.[25]

Federal Aviation Administration

The Federal Aviation Administration, as part of the DOT, is focused on protecting air travel infrastructure. One organization set up to do this is the Air Traffic Organization (ATO) Cybersecurity Group, whose mission is to secure ATO systems from cybersecurity threats. Within the ATO are various programs, one of which is the Aviation Cyber Initiative. This was set up to reduce risks to cybersecurity related to aviation, and to improve the resilience of aviation's cyberspace in case of an event. In all, they work to keep the nation's aviation safe, secure, and efficient. Also located within the ATO is the Cyber Enterprise Architecture group. This group works with stakeholders, both internal and external, to coordinate a comprehensive approach to risks, develop best practices, and identify technology to improve the security of aviation. They also identify architectural trends that may help with cybersecurity. Related to this is the Cybersecurity Engineering Group. This organization's mission is to develop tools that will help to increase cybersafety in the airlines.[26]

Federal Communications Commission

The Federal Communications Commission (FCC) has the responsibility to regulate interstate and international communications in the U.S. through radio, television, wire, satellite, and cable. They work to create networks that are sustainable during emergencies or natural disasters. Officials within the FCC have the ability to adopt, administer, and enforce rules related to cybersecurity of the communications critical infrastructure. According to FCC regulations, communications providers are required to report on the security of their infrastructures, including any service disruptions or outages that may impact the safety of the public or the ability for Emergency Response.

To assist with their task, the FCC works with other federal agencies and partners, others in the communications sector, and private owners and operators to ensure continuous and secure communications systems. They have instituted advisory committees such as the Communications, Security, Reliability, and Interoperability Council which seeks to advise the FCC on actions that can be taken to improve the security of both commercial and public safety communications systems. They also make recommendations concerning technical standards and gaps pertaining to the emergency communications system (i.e. the 911 system).[27] The FCC regulates the 911 or Emergency Alert System, which is a critical component of the nation's disaster preparedness system. The FCC must maintain this network to ensure it is secure and reliable if an event occurs or be able to restore it quickly if it is damaged so that emergency responders are able to communicate to help victims. See Table 5.2 for more information on the Emergency Alert System.

Table 5.2 Emergency Alert System

Emergency Alert System: a national public warning system requiring broadcasters, cable television operators, satellite digital audio radio providers, and direct broadcast satellite operators to provide communications capability for the president to address the nation during a national emergency. The public needs timely and accurate emergency alerts and warnings about impending disasters and other emergencies. The FCC and its federal partners are working toward a comprehensive alerting system that utilizes multiple communications technologies to reach the public quickly and effectively. The FCC is working toward the goals of operable and interoperable public safety communications systems for its spectrum users through its Commercial Mobile Alert System. https://www.fcc.gov/general/emergency-communications

FCC is also required, under presidential directive, to work with the DHS and other federal departments and agencies to identify and prioritize communications critical infrastructure and any vulnerabilities they face. The FCC is to work with stakeholders to address those vulnerabilities. They are also asked to cooperate with foreign governments and international organizations to increase the security and resilience of critical infrastructure within the communications sector.[28]

One agency within the FCC is the Public Safety and Homeland Security Bureau, which oversees policies to protect the safety of the public, homeland security, emergency management, cybersecurity, and disaster preparedness activities at the FCC. They ensure the public's access to the 911 system as well as emergency alerts. Equally as important is ensuring the reliability of communications systems for first responders. The bureau also collaborates closely with federal government partners and private owners who are responsible for protecting the nation's communications infrastructure.[29]

The mission of the Communications Security, Reliability, and Interoperability Council (CSRIC) is to make recommendations to the Commission to promote the security, reliability, and resiliency of the nation's communications systems. This includes the 911 system, emergency alerts, and national preparedness. They also focus on the availability of emergency communications for law enforcement Every two years, since 2009, CSRIC is assigned to address new issues if needed by the FCC's Chairman.[30]

Through its Cyberplanner program, the FCC assists small business improve their cybersecurity. Often, these businesses do not have the resources or knowledge base to have a complete plan to protect their networks and become easy targets for an attack. The FCC helps through their Small Biz Cyber Planner 2.0, which provides online resources to small businesses to help them create more complete cybersecurity plans.[31]

Another way the FCC helps the public improve their cybersecurity is through the yearly Cyber-Security Awareness Month. This is in October of each year. The FCC, along with other agencies, provides information to citizens on how they improve their network security and protect themselves from a cyberattack while online.[32] The FCC also developed websites and other educational materials to provide hints to people and small businesses to help prevent cyberattacks.[33]

Environmental Protection Agency

PDD-63 designates the EPA to serve as the lead agency and sector liaison for protecting the nation's water supply (drinking water and water treatment systems). Moreover, Presidential Decisions Directives 39, 62, and 63 mandate that EPA officials take part in a federal response program aimed at preparing for and responding to terrorist incidents so that clean water is available as quickly as possible after an event. The Office of Water is the lead EPA office in fulfilling EPA's national critical infrastructure protection responsibilities. Their main responsibility is to ensure safe drinking water for all citizens of the U.S., but they also maintain oceans, watersheds, and other aquatic systems for economic and recreation.

After the terrorist attacks of 2001, the Office of Water expanded its responsibilities to focus on providing technical and financial assistance to those who are carrying out vulnerability assessments. They also provide assistance for Emergency Response planning for drinking water and wastewater utilities. Another role of the Office of Water is to develop new technologies that will help water utilities protect their assets and public health. To do this, they carry out research programs in cooperation with other federal agencies and nongovernmental organizations. The Office of Water also works to increase the communication processes between privately owned or locally owned utilities and government officials regarding preparedness and response activities in the water sector. The office is also developing plans regarding the interdependencies between the water sector and others including energy and transportation.

Another agency within the EPA is the Office of Homeland Security, which oversees the EPA's programs to prevent, protect, respond to, mitigate, and recover from an intentional event or natural disaster. The Office was established to carry out the roles outlined in various executive orders, directives, and PPDs. The agency coordinates the EPA's responses to events that occur and helps to create plans for preventing future events. The office provides advice to the EPA administrator and other leadership related to national security. They also provide coordination for activities related to prevention and protection, and coordination for impacted regions for mitigation, response, and recovery efforts.[34]

In 2022, the EPA created Environmental Finance Centers to assist local communities apply for federal funding made available for projects that are intended to improve water infrastructure projects, among other things.[35]

That same year, the EPA announced a new initiative (the Industrial Control Systems Cybersecurity Initiative—Water and Wastewater Sector Action Plan) to guard the nation's water systems from a cyberattack. The plan identified actions that could be taken to protect water resources within the water sector by improving cybersecurity. The plan results from President Biden's Industrial Control Systems Initiative, which was established after National Security Memorandum 5, Improving Cybersecurity for Critical Infrastructure Control Systems. This is a program to implement technologies that identify threats and provide warnings as a way to create a coordinated approach to protecting the nation's water systems and support strategies for early detection of cyber threats so action can be taken.[36]

The Action Plan has the following goals

- Establishing a task force of water sector leaders;
- Implementing pilot projects to demonstrate and accelerate the adoption of incident monitoring;
- Improving information sharing and data analysis;
- Providing technical support to water systems.

Department of Interior

The Department of the Interior is the sector-specific agency for national monuments and icons, such as the Statute of Liberty (which ranks among the top ten of the department's critical infrastructure assets). The National Monuments and Icons Sector-Specific Plan promotes collaborative efforts at all levels of government to increase cooperation and improve the protection of assets. The goal of the department is to "protect and provide access to our nation's natural and cultural heritage and honor our trust responsibilities to Indian Tribes and our commitments to island communities." Recently, the department has recognized that information security is essential to protect systems from unauthorized access. One example of this is the requirements that records are maintained regarding visitors to assets. All visitors must record the purpose of their visit and who accompanies them. If a person attempts to get into the asset or can access a site without authorization, that must be reported.[37]

The department's safety plans are implemented by the U.S. Park Police and the National Park Service. The National Monuments Sector oversees a variety of assets located throughout the U.S. and includes monuments, physical structures, or objects that are recognized both nationally and internationally as a symbol that represents the nation's history, traditions, and values. These structures often have national, cultural, religious, historical, or political significance. They may also be memorials to people or events.[38] A ranking system helps to determine if a monument or icon is one of National Critical, National Significant, Regional Critical, or Local Significant asset. National assets include Flaming Gorge, Folsom, Shasta, Grand Coulee, Hoover, and Glen Canyon.[39] Further, assets are found in the nation's capital and the Statue of Liberty, Independence Hall, the Liberty Bell, and Mount Rushmore National Monument.

An agency within the Department of the Interior is the Bureau of Reclamation. Their mission is "to manage, develop, and protect water and related resources in an environmentally and economically sound manner in the interest of the American public." The bureau seeks to identify critical assets among its facilities. This includes items such as switchyards, transmission lines, and equipment that comprise the Bulk Electricity System that are owned and operated by the Bureau of Reclamation. This would include hydro-generation control centers, power plants, and transmission centers. They also perform a risk assessment on these facilities as a way to address vulnerabilities and improve security.[40]

Officials inside the bureau have identified Critical Cyber Assets within its critical assets to identify all cyber-based assets. This will help them identify which ones are "essential."

The Bureau of Reclamation, in cooperation with Oak Ridge Associated Universities (ORAU), conducts emergency exercises at sites that have been identified as being critical infrastructures. ORAU is an organization made up of academic institutions that work with agencies to further research and education. This will help officials know, first, if the dams could withstand an event, and how to improve security at these sites. The exercises involve planning, tabletop exercises, and workshops.[41]

In February 2023, the Office of the Inspector General for the DOI completed an investigation of the department's risk management practices to determine any security weaknesses that increase the risk of a compromise to their network. The analysis showed that the agency did not analyze and monitor security weaknesses so their systems were at an increased risk of attack. About a quarter of the systems did not complete annual audits and about half did not implement required security measures. Moreover, half of the agencies did not remediate identified security weaknesses, and over 60% did not have documented privacy controls. Recommendations were made to correct these problems.[42]

Department of Agriculture/Department of Health and Human Services

The Department of Agriculture (USDA), along with the Department of Health and Human Services, has been assigned to be the sector-specific agency for Food and Agriculture (meat, poultry, and egg products). This sector is important because most of the assets are privately owned (i.e. restaurants, farms, food factories, and stores).

In June 2022, the USDA announced a framework to protect the food supply chain that existed during the COVID-19 pandemic and the war in Ukraine. The plan was designed to increase access to food and other supplies. The focus is also on building a more resilient food chain that focuses on local food production.[43]

The 2022 Budget document for the department indicated an allocation of $101 million to fund activities to enhance cybersecurity. The money will be spent to increase cyber protections for the agency's network to protect the agency's data.[44]

Department of Energy

The electric power grid is critical to providing energy. It also supports national defense and emergency services, protects critical infrastructure, and supports the economy. There have been an increasing number of attacks on the nation's power grid in recent years. The Department of

Energy (DOE) oversees the federal government's policies to ensure the security of the cybersecurity surrounding the energy grid, and the oil and gas infrastructure, and ensure that there is resilience and that cyberattacks do not devastate the power grid. It is the lead agency for sector liaison for energy, including the production refining, storage, and distribution of oil and gas, and electric power except for commercial nuclear power facilities. Many energy assets are privately owned, so it is important for officials in this department to work closely with those owners and maintain strong partnerships to develop strategies to protect those assets. The DOE focuses on all stages of critical infrastructure protection: preparedness, incident response, and recovery. If an event occurs, DOE investigates the event by collecting evidence and analyzing it. The agency then shares pertinent information with others to ensure a quick response and recovery effort.

In May 2020, President Trump signed Executive Order 13920, Securing the U.S. Bulk-Power System. Trump assigned the DOE to work collaboratively with other federal government agencies (Departments of Commerce, Defense, Interior, and HS) to ensure that any future policies incorporate the security of the energy grid and cybersecurity needs.

The public–private relationships are important to the DOE because many of the critical infrastructure assets are owned privately. The DOE often reaches out to these owners for their input about policies. As threats evolve, they work with partners to ensure that citizens have knowledge about the threat and the best method to prevent or mitigate possible harm.

The DOE has established the Integrated Joint Cybersecurity Coordination Center (IJC3) within the Office of the Chief Information Officer. This agency works to reduce cyberthreats and risks to the DOE and seeks to enhance safe communication and sharing of information, expanding the protection of critical infrastructure. In the end, their goal is to improve policies for preparedness.[45]

The Office of Electricity works to increase the reliability and resilience of critical energy infrastructure. The lack of electricity can negatively impact the economy and public health. The office has several priorities, including protecting the energy infrastructure from cyber and physical threats as they continue to evolve. To do this, they have created the North American Energy Resilience Model to help with planning, identification of threats, and mitigating impacts while increasing resilience. They also protect the Defense Critical Electric Infrastructure to assure the U.S. government's role of protecting the country and operating the military. They work to support sharing of information and collaboration efforts.

The Office of Cybersecurity, Energy, Security, and Emergency Response (CESER) identifies threats that threaten to interrupt the flow of power to homes and businesses. They focus on improving the security of the energy infrastructure. They focus on preparedness, and creating a stronger, more secure, energy system. In part, this is done by the Cybersecurity Risk Information Sharing Program. They also work closely with the Electricity Subsector Coordinating Council and the Oil and Natural Gas Subsector Coordinating Council. CESER helps with training exercises such as GridEx.[46]

The 2018–2020 Cybersecurity Strategy for the DOE emphasizes that efforts to increase cybersecurity are essential for the department and that those efforts must involve both government employees and private citizens. Their strategy stresses the importance of sharing cyberthreat data with others to mitigate threats that are identified. They have methods for increasing cooperation with other federal agencies to implement best practices for cybersecurity. They also support ideas to build a system for increasing security for cloud data storage and a secure internal communication system for sharing sensitive data. They plan to design a framework for cyberrisk management to improve response to threats as they continue to identify, investigate, and mitigate threats.[47]

Department of Treasury

The Department of the Treasury is the lead agency for sector liaison for banking and finance. Because banking and finance play such a critical role in the economy, financial institutions are subject to significant regulation and examination standards. Many financial firms have chosen to collaborate with government officials through public–private partnerships. One of those is the Financial Services Sector Coordinating Council for Critical Infrastructure Protection and Homeland Security, LLC, and another is the Financial Services Information Sharing and Analysis Center. These agencies seek to inform financial firms on different methods to strengthen their cybersecurity and reduce the possibility of attack.

The Department of Treasury created an Office of Cybersecurity and Critical Infrastructure Protection. They seek to educate the public about cybersecurity risks so they can take steps to protect themselves.[48]

A secure cyber network is essential for the Department of the Treasury because they rely on sensitive information to conduct business. The Department's Cybersecurity program must protect information, services, and assets from a possible cyberattack. The department must work continually to maintain its network security. They will often coordinate with other federal agencies and private sector owners and operators. They monitor their systems to identify unauthorized access and report those to federal officials, who will then share them with appropriate other federal agencies and private organizations. They will also conduct program reviews regularly to help strengthen cybersecurity in the department.[49]

Conclusion

Many federal agencies have a role in protecting critical infrastructure across the nation. The goal is not only to protect the infrastructure from an attack but also to recover quickly if that infrastructure is damaged. These departments and agencies must work with each other but also in cooperation with private owners and operators to increase the security of their assets and reduce the risk of an attack.

Key Terms

Bureau of Counterterrorism
Bureau of Cyberspace and Digital Policy
Computer Crime and Intellectual Property Section (of Department of Justice)
Cyber-Digital Task Force
Ransomware and Digital Extortion Task Force
InfraGard
National White Collar Crime Center
Internet Crime Complaint Center (IC3)
Virtual Assets Unit (VAU)
National Cyber Investigative Joint Task Force
Domestic Alliance Security Council
Telecommunications Authorization Act
First Responder Network Authority
National Institute of Standards and Technology
Pipeline and Hazardous Materials Safety Administration
Office of Intelligence, Security, and Emergency Response
Air Traffic Organization Cybersecurity Group
Public Safety and Homeland Security Bureau
Office of Water
National Monuments and Icons Sector-Specific Plan

Review Questions

1 How is the Department of State involved with critical infrastructure protection?
2 Why is it important for federal agencies to work in conjunction with private owners of critical infrastructure?

3 What is the role of the Department of Justice (and FBI) in critical infrastructure protection?
4 What is InfraGard?
5 How does the Department of Commerce help to protect critical assets?

Notes

1 Sullivan, Becky (2023). "What to Know about the Train Derailment in East Palestine, Ohio." *NPR*, February 16. https://www.npr.org/2023/02/16/1157333630/east-palestine-ohio-train-derailment
2 Eaton, Sabrina (2023). "Norfolk Southern to Reimburse East Palestine Residents for Property Value Losses." Cleveland.com, May 10. https://www.cleveland.com/news/2023/05/ Norfolk-southern-to-reimburse-east-palestine-residents-for-property-value-losses.htm
3 Norton, Tom (2023). "Fact Check: Has Joe Biden Acknowledged East Palestine, Ohio, Derailment?" *Newsweek*, February 22. https://www.newsweek.com/fact-check-has-joe-biden-acknowledged-east-palestine-ohio-derailment-1783052
4 US Department of State, *FY 2014 Agency Financial Report*, November 2014. http://www.state.gov/s/d/rm/index.htm#mission
5 US Department of State, "Diplomacy in Action." http://www.state.gov/r/pa/map/index.htm
6 www.rewardsforjustice.net
7 US Department of State, "About Us—Bureau of Counterterrorism." https://www.state.gov/about-us-bureau-of-counterterrorism/
8 https://www.state.gov/supply-chain-ministerial/
9 https://www.justice.gov/about
10 "Computer Crime and Intellectual Property Section." https://www.justice.gov/criminal-ccips
11 US Department of Justice, "Cybersecurity Unit." https://www.justice.gov/criminal-ccips/cybersecurity-unit
12 *Report of the Attorney Generals' Cyber Digital Task Force*, July 2, 2018. https://www.justice.gov/media/1116921/dl?inline=
13 US Department of Justice, "Memorandum for all Federal Prosecutors," June 3, 2021. https://www.justice.gov/dag/page/file/1401231/download
14 Department of Justice, "Comprehensive Cyber Review," July 2022. https://www.justice.gov/media/1232936/dl?inline=
15 Department of Justice, Federal Bureau of Investigation, Office of the Private Sector. "Infragard." https://www.infragard.org/Files/InfraGard_Factsheet_2-24-2022.pdf
16 Source: NW3C Home. https://www.nw3c.org/
17 US Government Accountability Office, "Critical Infrastructure Protection: More Comprehensive Planning Would Enhance the Cybersecurity of Public Safety Entities' Emerging Technology," GAO-14-125, January 2014. http://www.gao.gov/assets/670/660404.pdf; "Internet Crime Complaint Center IC3." https://www.ic3.gov/
18 FBI (2021). *Internet Crime Report*. https://www.ic3.gov/Media/PDF/AnnualReport/2021_IC3Report.pdf
19 "The FBI Establishes a New Virtual Assets Unit," March 15, 2022. https://www.fbi.gov/news/press-releases/the-fbi-establishes-new-virtual-assets-unit
20 https://www.fbi.gov/investigate/cyber/national-cyber-investigative-joint-task-force
21 Cordesman, Anthony H., and Cordesman, Justin G. (2002). *Cyber-Threats, Information Warfare, and Critical Infrastructure Protection: Defending the U.S. Homeland*. Westport, CT: Praeger
22 US Department of Commerce, "First Net Authority." https://www.firstnet.gov/network
23 US Department of Transportation, "Pipeline and Hazardous Materials Safety Administration." https://www.phmsa.dot.gov/about-phmsa/phmsas-mission
24 US Department of Transportation, "Intelligence, Security and Emergency Response." https://www.transportation.gov/mission/administrations/intelligence-security-emergency-response
25 "4th Annual National Cybersecurity Summit." https://www.cisa.gov/cybersummit2021
26 https://www.faa.gov/air_traffic/technology/cas/aci
27 US Federal Communications Commission, "Communications Security, Reliability and Interoperability Council." https://www.fcc.gov/about-fcc/advisory-committees/communications-security-reliability-and-interoperability-council-0
28 US Government Accountability Office, "Critical Infrastructure Protection: More Comprehensive Planning Would Enhance the Cybersecurity of Public Safety Entities' Emerging Technology," GAO-14-125, January 2014. http://www.gao.gov/assets/670/660404.pdf

29 Federal Communications Commission, "Public Safety and Homeland Security." https://www.fcc.gov/public-safety-and-homeland-security
30 https://www.fcc.gov/CSRICReports; US https://www.fcc.gov/about-fcc/advisory-committees/communications-security-reliability-and-interoperability-council-0
31 Federal Communications Commission, "Cybersecurity for Small Business," March 30, 2022. https://www.fcc.gov/communications-business-opportunities/cybersecurity-small-businesses
32 https://www.fcc.gov/news-events/blog/2010/10/15/home-wi-fi-network-security
33 "FCC Launches New Consumer Webpages Dedicated to Online Safety and Security to Help Kick-off National Cybersecurity Awareness Month." https://www.fcc.gov/news-events/blog/2010/10/08/fcc-launches-new-consumer-webpages-dedicated-online-safety-and-security
34 US Environmental Protection Agency, "Homeland Security." https://www.epa.gov/homeland-security
35 US Environmental Protection Agency, "Environmental Finance Centers." https://www.epa.gov/waterfinancecenter/efcn
36 US Environmental Protection Agency, "EPA Announces Action Plan to Accelerate Cyber-resilience for the Water Sector." https://www.epa.gov/newsreleases/epa-announces-action-plan-accelerate-cyber-resilience-water-sector#:~:text=The%20Water%20and%20Wastewater%20Sector,to%20expedite
37 US Department of the Interior, Bureau of Reclamation, "Physical Security Plan XXXX Dam and Power-plant," March 15, 2012. www.usbr.gov
38 US DHS, Department of the Interior, "National Monuments and Icons Sector-Specific Plan," 2010. https://www.hsdl.org/?view&did=691263
39 U.S. Department of the Interior, Bureau of Reclamation, "Advances in Security Technology and Procedures to Safeguard Reclamation SCADA Systems," June 6, 2015. http://www.usbr.gov/resaerch/projects/detail.cfm?id=2331
40 U.S. Department of the Interior, Bureau of Reclamation, "Critical Asset Identification Methodology," September 9, 2010. http://www.usbr.gov/recman/temporary_releases/irmtrmr-35-AppA.pdf
41 US Department of the Interior, Bureau of Reclamation, "ORAU Helps Department of Interior Conduct Full-scale Emergency Exercises at Major US Dams," 2015. https://www.orau.org/national-security-emergency-management/success-stories/exercises-planning/bureau of reclamation.aspx
42 US Department of the Interior, "The US Department of Interior's CyberRisk Management Practices Leave its Systems at Increased Risk of Compromise." https://www.doioig.gov/sites/default/files/2021-migration/Final%20Evaluation%20Report_DOI%20Cyber%20Risk%20Management_Public.pdf
43 US Department of Agriculture, "USDA Announces Framework for Shoring up the Food Supply Chain and Transforming the Food System to be Fairer, More Competitive, More Resilient." https://www.usda.gov/media/press-releases/2022/06/01/usda-announces-framework-shoring-food-supply-chain-and-transforming
44 US Department of Agriculture, "FY 2022 Budget Summary." https://www.usda.gov/sites/default/files/documents/2022-budget-summary.pdf
45 US Department of Energy, "Office of the Chief Information Officer—CISO Programs." https://www.energy.gov/cio/office-chief-information-officer-ciso-programs
46 "Office of Cybersecurity, Energy Security, and Emergency Response." https://www.energy.gov/ceser/office-cybersecurity-energy-security-and-emergency-response
47 "DOE Cybersecurity Strategy 2018–2020." https://www.energy.gov/sites/default/files/2018/07/f53/EXEC-2018-003700%20DOE%20Cybersecurity%20Strategy%202018-2020-Final-FINAL-c2.pdf
48 https://home.treasury.gov/news/press-releases/jy1159
49 https://home.treasury.gov/about/offices/management/chief-information-officer/cyber-security

6 Public–Private Partnerships

Chapter Outline

Introduction

In late 2019, a major software firm based in Oklahoma, SolarWinds, was hacked and malicious code was added to their software system called Orion. This attack was particularly damaging because it was not discovered for many months, allowing hackers, thought to be from Russia, to spy on thousands of customers. It is estimated that there were around 18,000 customers affected, both from the public and private sectors. The victims included Microsoft, Intel, Cisco, and U.S. government agencies such as the Department of Homeland Security and the Treasury Department. The hack, one of the largest of its kind, caused long-term damage to multiple companies.

In response, the Cybersecurity and Infrastructure Security Agency (CISA—see Chapter 4), the Federal Bureau of Investigation (FBI), and other federal agencies, created a task force to manage the investigation and response to the attack. The task force, called the Cyber Unified Coordination Group (UCG), was asked to explore the scope of the hack and recommend a response. They are also tasked with identifying victims. Their findings will be shared with private and public sector agencies to inform them of evidence and future actions to prevent another similar attack.[1] See Table 6.1 for the press release from CISA announcing this initiative.

DOI: 10.4324/9781003434887-6

Table 6.1 CISA Press Release Announcing Task Force

Joint Statement by the Federal Bureau of Investigation (FBI), the Cybersecurity and Infrastructure Security Agency (CISA), the Office of the Director of National Intelligence (ODNI), and the National Security Agency (NSA)

Released January 05, 2021

On behalf of President Trump, the National Security Council staff has set up a task force construct known as the Cyber UCG, composed of the FBI, CISA, and ODNI with support from the NSA, to coordinate the investigation and remediation of this significant cyber incident involving federal government networks. The UCG is still working to understand the scope of the incident but has the following updates on its investigative and mitigation efforts.

This work indicates that an Advanced Persistent Threat actor, likely Russian in origin, is responsible for most or all of the recently discovered, ongoing cyber compromises of both government and non-governmental networks. At this time, we believe this was, and continues to be, an intelligence-gathering effort. We are taking all necessary steps to understand the full scope of this campaign and respond accordingly.

The UCG believes that, of the approximately 18,000 affected public and private sector customers of Solar Winds' Orion product, a much smaller number have been compromised by follow-on activity on their systems. We have so far identified fewer than ten U.S. government agencies that fall into this category and are working to identify and notify the nongovernment entities who also may be impacted.

This is a serious compromise that will require a sustained and dedicated effort to remediate. Since its initial discovery, the UCG, including hardworking professionals across the U.S. government, as well as our private sector partners have been working non-stop. These efforts did not let up through the holidays. The UCG will continue taking every necessary action to investigate, remediate, and share information with our partners and the American people.

As the lead agency for threat response, the FBI's investigation is presently focused on four critical lines of effort: identifying victims, collecting evidence, analyzing the evidence to determine further attribution, and sharing results with our government and private sector partners to inform operations, the intelligence picture, and network defense.

As the lead for asset response, CISA is focused on sharing information quickly with our government and private sector partners as we work to understand the extent of this campaign and the level of exploitation. CISA has also created a free tool for detecting unusual and potentially malicious activity related to this incident. In an Emergency Directive posted on December 14, CISA directed the rapid disconnect or power-down of affected SolarWinds Orion products from federal networks. CISA also issued a technical alert providing technical details and mitigation strategies to help network defenders take immediate action. CISA will continue to share any known details as they become available.

As the lead for intelligence support and related activities, ODNI is coordinating the Intelligence Community to ensure the UCG has the most up-to-date intelligence to drive United States Government mitigation and response activities. Further, as part of its information-sharing mission, ODNI is providing situational awareness for key stakeholders and coordinating intelligence collection activities to address knowledge gaps.

Lastly, the NSA is supporting the UCG by providing intelligence, cybersecurity expertise, and actionable guidance to the UCG partners, as well as National Security Systems, Department of Defense, and Defense Industrial Base system owners. NSA's engagement with both the UCG and industry partners is focused on assessing the scale and scope of the incident, as well as providing technical mitigation measures.

The UCG remains focused on ensuring that victims are identified and able to remediate their systems and that evidence is preserved and collected. Additional information, including indicators of compromise, will be made public as they become available.

Source: https://www.cisa.gov/news-events/news/joint-statement-federal-bureau-investigation-fbi-cybersecurity-and-infrastructure

The need for private and public agencies to work jointly to investigate attacks and propose changes to prevent future attacks became clear immediately after the terrorist events of September 11, 2001. At that time, most critical infrastructure protection efforts focused on action by the federal government. However, people quickly realized that it was essential to include the private sector in the planning process because a large portion of the critical infrastructure is owned and operated by the private sector. Moreover, the infrastructure asset is often located in communities that also can play a critical role in planning and recovery from an event. Over the years, partnerships have developed between government agencies and the owners and operators of the nation's infrastructure, with the Department of Homeland Security (DHS) taking the lead in coordinating critical infrastructure protection activities. These have increased the sharing of information relating to possible threats, vulnerabilities, and even interdependencies. This chapter describes the relationships that have developed between the public and private sectors for risk assessment and homeland security issues.

Private versus Public Sectors

The term "public sector" agency refers to organizations or agencies that are owned or operated by the government. Examples of these offices are federal, state, or local departments; law enforcement; public education; and the military. These groups often get financial support from the government through taxes and are regulated by Congress or the executive branch. They are responsible for protecting people and property in their jurisdiction. Unfortunately, state and local governments are often not able to keep up with changes in critical infrastructure as they do not have the funding and expertise to keep on par with the private sector.[2]

The term "private sector" critical infrastructure refers to any unit that is not operated by the state, federal, local, or tribal government. This can include privately owned banks, television or radio stations, oil and gas companies, electric companies, transportation firms, private education facilities, power companies, non-profit agencies, or nongovernmental organizations. Their goal is to make a profit or generate wealth for the owner. Therefore, the actions they take are geared toward minimizing any financial risk while maximizing profit. These organizations may attempt to influence government policy through the legislative process, but they do not have the formal authority to set policy. Those who work in the private sector must respond to authorities that are outside of the government such as parent companies, business partners, or even outside financial institutions.[3]

In the U.S., the private sector owns a large portion of the nation's critical infrastructure, up to 85%.[4] These owners and operators are the people who have the most information and understand the threats and vulnerabilities of that sector better than any other person or group. They will often know the best or most appropriate action that should be taken if something should occur. These organizations are typically regulated by government policies and regulations but make most policies regarding their companies internally.

The concept of a public–private partnership refers to an agreement that is worked out between a public agency and a private sector agency. These public–private relationships require a robust working connection and are essential to fully protecting critical infrastructure. The goal of the agreement is to draw on the skills and resources of each separate group so that a task can be accomplished efficiently. Developing a working partnership can result in an efficient and effective delivery of services that help with preparing for an event or recovering from an event. A great example is maintaining energy security and the power grid, an essential component of the American critical infrastructure and has been the subject of multiple attacks, both physical and cyber. Protecting the power grid requires both public and private action and cooperation between them. Federal agencies such as the DHS, the Department of Energy, and the Department of Defense each play a role, as do the private owners of the power companies involved. Protecting critical infrastructure is difficult as each has different characteristics, so strong cooperation between agencies is essential.

Information Sharing

Both public and private groups have a knowledge base and expertise that stems from their operations. Private owners have unique concerns and information about their organizations, while government operators have a different set of concerns. Private owners are aware of the day-to-day issues that arise from operating a business that is deemed to be critical infrastructure. They also are concerned with balancing their goal of making a profit with government regulations that may restrict their ability to do business. On the other side is the public sector, which has a different approach. Government agencies are less concerned with making a profit. They may have intelligence regarding a cyber threat or physical threat that may impact critical infrastructure that is unknown to private owners. At times, public sector groups have more funds that can be spent on critical infrastructure protection that can be shared with private organizations. Governments may also be more concerned with regulating businesses to protect the environment or the privacy rights of individuals.

Both groups have the same goal—to protect critical infrastructure—but different viewpoints. For the nation's policies for protecting critical infrastructure to be complete and effective, both groups must be able to express these unique concerns and knowledge with the other. Critical infrastructure protection must be a cooperative effort between the public and the private sector. There must be an open exchange of knowledge, interests, and concerns not only when new rules, policies, and laws are being made but every day as we carry out those laws to protect infrastructure. It is essential that employees in both the private and public sectors readily communicate with each other on a regular basis to exchange information regarding threats, mitigation efforts, or response and recovery concerns.[5]

When information is not shared, it may lead to an inaccurate assessment of an asset's vulnerability to risk, or to being attacked, leaving an asset at a higher risk. If an asset is undervalued and the risk is not clearly identified, that asset may be harmed to a greater extent if attacked. This can cause harm not only to the asset but to citizens as well. It will not be protected enough, making it more vulnerable to attack. On the other hand, if the identified risk associated with an asset makes it appear more significant than it is, the organization may waste resources in protecting it.

The private and public sectors can both have a sense of mistrust of the other and be hesitant to share information with each other for many reasons. Many private owners and operators are concerned that the information they are being asked to share might, if made available to the public or other competitors, result in damage to their company or its image, which may in turn affect their competitive position within the industry. The private sector may be concerned with confidentiality, or keeping the company's secrets, as they fear that their corporate secrets may be given to the public or to competing owners. They may also be concerned with the integrity of that information and seek assurances that the information will not be changed. In addition, the private sector may not want to share information about their asset because of the fear that the information may be used against them by government officials as part of a regulatory enforcement action. In other words, they may be afraid that, by sharing information, they are exposing themselves to increased or unwanted government oversight. In some cases, the company in question may be afraid that the information is made public, exposing losses to the public and possibly affecting its reputation or brand. Another fear is that the information released may expose them to investigations and increase liability or the information released may be used by potential terrorists and others who are intent on disrupting their infrastructure.

On the other hand, the government may be unlikely to share information because it is sometimes critical to keep methods and procedures for protecting critical infrastructure protection plans confidential to prevent an attack. There may be a need to hide certain information or policies geared toward protecting an asset or how that is done. A release of information may compromise intelligence activities or investigations. For many, there is a lack of trust that any information shared will remain restricted.

To overcome this hesitancy, Congress in November 2002 passed the Critical Infrastructure Information Act (S 3946; PL 107-296), which established the Protected Critical Infrastructure Information Program. This protects private or proprietary information that is shared by private owners with the government from being made public. The goal was to increase communication among government agencies and the private sector regarding infrastructure protection.

Executive Order 13010

Many documents regarding critical infrastructure protection, including PPD-63 and HSPD-21, note the importance of including the private sector in the planning process. In July 1996, President Clinton addressed this concern and established the President's Commission on Critical Infrastructure Protection in Executive Order 13010. He asked the members of the commission, who represented multiple departments (i.e. the Departments of Treasury, Justice, Defense, Commerce, Transportation, Energy, the CIA, Federal Emergency Management Agency (FEMA), FBI, and National Security Agency)) to assess the vulnerabilities of the country's critical infrastructures and then create a new plan to protect them. The committee's report was made public in 1997. In it, the commission members noted that increased information exchange was needed between all participants in critical infrastructure protection so that the government could analyze that information to determine vulnerabilities and to predict or prevent an attack. The commission indicated that it was necessary to develop methods for two-way sharing of information within each infrastructure but also across different sectors, and between sectors.

The PCCIP proposed Information Sharing and Analysis Centers (ISAC) that would be comprised of representatives from both government and the private sector. The ISAC would provide a forum in which information from all sources could easily be shared and then analyzed to identify vulnerabilities related to that asset. Information could also be shared after an incident to determine why that event happened and how to make changes to prevent a similar event in the future. However, the PCIP soon realized that this information, while ideal in theory, would be hampered by fears held by business owners that corporate secrets would be made public. Thus, it was necessary that those in the private sector have assurances that any confidential information they shared with others would be protected.

Soon after this report was published, President Clinton released Presidential Decision Directive No. 63 (PDD-63). In this document, Clinton requested that government officials, in particular the National Coordinator for Security, Infrastructure Protection and Counterterrorism, begin to cooperate more with private owners and operators of critical Infrastructures and begin to share information pertaining to critical infrastructure. He asked the National Coordinator to investigate possible liability issues that could emerge as a result of private companies sharing more information. Once those were identified, Clinton sought proposals for removing those obstacles. Clinton also wanted to have recommendations for ways to ensure that any confidential information shared by businesses would remain confidential.

As PDD-63 was implemented, many owners and operators of critical infrastructures formed ISACs, with the intent of sharing more information with federal officials. A report by the General Accounting Office in 2001 found that there was very little information exchanged between private agencies and the federal government. According to the Director of the National Infrastructure

Protection Center, the reason for this was the concern on the part of the private companies that the information they shared would remain confidential. Moreover, officials in the Partnership for Critical Infrastructure Security stated that it was not clear that any of the Freedom of Information Act exemptions would ensure that information would be protected.

Information Sharing and Analysis Centers

Formed in 2003, ISACs are sector-specific groups that facilitate interaction and communication between and among members. In short, they enable information sharing between the government and the private sector. Examples of Information an Analysis Centers (ISACS) include a Supply Chain ISAC, a Public Transportation ISAC, a Water ISAC, an Electricity Sector ISAC, Emergency Management and Response ISAC, Information Technology ISAC, Energy ISAC, Chemical Sector ISAC, Healthcare Services ISAC, Highway ISAC, Food and agriculture ISAC, Multistate ISAC, a Real Estate ISAC, a Research and Educational Networking ISAC, a Biotechnology and Pharmaceutical ISAC, and a Maritime ISAC. See Table 6.1 for information on different ISAACs.

The goal of these groups is to advance physical and cyber protection of critical infrastructure throughout the country through sharing information. According to PDD-63, an ISAC would be responsible for collecting, analyzing, and sharing incident and response information among its members. They were intended to increase the exchange of information among government agencies and the private sector. This helps the owners of critical infrastructure protect their facilities, employees, and customers from a disruption of service. The original idea was to have one ISAC, but that was later changed so that each sector would have its own center. It was thought that ISACs would be a place where facts from an attack or other incident would be reported, analyzed, and shared with others. They became a clearinghouse of information about any possible threats, mitigation efforts, or recovery efforts.[6]

In addition to sharing information, ISACs are involved in critical infrastructure protection in other ways. ISACs are constantly monitoring events and oversee an early threat and detection warning system. They work closely with Sector Coordinating Councils (SCCs) to provide detailed analysis of events or threats to their membership. If an event occurs, the ISAC will help coordinate the response and share relevant information with others in that sector and in other sectors, and among public and private agencies. They also help to plan, coordinate, and carry out training exercises.[7]

The membership and scope of every ISAC are different, but they share some similar characteristics. For example, each of the groups promotes sector-specific information and intelligence sharing regarding all hazard threats, vulnerabilities, and any incidents that may occur. They work with other agencies concerning mitigation efforts to reduce the risk of an attack. They also create a dialogue between related organizations and government agencies to identify best practices. In doing so, they seek to educate others and make them more aware of the environment in a safe and secure way.[8] While they do all of this, the ISAC team ensures that any proprietary or sensitive information is protected.

Some of the material that is made available through an ISAC includes information from U.S. and foreign government sources that is not publicly available; National and International Computer Emergency Response Team (CERT) information that is not publicly available; information from law enforcement and public safety agencies; manufacturing information from hardware and software vendors; research by independent researchers and sector experts; and geospatial analysis of threats.

The ISACs are overseen by the ISAC Council, which operates to advance the physical and cyber security of critical infrastructures across the U.S. They seek to advance interactions between and among the ISACs and with the government.

Table 6.2 Examples of ISAACS

AMERICAN CHEMISTRY COUNCIL: represents diverse companies that pertain to chemistry. Their mission is to support advocacy, member engagement, political advocacy, information sharing, communications, and scientific research. This allows companies to address common IT, cyber security, and security issues. www.americanchemistry.com/

AVIATION ISAC: supports information sharing and analysis of aviation issues to help protect global aviation businesses. Their vision is to provide a safe, secure, efficient, and resilient global air transportation system by analyzing and sharing timely, relevant, and actionable cyber security information to members. www.a-isac.com

COMMUNICATIONS ISAC: Also known as the DHS National Coordinating Center, the goal is to prevent and mitigate the impacts of attacks on the telecommunications infrastructure to ensure the communication networks remain operational. This ISAC collects, analyzes, and disseminates information on vulnerabilities, threats, and intrusions to carriers, ISPs, satellite providers, broadcasters, vendors, and other stakeholders. www.dhs.gov/national-coordinating-center-communications

ELECTIONS INFRASTRUCTURE ISAC: comprised of election officials and staff, vendors, federal agencies, and cybersecurity experts, they seek to protect the U.S. elections infrastructure. They work to share threats, establish educational opportunities, and carry out cybersecurity controls. https://www.cisecurity.org/ei-isac/

ELECTRICITY ISAC: aids in situational awareness, incident management, coordination, and communication capabilities within the electricity sector. This group is the primary security communications organization for the electricity sector and seeks to improve the sector's ability to prepare for and respond to cyber and physical threats, vulnerabilities, and incidents. www.eisac.com

EMERGENCY MANAGEMENT AND RESPONSE ISAC: has the mission of collecting, analyzing, and disseminating information about critical infrastructure protection and resilience information related to Emergency Services Sector departments and agencies. www.usfa.dhs.gov/emr-isac

FINANCIAL SERVICES ISAC: seeks to protect the resilience and security of the global financial services infrastructure by sharing information on threats and vulnerabilities. They also carry out coordinated contingency planning exercises, manage rapid response communications, conduct education and training programs, and foster collaborations with other sectors and government agencies. www.fsisac.com

HEALTHCARE READY: seeks to strengthen healthcare supply chains and enhance resiliency by supporting collaboration between public and private agencies and by assisting before, during, and after disasters. They attempt to develop solutions to critical problems and share best practices for healthcare preparedness and response.www.healthcareready.org

INFORMATION TECHNOLOGY ISAC: Involves actors in the information technology sector who seek to assist corporations protect the IT infrastructure. They provide a way for experts from leading technology companies to collaborate with others by supporting the automated sharing of threat information and analysis. They provide members with thousands of threat indicators each week so they can manage risks through analysis and collaboration with others. www.it-isac.org

MARITIME ISAC: serves ocean carriers, cruise lines, port facilities, and terminals, and others by increasing the security of the maritime community. They often serve as liaisons between the industry and government. They also disseminate information to members as well as develop new technologies and provide educational and informational conferences. www.maritimesecurity.org

NATIONAL DEFENSE ISAC: the national defense sector's ISAC that was formed to enhance the security and resiliency of the defense industry. They provide a forum for sharing information on cyber and physical threat information, best practices, and mitigation strategies. This includes all-hazards threat sharing, alerts, and warnings. www.ndisac.org

RETAIL AND HOSPITALITY ISAC: provides for sharing cyber security information and intelligence to those in the industry with the goal of increasing security for the retail and hospitality industries. They assist companies in retail, hospitality, gaming, travel, and other related entities. www.rhisac.org

SPACE ISAC: facilitates collaboration within the space industry. Their goal is to enhance the ability of organizations to prepare for and respond to vulnerabilities, incidents, and threats and to disseminate information among members. https://s-isac.org/

WATER ISAC: the information sharing and operational arm of the water and wastewater sector. They assist members to strengthen their physical and cyber security and to recover from natural and man-made disasters. www.waterisac.org

Source: National Council of ISACS, https://www.nationalisacs.org/member-isacs-3

The individual ISACs are coordinated by the National Council of ISACs (NCI). Twenty-six ISACs are part of this organization. This provides a means for the ISACs to communicate with other groups but also with government agencies. They can share information on threats or mitigation strategies or can even coordinate responses if need be. The NCI coordinates this communication with daily reports, monthly meetings, and other actions. They also provide training for its members to help prepare for events. Examples of ISACs are found in Table 6.2.).[9]

Cyberspace Policy Review

In May 2009, President Obama established a committee to review the nation's cyberspace security policy. The committee, called the Cyberspace Policy Review Committee, made recommendations to the president about ways to improve cyber security. One of the recommendations the committee made was for the president to appoint an Executive Branch Cybersecurity Coordinator, which he did. The committee also recommended that the Executive Branch work more closely with all key players who are involved in U.S. cybersecurity policy, including those in state and local government officials and the private sector. This would help to ensure a more organized and unified response to future cyber incidents and strengthen existing public/private partnerships to find solutions to ensure cyber security.[10]

Fusion Centers

State-owned and operated fusion centers are one important way for federal, state, local, tribal, and territorial agencies can facilitate the sharing of information and intelligence. Many states and large cities have established fusion centers to communicate with each other and share information and intelligence within their own jurisdictions and with the federal government. The fusion centers collect information on threats and share that with other groups. They ensure that both classified and unclassified information can be shared among the group, with expertise at all levels sharing information.[11] The information is shared by stakeholders and experts in the particular subject matter.[12]

Fusion centers have access to data resources that are necessary to carry out analysis. They also support protection-related exercises that are planned by federal, state, and regional officials. The analysis in the fusion centers work alongside other analysts, law enforcement officials, public safety officials, and private sector representatives to analyze information. Some of the information that can be exchanged within a fusion center includes site-specific security risks, interdependencies with other sectors, any suspicious activity reports, adversary techniques, best practices in asset protection and resiliency, standard operating procedures for incident response, and emergency communications capabilities.

Joint Terrorism Task Forces

Led by the FBI, Joint Terrorism Task Forces (JTTFs) help to share information on threats to homeland security and critical infrastructure. JTTFs assist with investigations of these threats or incidents as they occur. Members of the JTTFs also provide valuable information on threats and activities for law enforcement and other activities. They often include state, local, and tribal law enforcement agencies.[13]

InfraGard

InfraGard is one of the nation's largest public–private partnerships. This an organization within the FBI between them and members of the private sector who seek to increase information sharing and analysis between the organizations regarding both cyber and physical critical infrastructure protection. The private sector members include businesses, academic institutions, state and local law enforcement agencies, and others. Each of the members shares key information and intelligence that is somehow related to the protection of the U.S. critical infrastructure from both physical and cyber threats. InfraGard chapters are formed throughout the country and are each linked to an FBI agent. This is described in more detail in Chapter 5.

State, Local, Tribal, and Territorial Government Coordinating Council

The State, Local, Tribal, and Territorial Government Coordinating Council was formed in 2007 in the National Infrastructure Protection Program. Since a large portion of the nation's critical infrastructure is located in these jurisdictions and is overseen by these governments, it plays a vital role in protecting the communities. This provides a way for officials in state, local, tribal and territorial governments to communicate with each other and with the federal government. They have developed working groups that specialize in particular areas. Those working groups include Critical Infrastructure and Cybersecurity. There are currently representatives from 26 jurisdictions in the organizations that cover all 10 CISA regions. This includes representatives from 14 states, 6 counties, 9 cities, 1 tribe, and 1 territory.

The organization shares information on best practices for protecting critical infrastructure and mitigating possible harm. They also get involved in supporting or opposing legislation in Congress on new policies under consideration. They host a webinar series to share critical information with stakeholders throughout the country.[14]

Homeland Security Information Network

The Homeland Security Information Network (HSIN) is a secure web-based system that was established by the DHS to increase information sharing and collaboration efforts between government agencies and the private sector that have a concern with protecting critical infrastructure. They have developed a way to share "sensitive but unclassified" information between federal, state, local, territorial, tribal, and international governments and private owners and operators of critical infrastructure. HSIN is comprised of Communities of Interest that allows users in all 50 states to share information with others in their communities through a safe environment in real time. It also allows members to discuss problems or seek information from other communities. Through HSIN, groups are able to convene meetings in a virtual meeting space or through instant messaging and document sharing.[15]

US-CERT

As another way to share more information, the DHS created the U.S. Computer Emergency Readiness Team (US-CERT), which was created in 2003. Employees of the CERT seek to improve the cybersecurity of the nation by sharing information about threats with key stakeholders in a 24/7 Operations Center. This agency makes information related to computer-related vulnerabilities and threats available to others. Members of the public can seek out information about cybercrimes on the agency's website, and training is provided for public and private infrastructure owners and operators. They also provide information about responses to an incident. US-CERT collects incident reports from others around the country and analyzes that information to look for patterns and trends in computer-based crime. If an event occurs, CERT experts will help with the investigation and recovery support. Officials here manage the National Cyber Alert System, which provides general information to any organization or individual who subscribes.[16]

Protective Security Advisors Program

Originally developed in 2004, the Protective Security Advisor (PSA) program is housed in CISA (part of the DHS). These advisors are experts who have knowledge of infrastructure protection and mitigation techniques. They will advise state, local, and private owners of critical infrastructure about how to adapt their current practices to improve the security of their assets. They will also hold field training exercises and outreach to owners who are hosting large events (i.e. a sporting event) in their jurisdiction.

The PSA program focuses on three areas: enhancing infrastructure protection; assisting with incident management; and facilitating information sharing. To enhance infrastructure protection, PSAs help owners and operators of critical infrastructure by providing training and vulnerability assessments when they are requested. They help in the process of identifying possible risks to critical infrastructure and then mitigating those risks.

The second area is assisting with incident management. PSAs will respond after an emergency or other disaster. During an event, they will cooperate with other officials to keep the DHS and other government offices up to date on assessments of damage to critical infrastructure. They can also help provide advice on recovery activities.

The third area is facilitating information sharing. Here, PSAs help the transfer of information among all levels of government and the private sector. During times when there are no emergencies, the directors and PSAs may hold briefings and meetings with critical infrastructure protection partners to disseminate information. In the case of an event, PSAs will communicate with others about critical infrastructure response and recovery. They will also provide information on conditions to the DHS and other federal agencies.[17]

Critical Infrastructure Partnership Advisory Council

The DHS established the Critical Infrastructure Partnership Advisory Council (CIPAC) in 2006 as a way to increase communication between federal programs and state, local, territorial, and tribal agencies that provide infrastructure protection. CIPAC is comprised of critical infrastructure owners and operators and the trade organizations that are members of the SCC. The council also includes representatives from federal, state, local, and tribal governmental groups who are members of the Government Coordinating Councils for each sector.

CIPAC helps to coordinate federal infrastructure protection programs with private sector owners and operators. It provides a forum in which they can discuss issues or participate in activities that will increase coordination of critical infrastructure protection. The membership often meets to discuss planning, security concerns, operational activities, incident response, or recovery. They may also discuss information about possible threats, vulnerabilities, or protective measures. They will advise the DHS or other policy makers regarding infrastructure protection policies and regulations.[18]

Office of Intelligence and Analysis

The Office of Intelligence and Analysis (I&A), located within the DHS, has developed strong ties with the private sector. They help private owners and operators have the information they need to protect their assets. I&A works with CISA, the FBI, Fusion Centers, NCIAC, and Infraguard, among others, to provide accurate and timely information to the private sector regarding emerging threats. I&A often holds events to increase discussion between the public and private sectors on different issues.

Public–Private Analytic Exchange Program (AEP)

The Public–Private Analytic Exchange Program is part of the DHS and I&A section. The agency provides a way for government experts and private sector stakeholders to work together to protect assets. This is a way to provide private owners with a better understanding of issues facing the country and critical infrastructure protection. Topics discussed include the impacts of 5G on cybersecurity, implications of severe weather events, U.S. port cybersecurity, and national security readiness.[19]

FEMA Grants

The DHS/FEMA sponsors an annual grant program to provide financial assistance for programs that seek to increase the nation's infrastructure protection and security of its assets. The program is overseen by the Grant Programs Directorate (GPD). The focus of GPD is to establish and promote communication with state, local, and tribal stakeholders and increase the nation's level of preparedness and its ability to protect against and respond to an attack or event. In all, the grant programs help to fund many activities related to homeland security and emergency preparedness. Some of the programs are about planning, organization, equipment purchase, training, exercises, and management and administration costs.[20]

Training and Exercise Support

The DHS, through the National Preparedness Directorate, has established many programs to help local communities train their first responders and others for an event, either natural or man-made. Through the National Domestic Preparedness Consortium (NDPC) Training Program, the DHS and FEMA officials provide training to emergency responders in the U.S. and its territories. Three levels of training are provided: awareness-level training, performance level training, and management-level training.

Awareness-level training programs are designed for a wide array of groups and introduce basic concepts to participants. They are intended to teach classes about key principles and policies related to infrastructure protection and security. These courses usually focus on basic topics related to the prevention of incidents and preparation for them (including mitigation techniques).

The second type of training course is Performance Level Training courses. In these, the DHS officials help to train participants so they can carry out specific tasks or have specific skills that are needed when responding to, or recovering from, an event. For example, some of the courses focus on how to deal with chemical spills or radioactive materials.

The third type of training course made available by the DHS is management-level training classes. These are meant to improve the leadership skills of those who are in management or in other supervisory positions.

If needed, the DHS will also provide communities and first responders with exercises so that any potential deficiencies in either personnel needs or response plans can be identified. The Homeland Security Exercise and Evaluation Program helps in designing exercises, conducting them, and then evaluating them. Exercises may include seminars, workshops, drills, games, tabletop exercises, functional exercises, or full-scale exercises.

Conclusion

Attacks on U.S. critical infrastructure continue to increase. In today's world of increased connectivity, vulnerabilities will always exist. The results from a breach can lead to mass injuries, harm, or even death. Each attack is different as new malware is developed and adapted by those seeking to do harm to others. It is critical that public and private agencies stay aware of threats to their infrastructure. This necessitates effective cooperation and sharing of information. Public–private partnerships such as the ones described here can help all agencies address the growing threats. Public and private agencies must collaborate to mitigate risks and develop resiliency plans to protect citizens.

Newly developed partnerships between the federal government and the private sector that emerged after the terrorist attacks in 2001 have improved the communication between the two groups. In the end, this helps for better planning for protecting the nation's critical infrastructure. In the future, improved relationships and more exchange of information will help protect the nation from harm even further.

Key Terms

Private Sector
Public Sector
Public–Private Partnerships
Information Sharing
Critical Infrastructure Information Act
Executive Order 13010
ISAC
ISAC Council
Cyberspace Policy Review Committee
Fusion Center
Joint Terrorism Task Forces
InfraGard
State, Local, Tribal, and Territorial Government Coordinating Council
Homeland Security Information Network (HSIN)
US-CERT
Protective Security Advisors Program
Critical Infrastructure Partnership Advisory Council
Office of Intelligence and Analysis
Public–Private Analytic Exchange Program
FEMA Grants
Training and Exercise Support

Review Questions

1 What are the advantages of sharing information about critical infrastructure protection? Why would some companies be hesitant to do so?
2 Describe an ISAC, and why it is important in protecting assets.
3 Explain what is meant by a fusion center and its role in critical infrastructure protection.
4 What does the HSIN do?
5 What are the three areas of concentration for the PSAP?
6 Why is the Critical Infrastructure Partnership Advisory Council important?

Notes

1 Data Guidance, "USA: CISA and Federal Agencies Create Cyber Unified Coordination Group in Response to SolarWinds' Orion Attack," January 7, 2021. https://www.dataguidance.com/news/usa-cisa-and-federal-agencies-create-cyber-unified
2 Cordesman, A.H., and Cordesman, J.G. (2002). *Cyber-Threats, Information Warfare, and Critical Infrastructure Protection: Defending the U.S. Homeland.* Westport, CT: Praeger.
3 Radvanovsky, R., and McDougall, A. (2013). *Critical Infrastructure*. Boca Raton, FL: CRC Press.
4 Stevens, Gina Marie, and Tatelman, Todd B. (2006). "Protection of Security-Related Information." Congressional Research Service, September 27. http://fas.org/sgp/crs/secrecy/RL33670.pdf; see also CISA, Partnerships and Collaboration, https://www.cisa.gov/topics/partnerships-and-collaboration#:~:text=The%20private%20sector%20owns%20and,critical%20infrastructure%20security%20and%20resilience

5 DHS, FEMA, "CIKR Awareness AWR-213, Participant Guide," September 2010, 1:16.
6 National Council of ISACs, "About ISACs. National Council of ISACs | About ISACs," 2002. nationalisacs.org
7 DHS, 2013 NIPP.
8 FEMA, DHS, "Advanced Critical Infrastructure Protection." MGT-414 Participant Guide, April 2013.
9 National Council of ISACs, "About NCI, National Council of ISACs | About NCI," 2022. nationalisacs.org
10 FEMA, DHS, "Advanced Critical Infrastructure Protection." MGT-414 Participant Guide, April 2013; see also White House Fact Sheet, "Fact Sheet: Cyberspace Policy Review. White House Fact Sheet on Cyberspace Policy Review (fas.org)," May 29, 2009. https://irp.fas.org/news/2009/05/cyber-fs.html.
11 FEMA, DHS, "Advanced Critical Infrastructure Protection." MGT-414 Participant Guide; see also DHS, "Fusion Centers," April 2013. https://www.dhs.gov/fusion-centers
12 US Department of Homeland Security, and US Department of Justice, Global Justice Information Sharing Initiative, "Critical Infrastructure and Key Resources, Protection Capabilities for Fusion Centers," December 2008. https://it.ojp.gov/documents/d/CIKR%20protection%20capabilities%20for%20fusion%20centers%20s.pdf
13 Department of Homeland Security, Fusion Centers and Joint Terrorism Task Forces. https://wwwdhs.gov/fusion-centers-and-joint-terrorism-task-forces#
14 CISA, "State, Local, Tribal, and Territorial Government Coordinating Council." https://www.cisa.gov/resources-tools/groups/state-local-tribal-and-territorial-government-coordinating-council#:~:text=Formed%20in%20April%202007%2C%20the,the%20National%20Infrastructure%20Protection%20Plan
15 US Department of Homeland Security, "Homeland Security Information Network," April 10, 2022. https://www.dhs.gov/homeland-security-information-network-hsin
16 US Department of Homeland Security, "US-CERT." https://www.cisa.gov/sites/default/files/publications/infosheet_US-CERT_v2.pdf
17 CISA, "Protective Security Advisor (PSA) Program." https://www.cisa.gov/resources-tools/programs/protective-security-advisor-psa-program
18 CISA, "Critical Infrastructure Partnership Advisory Council (CIPAC)." https://www.cisa.gov/resources-tools/groups/critical-infrastructure-partnership-advisory-council-cipac
19 US Department of Homeland Security, "2023 Public-Private Analytic Exchange Program." https://www.dhs.gov/sites/default/files/2022-12/2022_1003_ia_2023-aep-program-overview-508.pdf
20 FEMA, "FEMA Grants," March 22, 2023. http://www.fema.gov/grants

7 Laws and Regulations

Chapter Outline

Introduction

Hurricane Katrina, a Category 5 hurricane, struck the city of New Orleans and other Gulf Coast states in late August 2005. For days before the storm, weather officials warned citizens about the imminent danger and asked residents to evacuate. When Katrina hit land, the levees protecting the city failed, resulting in widespread flooding. Over 1,800 people were killed and the city saw $100 billion in property damage.[1]

The city and many of its residents found themselves in dire need of government assistance for basic needs—food, water, shelter, and healthcare. Unfortunately, the government's response was slow and inadequate. Many residents remained stranded for days, receiving little or no help. Firefighters and first responders who travelled to New Orleans to help were forced to watch presentations about sexual harassment and the history of the Federal Emergency Management Agency (FEMA) before being allowed to help. Law enforcement agencies seemed to be involved in a "turf war" over who should take the lead in providing a response. The then-director of FEMA, Michael Brown, who later resigned, was criticized for lack of action, and instead blamed politics for a limited response. The government chose not to waive the Stafford Act, which required local governments to help pay for clean-up projects, for months after the hurricane.[2] FEMA did not work in conjunction with non-government agencies, such as the American Red Cross, which was denied access to the city and its residents.[3] The head of the Department of Homeland Security (DHS) at

DOI: 10.4324/9781003434887-7

the time, Secretary Chertoff, faced criticism for not reaching out to other Cabinets to coordinate assistance and appeared to be slow in providing a response. There seemed to be disagreement over who was responsible for providing basic assistance to victims and repairing the levees. Despite previous trainings, the city seemed unprepared for such an event.[4]

In response to these (and other) problems, members of Congress carried out an investigation as to the causes of the delayed response and the gaps that occurred that forced its residents to go without assistance. Congress eventually proposed and passed new legislation to prevent similar non-response in the future after an event like Hurricane Katrina. The resulting legislation was the Post Katrina Emergency Management Reform Act of 2006 (PKEMRA), signed by President George W. Bush in 2006. In passing this legislation, Congress reformed how the government responds to disasters, and in particular how FEMA should respond. The law changed the roles and responsibilities of FEMA, making it easier for the agency to respond. Among other things, PKEMRA established the National Emergency Family Registry and Locater System to help reunite family members after an emergency; created the position of Disability Coordinator who will oversee the inclusion of individuals with disabilities in a disaster relief plan; and provides assistance for evacuation plans in advance of an event, including provision of transportation for residents who need it.[5]

Over the years, Congress has proposed and passed many laws that authorize new policies or methods to protect the country's critical infrastructure (CI). While many proposals were not passed, the debate that ensued during the process helped to educate the public and raise awareness about CI protection and cybersecurity. This chapter includes information on laws that were passed by Congress for protecting critical assets, followed by an appendix that shows what types of proposals are introduced, maybe even debated, but not passed into law.

106th Congress (1999–2000)

Even before the terrorist attacks of 2001, Congress and other officials were concerned about protecting the nation's CI. In the 106th Congress, the members passed HR 4205 (PL 106-398), the National Defense Authorization Act. In Section 1033 of this bill, the Defense Secretary was authorized to guarantee up to $10 million of the payment of a loan that was made to a commercial firm for the purpose of improving the protection of CI assets, and even to refinance any improvements that were previously made for such protection. To do this, the Congress gave the Secretary authorization to use up to $500,000 each year from the Department of Defense budgets. In another part of the bill, Section 1053, the Congress asked the President to provide them with a report on the specific steps that the Federal Government was taking to develop CI protection strategies.[6]

107th Congress (2001–2002)

The terrorist attacks of September 11, 2001, occurred in the middle of the 107th Congressional session. As a result, there was more attention focused on protecting CI and the nation's ability to respond to any future similar event. Congress passed many new laws as a direct response to these events—particularly within the transportation sector and border security.

Congress amended the Merchant Marine Act of 1936 with the Maritime Transportation Security Act of 2002 (PL 107-295). In short, the new law increased the security of ports and waterways in the country to counteract the threat of terrorism. The new law authorized the hiring of new security officers, the purchase of new screening equipment, and the building of security infrastructure at seaports. Additionally, the bill tasked the Secretary of Transportation to identify the types of vessels and the ports that were a high risk of being involved in a security incident and to assess the vulnerability of these vessels and ports. The Secretary of Transportation was to submit a Security

Plan for deterring incidents and responding one shall it occur that will minimize damage. In addition, the owner or operator of a vessel or facility was required to prepare a security plan for the vessel or facility and submit it to the Secretary. That meant that chemical facilities, which are often located at or near ports, were required to prepare a plan to keep that facility safe.

Another law passed after 9/11 was the Aviation and Transportation Security Act (PL 107-71). This act established a new agency called the Transportation Security Administration (TSA) to be housed in the Federal Aviation Administration. The new agency was later transferred to the DHS. The TSA was made responsible for the safety of air travel, including pre-flight screenings of passengers. Prior to this, passenger screening was the responsibility of airlines.

USA PATRIOT Act

A major change resulting from the new legislation this year was the creation of the DHS through the USA PATRIOT Act, or the Uniting and Strengthening America by Providing Appropriate Tools Required to Intercept and Obstruct Terrorism Act of 2001 (HR 3162; PL 107-107). This was a complex bill that had many parts. In general, the PATRIOT Act allows for more information sharing by law enforcement and intelligence agencies so that critical information can help officials to disrupt potential terrorist plots and other criminal activity.

When it came to CI, this bill required the Director of the U.S. Secret Service to develop a national network of electronic crime task forces that would be located throughout the U.S. These task forces would work to prevent, detect, and investigate various forms of electronic crimes, including potential terrorist attacks against CI and financial payment systems.

The most attention to CI in the PATRIOT Act was the Critical Infrastructure Protection Act of 2001 (§5195c). The portion of the bill defines CI as "systems and assets, whether physical or virtual, so vital to the U.S. that their incapacity or destruction would have a debilitating impact on security, national economic security, national public health or safety, or any combination of those matters." The law mandated that it was U.S. policy that any physical or virtual disruption of the operation of CI in the country will be rare, brief, geographically limited in effect, manageable, and minimally detrimental to the economy, human and government services, and U.S. national security. Further, to achieve this, a public–private partnership involving corporate and nongovernmental organizations should be established. It also required that a comprehensive and effective program that will ensure the continuity of CI be created. To do these things, the new law mandated that a National Infrastructure Simulation and Analysis Center be established that will address the need for CI protection and continuity.

In Section 105 of the USA PATRIOT Act, the Director of the U.S. Secret Service was required to develop a national network of electronic crime task forces throughout the U.S. that would work to prevent, detect, and investigate different types of electronic crimes, including potential terrorist attacks against CI and financial payment systems.

Another bill passed during this session of Congress was HR 5005 (PL 107-296), the Homeland Security Act of 2002. This new law outlined the DHS's responsibilities for protecting the Nation's CI. It established the DHS mission, which was to reduce the nation's vulnerability to terrorist attacks, major disasters, and other emergencies. The organization was also given the responsibility to evaluate potential vulnerabilities to assets and then create steps to protect them. Another critical part of the new law was entitled the Critical Infrastructure Information Act of 2002. In this part of the bill, CI information was defined as information that is not customarily in the public domain and related to the security of CI or protected systems.

In this part of the bill, the president and the Secretary of the DHS were asked to exempt any CI information that was voluntarily submitted to the federal government from FOIA disclosure requirements. The reason for this is many private groups share confidential or proprietary information with federal agencies in an effort to create more comprehensive plans to protect assets. Private companies are sometimes hesitant to share that information unless they have guarantees that it will not be shared. This new law ensures that shared information will not be disclosed, even if someone files an FOIA request. In this new law, the Secretary of the DHS was asked to establish procedures for the receipt, care, and storage of critical information that is shared to protect its confidentiality.

Provisions within the CIIA detailed the circumstances under which the DHS may obtain, use, and disclose CI information as part of a CI protection program. CIIA created several limitations on the disclosure of CI information voluntarily submitted to the DHS (Section 214). If someone shared the protected information, they would face possible criminal penalties including a possible prison term of up to one year and fines, or both. In addition, they could be removed from their job.[7]

Section 214(e) of the bill requires that the Secretary of the DHS create methods for the receipt and storage of CI information. To do this effectively, the Secretary of the DHS was to work collectively with the National Security Council and the Office of Science and Technology Policy to establish uniform procedures.[8] The bill also had provisions to authorize the Federal Government to issue advisories, alerts, and warnings to relevant companies, targeted sectors, other governmental entities, or the general public regarding potential threats to CI.

A new position, called an Under Secretary for Information Analysis and Infrastructure Protection was made in the new law. This office was given the responsibility for:

1 Receiving and analyzing information and intelligence from law enforcement officials to help them understand the nature and scope of the terrorist threat made against the U.S. homeland and to detect and identify potential threats of terrorism within the U.S.;
2 Assessing the vulnerability of key resources and CIs found in the U.S.;
3 Integrating relevant information and intelligence analyses, along with vulnerability assessments to set priorities and fund protective measures;
4 Developing a comprehensive national plan for protecting critical resources and infrastructures;
5 Taking needed measures to protect critical resources and infrastructures;
6 Administering the Homeland Security Advisory System, being responsible for threat advisories, and providing specific warning information to state and local governments and the private sector, as well as providing suggestions regarding necessary protective actions and threat countermeasures; and
7 Reviewing, analyzing, and recommending improvements in the policies that regulate sharing of law enforcement intelligence, and other information related to homeland security within and between federal agencies and state and local governments.

108th Congress (2003–2004)

The Congress passed the Intelligence Reform and Terrorism Prevention Act of 2004 (PL 108-458) during the 108th Congress. Among other things, this bill established the Director of National Intelligence who was to oversee the National Counterterrorism Center. The Director was to promote intelligence information sharing among the Intelligence Community, but also address transportation security, border surveillance, and immigration concerns.

109th Congress (2005–2006)

In the 109th Congress, members passed the Post Katrina Emergency Management Reform (PKEMR) Act of 2006. This new law (PL 109-295) amended the Homeland Security Act of 2002 to make changes to the role of FEMA and their ability to strengthen emergency response services to those affected by natural events and emergencies. Members of Congress debated where FEMA should be located, but in the end, they chose to keep FEMA as part of the DHS. Some provisions in the law clarified FEMA's mission, which included:

1 Leading the nation's efforts to prepare for, respond to, recover from, and mitigate the risks of, any natural and man-made disaster, including catastrophic incidents;
2 Implementing a risk-based, all-hazards plus strategy for preparedness; and
3 Promoting and planning for the protection, security, resiliency, and post-disaster restoration of CI and key resources, including cyber and communications assets;

There is also specific language in the PKEMR Act regarding the role, qualifications, authority, and responsibilities of the administrator of FEMA, who should:

4 Have not less than five years of executive leadership and management experience, significant experience in crisis management or another relevant field, and a demonstrated ability to manage a substantial staff and budget;
5 Report to the Secretary of HS without being required to report through any other DHS official;
6 Be the principal emergency preparedness and response advisor to the President, the Homeland Security Council, and the Secretary;
7 Provide federal leadership necessary to mitigate, prepare for, respond to, and recover from a disaster;
8 Develop a national emergency management system capable of responding to catastrophic incidents; and
9 Develop and submit to Congress each year an estimate of the resources needed for developing the capabilities of federal, state, and local governments necessary to respond to a catastrophic incident.

The law created two new positions within FEMA: a Director for Preparedness and a Director for Response and Recovery. The changes also mandated the creation of ten Regional Offices and area offices for the Pacific, for the Caribbean, and for Alaska. The personnel in these new offices were in turn asked to establish multiagency strike teams that would be prepared to respond to disasters, including catastrophic incidents, at a moment's notice. In addition, a National Advisory Council on Preparedness and Response was created, whose task was to advise the FEMA Administrator on preparedness and emergency management. Members of the Council would be appointed by the Administrator and should represent emergency management and law enforcement, general emergency response providers from state, local, and tribal governments, private sector, and nongovernmental organizations.

Congress also passed the Tsunami Preparedness Act (PL 109-424). This bill provided for improvements in detecting tsunamis, preparedness, and mitigation of these events to protect the lives of people living in affected areas. As part of the law, there would be a community-based tsunami hazard mitigation program established to improve preparedness plans. This also included improved communications systems to provide for warnings.

Another law was passed by Congress in this session was intended to protect the nation's ports. PL 109-347, the Safe Port Act, which stands for Security and Accountability for Every Port, required the Secretary of Homeland Security to establish a Port Security Training Program that would help facilities submit plans to prevent, prepare for, respond to, mitigate against, and recover from acts of terrorism, natural disasters, or other emergencies. The Program would also test the ability of governments and personnel in the private sector prepare for, mitigate again, respond to, and recover from acts of terrorism or natural acts that affect a seaport. Any port that was identified as "high risk" would be required to conduct training exercises to test their ability to respond to, or recover from, threats and incidents.

110th Congress (2007–2008)

The members of the 110th Congress continued to be concerned about the protection of CI. In this session, they passed HR 1 (PL 110-53), or the Implementing Recommendations of the 9/11 Commission Act of 2007. In this law, new policies were established to protect the nation's critical assets. Section 101 of the bill amended the Homeland Security Act of 2002 to create a Homeland Security Grant Program (specifically the Urban Area Security Initiative and a State Homeland Security Grant Program). The Secretary of Homeland Security and the FEMA Administrator were given the responsibility to decide who would be awarded the grants. The grant funding could be used for training, exercises, protecting CI, and/or purchasing equipment, among other reasons. The considerations for awarding the grants were provided in the bill, and they included things like the risk posed to the state, the degree of threat, vulnerability, and consequences related to CI or key resources, and the anticipated effectiveness of the state's proposed use of grant funds to improve interoperability.

One section of the bill, Section 409, made amendments to the Homeland Security Act so that the administrator of FEMA would establish model standards for credentialing CI workers. The administrator would then be asked to provide the standards, along with technical assistance to state, local, and tribal governments to aid them in protecting their assets.

Title X of the new law, Section 1001, was entitled "Improving Critical Infrastructure Security." In this section, the Secretary of the DHS was asked to establish, maintain, and update a National Asset Database of each asset that the DHS determines to be a vital asset. Those assets were defined as those assets and services, if lost, interrupted, incapacitated, or destroyed, would have a negative or debilitating effect on the economic security, public health, or safety of the U.S., any state, or any local government. To help with this task, the bill required a National Infrastructure Protection Consortium be established that would advise the DHS on the best way to identify assets and maintain the database. It was noted, however, that the Secretary should ensure that levees were included in the list of CI assets. The Secretary was also to maintain a classified and prioritized list of assets that the Secretary determines would, if destroyed or disrupted, cause national or regional catastrophic effects. The Secretary was asked to present annual reports on the database and the list to the Homeland Security Committees in both the House and Senate.

In Section 1519 of the law, the DHS was asked to assess the methods that would likely be used in a deliberate terrorist attack against a railroad tank car used to transport toxic-inhalation-hazard materials. Included in that analysis was to be an estimate of the degree to which that method could cause death, injury, or serious health effects to victims, the environment, CI, national security, the national economy, or public welfare. Moreover, the DHS was asked to conduct an analysis, with the assistance of the National Infrastructure Simulation and Analysis Center, on the analysis of a possible release of toxic-inhalation-hazard materials that might occur from a terrorist attack on a loaded railroad tank car that was loaded with such materials in urban and rural environments. This would help with the planning process for such an attack.

111th Congress (2009–2010)

The members of Congress during the 111th Session passed the Reducing Overclassification Act that was intended to, in part, increase communication of information pertaining to CI protection between public and private organizations. According to this new law, the Secretary of Homeland Security was required to develop a way to prevent overclassification of materials related to homeland security, cyber threats, and other relevant information so that it could be easily shared with organizations as they develop mitigation strategies. This became Public Law 111-258.

In another Congressional action, members passed the Department of Homeland Security Appropriations Act on October 28, 2009. Title 3 of the bill (HR 2892), called Protection, Preparedness, Response, and Recovery, provided funds for the National Protection and Programs Directorate for infrastructure protection and information security programs. It also provided funds for FEMA grants for state and local programs, emergency management, disaster relief, and other programs. This became PL 111-83.

Members focused on protecting the safety of the nation's food when it passed HR 2751, Improving Capacity to Prevent Food Safety Problems (PL 111-353). This law requires the FDA to allocate funds to inspect food facilities and imported food to ensure that it meets safety standards.

112th Congress (2011–2012)

The Coast Guard and Maritime Transportation Act of 2012 (PL 112-213), passed during this Congressional session, authorized the head of the Coast Guard to establish an Integrated Cross-Border Maritime Law Enforcement Operations Program to strengthen border security and respond to terrorism and border-related violations of law, among other things. It also provided authority for joint training with Canada and coordination with DHS border security and antiterrorism programs.

113th Congress (2013–2014)

One bill passed in the 113th Congress was HR 2952 (PL 113-246), also known as the Cybersecurity Workforce Assessment Act. Under this law, the Secretary of Homeland Security would conduct a yearly assessment of the cybersecurity workforce in DHS. For example, it would analyze the readiness of employees to meet its cybersecurity mission; the location of the cybersecurity workforce within DHS; a description of which positions are carried out by permanent full-time DHS employees, by independent contractors, by individuals employed by other federal agencies, or which positions are not filled; the percentage of individuals within certain specialty areas who have received essential training to perform their jobs; and any challenges that were confronted for that training.

Congress in this session also passed S 2510 into law (PL 113-282). Called the National Cybersecurity Protection Act, this law amended the Homeland Security Act of 2002 to establish a National Cybersecurity and Communications Integration Center in the DHS. The new agency would be responsible for overseeing CI protection, cybersecurity, and related DHS programs. The agency would be a clearinghouse to collect and share information on cybersecurity risks, incidents, analysis, and warnings. The Center was also tasked with facilitating cross-sector coordination to address risks and incidents that could have consequential impacts across multiple sectors. They should also provide technical assistance, risk management, and recommendations for increasing security. Congress would also receive reports with recommendations on how to expedite the information sharing processes between organizations.

PL 113-274, the Cybersecurity Enhancement Act of 2014, included many provisions to increase collaboration on cybersecurity concerns between public and private groups. The law gave the Secretary of Commerce the authority to facilitate the development of a voluntary, consensus-based, industry-led set of procedures to reduce cyber risks to CI. The Secretary should, according to the law, coordinate with private sector personnel, CI owners and operators, ISACs, and others to identify measures that could be voluntarily adopted by owners and operators to identify and manage cyber risks.

In addition, multiple federal agencies were required to develop and regularly update a strategic plan to increase cybersecurity. Some of these agencies were the Department of Agriculture, Department of Commerce, Department of Defense, Department of Education, Department of Energy, the EPA, and others. The report was to include methods to guarantee privacy, verify software, and protect information.

Workforce development was also part of this law. It directed DHS and the Department of Commerce to support competitions and challenges that help to recruit new people. The Office of Personnel Management (OPM) was asked to support internships or other types of work experience for the winners of the competitions and challenges. The Cyber Scholarship-for-Service program was also encouraged.

Public Law 113-254 was passed in December of 2014 and was titled Protecting and Securing Chemical Facilities from Terrorist Attacks Act of 2014. This new law amended the Homeland Security Act of 2002 to reestablish the Chemical Facility Anti-Terrorism Standards Program in the DHS. Through this program, the Secretary of Homeland Security is required to identify chemical facilities and ask each facility to submit information so that a vulnerability analysis of the facility's security risks can be completed, along with a security plan. The Secretary can then approve or disapprove each site security plan and suggest an alternative security program if needed.

In Public Law 113-5 (HR 307), the Pandemic and All-Hazards Preparedness Reauthorization Act of 2013, the Congress sought to strengthen national preparedness and response for public health emergencies. This law required the Secretary of Health and Human Services to submit the National Health Security Strategy. The Strategy should include methods for increasing the preparedness, response capabilities, and surge capacity of ambulatory care facilities, dental health facilities, and critical care service systems. Provisions should be taken into account the unique needs of individuals with disabilities, and should include strategic measures to diagnose, mitigate, prevent, or treat harm from any biological agent or toxin or any chemical, radiological, or nuclear agent. The Secretary is required to report to Congress about coordination with the Department of Defense (DOD) regarding countermeasure activities to address chemical, biological, radiological, and nuclear threats.

The Assistant Secretary should oversee research, development, and procurement of countermeasures for a pandemic. Gaps and duplication in public health preparedness and response activities should be identified. The Assistant Secretary for Preparedness should take responsibility for emergency preparedness and response policy and coordination. This person will also have authority over and responsibility for the Biomedical Advanced Research and Development Authority (BARDA) as well as grants and related authorities related to trauma care.

One provision of the law reauthorizes the National Disaster Medical System. Found within DHS, this organization will provide state and local governments with extra emergency medical personnel and supplies during and after an emergency.

114th Congress (2015–2016)

Public Law 114-111 (HR 6381) was passed on December 18, 2015, and was known as the Emergency Information Improvement Act of 2015. Section 2 of the law amends the Robert T. Stafford Disaster Relief and Emergency Assistance Act to include broadcasting facilities within the definition of a "private nonprofit facility" that provides essential services to the general public. This also includes broadcast and telecommunications within the definition of a "critical service" provided by such a facility so that it is eligible for disaster assistance.

A new law was created when Public Law 114-183 passed Congress on June 22, 2016. Called Protecting Our Infrastructure of Pipelines and Enhancing Safety Act of 2016 or the PIPES Act of 2016, the Department of Transportation was tasked with creating a voluntary information sharing system that would support collaborative efforts to improve information sharing regarding hazardous liquid pipeline integrity.

HR 636, or the FAA Extension, Safety and Security Act of 2016, became Public Law 114-190. Upon implementation of this law, the FAA would be required to report a cybersecurity plan to Congress that was built on NIST standards. The FAA was also required to establish a research and development plan for the national airspace to keep it safe for travelers.

115th Congress (2017–2018)

HR 3359, the Cybersecurity and Infrastructure Security Agency Act amended the Homeland Security Act of 2002. Here, the National Protection and Programs Directorate in the DHS was renamed the Cybersecurity and Infrastructure Security Agency. This became law on November 16, 2018 (PL 115-278). Under this new legislation, the agency would be headed by the Director of National Cybersecurity and Infrastructure Security. It would be responsible for leading cybersecurity and CI security programs and operations. It would also oversee the implementation of DHS's responsibilities regarding chemical facility antiterrorism standards. It also transferred the Office of Biometric Identity Management to the Management Directorate.

Another new law was Public Law 15-236 (S770) which was signed into law on August 14, 2018. Known as the NIST Small Business Cybersecurity Act, this new law required NIST to take the interests and needs of small businesses into account when developing voluntary guidelines for reducing cybersecurity risks to CI. Moreover, NIST must publish available resources for small businesses online to help them identify, assess, and manage risks.

America's Water Infrastructure Act of 2018, or Public Law 115-270 (S 3021), makes the Corps of Engineers must notify nonfederal stakeholders and allow them to provide input when it develops policy. The corps must improve the reliability and the maintenance of existing infrastructure and improve such infrastructure's resilience to cyber-related threats.

Finally, HR 7327, or Public Law 115-390 (signed on December 21, 2018) was called Strengthening and Enhancing Cyber-capabilities by Utilizing the Risk Exposure Technology Act or the SECURE Technology Act. This bill directed the secretary of DHS to create a system for individuals or organizations to report security vulnerabilities to DHS, and then develop a method to mitigate the vulnerabilities that were reported. In addition, DHS was asked to establish a "bug bounty" program through which individuals or organizations can receive compensation for reporting vulnerabilities.

116th Congress (2019–2020)

A new bill called Safeguarding Tomorrow Through Ongoing Risk Mitigation Act (STORM Act) became a new law when Congress passed it this session (S 3418/PL 116-284). This authorized FEMA to enter into agreements with any state or tribal government to make grants to establish hazard mitigation loan funds. The funds will provide funding assistance to local governments to carry out projects to reduce disaster risks for homeowners, businesses, nonprofit organizations, and communities to decrease the loss of life and property, the cost of insurance claims, and federal disaster payments.

Public Law 116-22 (S 1379) was called the Pandemic and All-Hazards Preparedness and Advancing Innovation Act of 2019. This bill established several programs to improve preparedness and response to a public health emergency. The new agencies created are the National Health Security Strategy, the Public Health Emergency Preparedness cooperative-agreement program, and the Hospital Preparedness Program. Another section of the bill establishes federal guidelines for regional healthcare facilities' response to public health emergencies, and more funds are made available to respond to public health emergencies.

Other sections of this law mandate an evaluation of the available workforce available to respond to public health emergencies and possible gaps in the workforce. Another section concerns an assessment of security threats that could result in a public health emergency. Third, the Strategic National Stockpile was reauthorized. Another section of the law required the Department of Health and Human Services to provide a report on international coordination for developing necessary products during health pandemics and to develop a plan for addressing cybersecurity threats to national health security.

117th Congress (2021–2022)

Congress passed the National Cybersecurity Preparedness Consortium Act, S 658, which the president signed on May 12, 2022 (PL 117-122). The new law allowed officials in DHS to cooperate with nonprofit groups to develop, update, and deliver cybersecurity training in support of homeland security.

A significant bill that passed Congress was the Infrastructure and Jobs Act (HR 3684/PL117-58). The bill was signed by President Biden in November 2021. This bill allocated funding for a wide variety of infrastructure projects, such as:

- Roads, bridges, and major projects;
- Passenger and freight rail;
- Highway and pedestrian safety;
- Public transit;
- Broadband;
- Ports and waterways;
- Airports;
- Water infrastructure;
- Power and grid reliability and resiliency;
- Resiliency, including funding for coastal resiliency, ecosystem restoration, and weatherization;
- Clean school buses and ferries;
- Electric vehicle charging;
- Addressing legacy pollution by cleaning up Brownfield and Superfund sites and reclaiming abandoned mines; and
- Western Water Infrastructure.

Another proposal made in the Senate, the Community Disaster Resilience Zones Act (S 3875; PL 117-255) was signed into law on December 20, 2022. This new law requires the President to continue to maintain a natural hazard assessment program that outlines the risks of natural hazards in the country. The assessment shows the risk of natural hazards and community resilience. This must be completed at least every five years.

Conclusion

Protection of the nation's CI and assets is a top priority for the members of Congress, and they have passed many new laws to ensure that threats will be investigated and in the case of an event, damage will be kept to a minimum. For this reason, Congress has passed laws that attempt to increase communication among and between groups, and that have provided funding for protection efforts. In the future, Congress will continue to pass new laws so protect assets and save property and lives.

Career Focus: President's National Infrastructure Advisory Council

The President's National Infrastructure Council (NIAC) was created in Executive Order 13231 issued by President George W. Bush in October 2001. The purpose of the Council is to advise the president on practical strategies for reducing risks to, and protecting, CI in industry and government.

The Council comprises 30 senior executives who have been appointed to the position. It volunteers their time to research problems and work cooperatively with other experts to come up with solutions to sometimes very issues. The Council's membership includes leaders from the private sector and from state and local governments. They come from an array of backgrounds with different expertise, which provides wide perspectives and solutions to issues. They advise the president and others in the administration about how to protect CI by reducing both physical risks and cyberattacks on the nation's assets. They also provide advice on how to make infrastructure more resilient if an event should occur. Another goal is to improve the communication between public and private owners and operators of CI. The NIAC is overseen by the U.S. Cybersecurity and Infrastructure Security Agency, which is part of the DHS.

The current head of the NIAC is Mr. Adebayo Ogunlesi. He is the chairman and CEO of Global Infrastructure Partners, an independent asset manager. He also worked at Credit Suisse for 23 years, served as an attorney, and clerked for Honorable Thurgood Marshall.

Some of the reports submitted by NIAC include topics such as:

- Improving intelligence information sharing across government and industry;
- Identifying and reducing complex cyber risks, particularly for cyber-physical systems that operate critical processes;
- How to better prepare for, and respond to, disruptions that can ripple across multiple infrastructure systems and paralyze services to entire regions;
- Facilitating cooperative decision-making among senior executives and federal leaders during imminent threats and disaster responses;
- Addressing the skills gaps and loss of institutional knowledge in key workforces.

Source: niac@cisa.dhs.gov

Key Terms

USA PATRIOT Act
Maritime Transportation Security Act
Post Katrina Emergency Management Reform (PKEMR) Act of 2006
Cybersecurity Workforce Assessment Act
Pandemic and All-Hazards Preparedness Reauthorization Act
PIPES Act
Cybersecurity and Infrastructure Security Agency Act
NIST Small Business Cybersecurity Act
SECURE Technology Act

Review Questions

1 What role did the USA PATRIOT Act play in critical infrastructure protection?
2 What was the goal of the Homeland Security Act?

3 In the 109th Congress, members passed the Post Katrina Emergency Management Reform (PKEMR) Act of 2006. What were the major elements of this bill as it pertains to critical infrastructure protection?
4 Describe some of the bills that were not passed by Congress pertaining to critical infrastructure protection.
5 If you were a member of Congress, what would you include in an infrastructure protection bill for the country?

Notes

1 Chris Edwards, Chris (2015). "Hurricane Katrina: Remembering the Federal Failures." Cato Institute, August 27. https://www.cato.org/blog/hurricane-katrina-remembering-federal-failures#:~:text=Perhaps%20the%20most%20appalling%20aspect,%E2%80%8Bof%E2%80%90%E2%80%8Bstate%20headquarters
2 "Ten Facts about the Katrina Response." *Politico*. https://www.politico.com/story/2012/10/10-facts-about-the-katrina-response-081957#:~:text=Within%20four%20days%20of%20Katrina's,additional%20%2451.8%20billion%20in%20aid
3 Chaudhuri, Debarshi. "Government: Response to Katrina. Massachusetts Institute of Technology." http://web.mit.edu/12.000/www/m2010/finalwebsite/katrina/government/government-response.html
4 "Four Overreaching Factions Contributed to the Failures of Katrina." http://www.disastersrus.org/katrina/senatereport/KatCon.pdf
5 FEMA. "Post-Katrina Emergency Management Reform Act." https://emilms.fema.gov/is_0230e/groups/21.html#:~:text=Provides%20transportation%20assistance%20for%20relocating,of%20survivors%20of%20major%20disasters
6 National Defense Authorization, Public Law 106-398. (October 30, 2000). https://www.govinfo.gov/content/pkg/PLAW-106publ398/pdf/PLAW-106publ398.pdf
7 Stevens, Gina Marie, and Tatelman, Todd B. (2006). "Protection of Security-Related Information." Congressional Research Service, September 27. http://fas.org/sgp/crs/secrecy/RL33670.pdf
8 Stevens, Gina Marie, and Tatelman, Todd B. (2006). "Protection of Security-Related Information." Congressional Research Service, September 27. http://fas.org/sgp/crs/secrecy/RL33670.pdf

Appendix

Proposed Bills Concerning Critical Infrastructure Protection

106	S 2448	Internet Integrity and CI Protection Act	Appointment of a Deputy AG for Computer Crime and Intellectual Property to advise federal prosecutors and law enforcement on computer crime. Coordinates national and international activities for combating cybercrime. Provides criminal penalties for those who access protected computers when the offense causes losses of at least $5,000 or when the offense causes the modification of medical records or when the offense causes a physical injury to any person or to the public's health or safety. Requires forfeiture of property used in committing such offenses or derived from proceeds. Requires the Director of the FBI to establish a National Cyber Crime Technical Support Center to provide technical assistance for computer-related criminal activities. Requires the Director to develop at least ten regional computer forensic laboratories, and to provide support for existing laboratories.
106	S 3188 HR 4246	Cyber Security Enhancement Act	Facilitates the protection of CI in the U.S. and to enhance the investigation and prosecution of computer-related crimes. Creates methods to ensure that information on CI and CI protection will not be made available under the Freedom of Information Act if the person submitting the information requests that the information not be made available.
107	HR 4	Energy Policy Act of 2002	Authorizes appropriations for a research and technology program pertaining to critical energy infrastructure protection; and provides funds to states for security to protect critical energy infrastructure facilities.
107	HR 1158; HR 4660	National Homeland Security Agency Act	Establishes a National Homeland Security Agency; transfers FEMA, Customs Service, Border Patrol, the Coast Guard, the Critical Infrastructure Assurance Office and the Institute of Information Infrastructure Protection of the Department of Commerce, and the National Infrastructure Protection Center and the National Domestic Preparedness Office of the FBI. Requires the Director to be an advisor to the National Security Council.
107	HR 2435 S 1456	Cyber Security Information Act	Prohibits the disclosure of cyber security information that is voluntarily provided to a federal entity. Such information shall be exempt from disclosure under FOIA; cannot be disclosed to a third party; and cannot be used in any civil action.
107	HR 2975	United and Strengthening America Act (USA Act)	**Information Sharing for Critical Infrastructure Protection**—Amends the Omnibus Crime Control and Safe Streets Act of 1968 to extend Bureau of Justice Assistance regional information sharing system grants to systems that enhance the investigation and prosecution abilities of participating Federal, State, and local law enforcement agencies in addressing multi-jurisdictional terrorist conspiracies and activities.
107	HR 4660	National Homeland Security and Combating Terrorism Act of 2002	Plans, coordinates, and integrates government border security, CI protection, and emergency preparedness activities.

(Continued)

(Continued)

107	HR 5710	Homeland Security Information Sharing Act (CI Information Act)	Allows a CI Protection program to be formed by the President or the Secretary of the DHS. Exempts from the FOIA any CI information that is voluntarily submitted to a federal agency for use in the security of CI when accompanied by a statement that such information is being submitted voluntarily in expectation of nondisclosure protection. Requires that CI information that was voluntarily submitted be stored so that it will not be made public. Sets criminal penalties for unauthorized disclosure of such information. Authorizes the federal government to issue advisories, alerts, and warnings to relevant companies, targeted sectors, other governmental entities, or the public regarding potential threats to CI.
107	S 1407	Critical Infrastructure Protection Act of 2001	Directs the National Infrastructure Simulation and Analysis Center (NISAC) to provide support for the activities of the President's Critical Infrastructure Protection and Continuity Board for simulation and analysis of any CI systems (cyber and/or physical) as a way to get a better understanding of their complexity and to facilitate changes that will mitigate any threats and to increase training and to enhance the stability of CI. Requires the Board to provide information on the CI requirements of each Federal agency.
107	S1419	Department of Defense Authorization Act	Authorizes funds for the Navy's critical infrastructure (CI) protection initiative.
107	S1510	Uniting and Strengthening American Act	Requires the Director of the U.S. Secret Service to develop a national network of electronic crime task forces throughout the U.S. to prevent, detect, and investigate various forms of electronic crimes, including potential terrorist attacks against CI and financial payment systems. Amends the Omnibus Crime Control and Safe Streets Act of 1968 to extend Bureau of Justice Assistance regional information sharing system grants to systems that enhance the investigation and prosecution abilities of federal, state, and local law enforcement agencies in addressing multi-jurisdictional terrorist conspiracies and activities.
107	S 1534	Department of National Homeland Security Act	Establishes the Department of National Homeland Security. Among other agencies to be transferred here are the Critical Infrastructure Assurance Office, the Institute of Information Infrastructure Protection of the Department of Commerce, the National Infrastructure Protection Center, and the National Domestic Preparedness Office of the FBI. Establishes within the Department: (1) separate Directorates of Prevention, Critical Infrastructure Protection, and Emergency Preparedness and Response; and (2) an Office of Science and Technology to advise the Secretary with regard to research and development efforts and priorities for such directorates. Requires the Secretary to establish mechanisms for the sharing of information and intelligence with U.S. and international intelligence entities.
107	S1593	Water Infrastructure Security and Research Development Act	Requires the Administrator of the EPA to establish a program of grants to research institutions to improve the protection and security of public water supply systems by carrying out projects concerning processes to address physical and cyber threats to water supply systems. The projects should assess possible security issues and protect systems from potential threats by developing technologies, processes, guidelines, standards, procedures, real-time monitoring systems, and educational and awareness programs.

107	S 1766	Energy Policy Act: Energy Science and Technology Enhancement Act	Sets forth a national energy research and technology program that will encourage partnerships with industry, national laboratories, and institutions of higher learning. Each program will include research into critical energy infrastructure protection. Authorizes the Secretary of Energy to establish security enhancement programs for critical energy infrastructure. Instructs the Secretary of the Interior to establish the Outer Continental Shelf Energy Infrastructure Security Program to 1 provide financial assistance for State security plans against threats to critical OCS energy infrastructure facilities, and provide support for activities needed to maintain the safety and operation of critical energy infrastructure activities.
107	S2077	Securing Our States Act of 2002	Requires the Director of FEMA to make grants to States to improve CI protection.
107	S2452	Energy Science and Technology Enhancement Act of 2002	Sets forth a national energy research and technology program that operates in partnership with industry, national laboratories, and institutions of higher learning. Includes within the program is a focus on critical energy infrastructure protection research and development. Directs the Energy Secretary to plan Government activities relating to border security, CI protection, and emergency preparedness and to act as the focal point regarding crises and emergency planning and response. Transfers the Critical Infrastructure Assurance Office of the Department of Commerce and the National Infrastructure Protection Center of the FBI to the Treasury Department. Establishes within the Department: (1) Directorates of Border and Transportation Protection, Critical Infrastructure Protection, and Emergency Preparedness and Response. Provides that the Director of OST assist the Directorate of Critical Infrastructure Protection.
107	S 2509/ S 2515	Transparent and Enhanced Criteria Act	Amends the Defense Base Closure and Realignment Act of 1990 to add the following to the selection criteria for the 2005 round of defense base closures and realignments: the costs and effects of relocating CI.
107	S Amdt 1812		Sets aside funds for the CI protection initiative of the Navy
107	S Amdt 2363		Sets aside funds for the CI protection initiative of the Navy
107	S Amdt 2514		Increases the amount provided for research for CI protection
107	S Amdt 4422		Sets aside $6,000,000 of operation and maintenance from the Navy for the CI Protection Program
107	S Amdt 4552		Identifies certain sites as key resources for protection by the Directorate of Critical Infrastructure Protection
108	HR 6	Energy Policy Act of 2003	Authorizes appropriations for a research and technology deployment program pertaining to critical energy infrastructure protection.
108	HR 10	9/11 Recommendations Implementation Act	Directs the Secretary of the Treasury to submit a report on public/private partnerships to protect critical financial infrastructure.
108	HR 3367 S 1230	To provide for additional responsibilities for the Chief Information Office of the DHS	Directs the Chief Information Officer of the DHS to establish a program to provide for efficient use of geospatial information. This will include providing geospatial information as necessary to implement the CI protection programs and providing leadership in managing databases used by those responsible for planning, prevention, mitigation, assessment, and response to emergencies and CI.

(Continued)

(Continued)

108	HR 4852	DHS Authorization Act for FY 2005	Establishes within the Directorate for Information Analysis and Infrastructure Protection a National Cybersecurity Office. Establishes the Liberty Shield Award for Innovation and Excellence in CI Protection. Urges the DHS Homeland Security Operations Center to increase on-site participation of representatives from private sector CI sectors. Directs the Secretary of the DHS to develop and distribute CI protection awareness and education materials for emergency response providers.
108	S 1229	Federal Employee Protection of Disclosures Act	Amends the Homeland Security Act of 2002 to prohibit disclosure of voluntarily shared CI information that is protected information.
108	S 2021	Domestic Defense Fund Act of 2004	Authorizes the Secretary of the DHS to award grants to States, units of local government, and Indian tribes for homeland security development. The grants can be used to improve cyber and infrastructure security. Requires 70% of grant funds to be allocated among metropolitan cities and urban counties based on the Secretary's calculations of infrastructure vulnerabilities and threats such as proximity to international borders, nuclear or other energy facilities, air, rail or water transportation, and national icons and Federal buildings. Allocates funds for discretionary grants for the protection of CI.
108	S Amdt 743		Sets aside funds for the Collaborative Information Warfare Network at the Critical Infrastructure Protection Center at the Space Welfare Systems Center.
108	S Amdt 1235		Makes available from Research, Development, Test, and Evaluation for the Critical Infrastructure Protection Center.
109	HR 867/S 394	Openness Promotes Effectiveness in our National Government Act (OPEN government Act	Requires the Comptroller General to annually report on the implementation of provisions to protect voluntarily shared CI information.
109	HR 1817	DHS Authorization Act for FY 2006	Requires the Comptroller General to annually report on the implementation of provisions to protect voluntarily shared CI information.
109	HR 4881	National Defense Critical Infrastructure Protect Act of 2006	Authorizes appropriations for the DHS for FY 2006, including for CI grants. Amends the Homeland Security Act of 2002 requires the Under Secretary for Information Analysis and Infrastructure Protection to disseminate information relevant to CI sectors. Requires the Under Secretary to assign Information Analysis and Infrastructure Protection functions to the Assistant Secretary for Information Analysis, the Assistant Secretary for Infrastructure Protection, and the Assistant Secretary for Cybersecurity. Subtitle D: Critical Infrastructure Prioritization—Directs the Secretary of the DHS to complete prioritization of the nation's CI according to the threat of terrorist attack; the likelihood that an attack would destroy or significantly disrupt such infrastructure; and the likelihood that an attack would result in substantial numbers of deaths and serious bodily injuries, a substantial adverse economic impact, or a substantial adverse impact on national security. Requires the Secretary of the DHS, in coordination with federal agencies, state, local, and tribal governments, and the private sector, to review plans for securing CI. Protects certain CI information generated, compiled, or disseminated by the DHS to be protected.

109	HR 5441	DHS Appropriations Act	Transfers to FEMA all functions of the Directorate of Preparedness except for the Office of Infrastructure Protection, the National Cybersecurity Division, and others. Establishes within FEMA a National Integration Center, a National Infrastructure Simulation and Analysis Center, a National Operations Center, and a Chief Medical Officer.
109	H Amdt 149		Requires the DHS to coordinate its activities regarding CI protection with other relevant Federal agencies
109	S 494 S 2361 S 2766	Federal Employee Protection Act	Amends the Homeland Security Act of 2002 to provide that, for purposes of provisions regarding the protection of voluntarily shared CI information, permissible use of independently obtained CI information includes any lawful disclosure an employee or applicant reasonably believes is credible evidence of waste, fraud, abuse, or gross mismanagement, without restriction as to time, place, form, motive, context, or prior disclosure.
109	S 2361	Honest Leadership and Accountability in Contracting Act	Amends the Homeland Security Act of 2002 to provide that, for purposes of provisions regarding the protection of voluntarily shared CI information, permissible use of independently obtained CI information includes any lawful disclosure an employee or applicant reasonably believes is credible evidence of waste, fraud, abuse, or gross mismanagement, without restriction as to time, place, form, motive, context, or prior disclosure.
109	S 3721	Post Katrina Emergency Management Reform Act of 2006	Sets forth provisions regarding FEMA's mission, which includes planning for the protection, security, resiliency, and restoration of CI and key resources, including cyber and communications assets. Directs the FEMA Administrator to establish at least two pilot projects to develop and evaluate strategies and technologies for capabilities in a disaster in which there is significant damage to CI.
110	1309	Freedom of Information Act Amendments of 2007	Requires the Comptroller General to report on the implementation of provisions for the protection of voluntarily shared CI information.
110	HR 1326 S 849	OPEN Government Act of 2007	Requires the Comptroller General to report on the implementation of provisions for the protection of voluntarily shared CI information.
110	S 4	Improving America's Security Act of 2007	Establishes the State Homeland Security Grant Program to assist state, local, and tribal governments in preventing, preparing for, protecting against, responding to, and recovering from terrorist acts. The bill provides lists of permissible uses of grants, including the payment of personnel costs for protecting CI and key resources identified in the CI List. Lists considerations in awarding grants, including the nature of the threat, the location, risk, or vulnerability of CI and key national assets, and the extent to which geographic barriers pose unusual obstacles to achieving, maintaining, or enhancing emergency communications operability or interoperable communications. Directs the Administrator to create model standards that states may adopt in conjunction with CI owners and operators and their employees to permit access to restricted areas in the event of a natural disaster, terrorist act, or other man-made disaster.

(*Continued*)

(Continued)

			Title XI: Critical Infrastructure Protection: Directs the Secretary to establish a risk-based prioritized list of CI and key resources that (1) includes assets or systems that, if successfully destroyed or disrupted through a terrorist attack or natural catastrophe, would cause catastrophic national or regional impacts; and (2) reflects a cross-sector analysis of CI to determine priorities for prevention, protection, recovery, and restoration. Requires the Secretary to include levees in the Department's list of CI sectors. Authorizes the Secretary to establish additional CI and key resources priority lists by sector. Directs the Secretary to prepare a risk assessment of the CI and key resources of the nation. Authorizes the DHS to rely on a vulnerability or risk assessment prepared by another federal agency that the DHS determines is prepared in coordination with other DHS initiatives relating to CI or key resource protection and partnerships between the government and private sector. Directs the Secretary to report to specified committees for each fiscal year detailing the actions taken by the government to ensure the preparedness of industry to (1) reduce interruption of CI operations during a terrorist attack, natural catastrophe, or other similar national emergencies. Directs the Secretary to provide sufficient financial assistance for the reasonable costs of the Information Sharing and Analysis Center for Public Transportation (ISAC) established to protect CI; (2) require public transportation agencies at significant risk of terrorist attack to participate in ISAC; and (3) encourage all other public transportation agencies to participate in ISAC.
109	S 274	Federal Employee Protection of Disclosures Act	Amends the Homeland Security Act of 2002 to provide that, for purposes of provisions regarding the protection of voluntarily shared CI, permissible use of independently obtained CI information includes any lawful disclosure an employee or applicant reasonably believes is credible evidence of waste, fraud, abuse, or gross mismanagement, without restriction as to time, place, form, motive, context, or prior disclosure.
110	S606	Honest Leadership and Accountability in Contracting Act of 2007	Amends the Homeland Security Act of 2002 to provide that, for purposes of provisions regarding the protection of voluntarily shared CI, a permissible use of independently obtained CI includes any lawful disclosure an employee or applicant reasonably believes is credible evidence of waste, fraud, abuse, or gross mismanagement, without restriction as to time, place, form, motive, context, or prior disclosure.
110	S 1804	National Agriculture and Food Defense Act of 2007	States that the Secretary of the DHS shall lead federal, state, local, tribal, and private efforts to enhance the protection of U.S. CI and key resources, including the agriculture and food system.
110	S 2215	Supporting America's Protective Security Advisor Act of 2007	Amends the Homeland Security Act of 2002 to establish the Protective Security Advisor Program Office within the Protective Security Coordination Division of the Office of Infrastructure Protection of the DHS. Requires the Office to have primary responsibility within the DHS for encouraging state, local, and tribal governments and private sector owners and operators of CI and key resources to operate within the risk management framework of the National Infrastructure Protection Plan. The Office should also coordinate national and intergovernmental CI and key resource activities with such governments, owners, and operators as well as conduct risk assessment analyses that enhance the preparedness of CI and key resources.

110	S Amdt 415		Amends title X, with respect to CI protection efforts by Federal departments and agencies
111	HR 6351	Strengthening Cybersecurity for Critical Infrastructure Act	Grants the Secretary of the DHS primary authority in the creation, verification, and enforcement of measures for the protection of critical information infrastructure, including practices applicable to critical information infrastructures that are not owned by or under the direct control of the federal government. It authorizes the Secretary to conduct audits to ensure that appropriate measures are taken to secure critical information infrastructure. Establishes a National Office for Cyberspace, headed by an Executive Cyber Director, who shall have authority to oversee interagency cooperation on security policies relating to the creation, verification, and enforcement of measures regarding the protection of critical information infrastructure.
111	S 946	Critical Electric Infrastructure Protection Act of 2009	Directs the Secretary of Homeland Security to investigate to determine if the security of federally owned programmable electronic devices and communication networks (including hardware, software, and data) essential to the operation of critical electric infrastructure have been compromised. Directs the Secretary to make ongoing assessments and provide periodic reports with respect to cyber vulnerabilities and cyber threats to CI, including critical electric infrastructure and advanced metering infrastructure. Directs the Federal Energy Regulatory Commission to establish mandatory interim measures to protect against known cyber vulnerabilities or threats to the operation of the critical electric infrastructure.
111	S 1462	American Clean Energy Leadership Act of 2009	Amends the Federal Power Act to require FERC to issue rules or orders to protect critical electric infrastructure from cybersecurity vulnerabilities. Makes such rules or orders expire when a standard is developed and approved to address the vulnerability. Authorizes the Secretary of Energy to require people to take action to avert or mitigate an immediate cybersecurity threat to critical electric infrastructure.
112	HR 1136	Executive Cyberspace Coordination Act of 2011	Requires the Director of the National Office for Cyberspace to review federal agency budgets relating to the protection of information infrastructures. Establishes in the Executive Office of the President the Office of the Federal Chief Technology Officer. The duties of the Officer are to advise the President and agency officials on information technology infrastructures, and strategy; and to establish public–private sector partnership initiatives. Grants the Secretary of Homeland Security primary authority for the protection of the CI infrastructure.
112	HR 2658	Federal Protective Service Reform and Enhancement Act	Revises provisions governing the Federal Protective Service. Declares FPS's mission to be to secure all facilities and surrounding federal property under its protection and to safeguard all occupants. Requires the Director of FPS to report to the Under Secretary responsible for CI. Directs the Secretary to submit a strategy for cooperation between the Under Secretary responsible for CI protection and the Under Secretary for Science and Technology regarding research, development, and deployment of security technology.

(*Continued*)

(Continued)

112	HR 3674	PRECISE Act of 2012	Amends the Homeland Security Act of 2002 to direct the Secretary of the DHS to perform necessary activities to facilitate the protection of federal systems and to assist CI owners and operators, upon request, in protecting their CI information systems. Directs the Secretary, in carrying out cybersecurity activities, to coordinate with relevant federal agencies, state and local government representatives, CI owners and operators, suppliers of technology for such owners and operators, academia, and international organizations and foreign partners; and develop and maintain a strategy that articulates DHS actions that are needed to assure the readiness, continuity, integrity, and resilience of federal systems and CI information systems. Directs the Secretary to make appropriate cyber threat information available to appropriate owners and operators of CI on a timely basis.
112	H Amdt 326		To coordinate federal information security policy through the creation of a National Office for Cyberspace, and establishing measures for the protection of CI from cyberattacks
112	S 1342	Grid Cyber Security Act	Amends the Federal Power Act to direct the Federal Energy Regulatory Commission to determine whether certain reliability standards are adequate to protect critical electric infrastructure from cyber security vulnerabilities and order the Electric Reliability Organization (ERO) to submit a proposed reliability standard that will provide adequate protection of critical electric infrastructure from cyber security vulnerabilities. Authorizes the Secretary of Energy to (1) require persons subject to FERC jurisdiction to take immediate action that will best avert or mitigate the cyber security threat if necessary to protect critical electric infrastructure. Applies specified disclosure restrictions to critical electric infrastructure information submitted to FERC or DOE or developed by a federal power marketing administration or the Tennessee Valley Authority, under this act to the same extent as they apply to CI information voluntarily submitted to the DHS. Authorizes FERC to require the ERO to develop and issue emergency orders to address vulnerabilities and take immediate action to protect critical electric infrastructure if necessary. Directs the Secretary of Energy to assess the susceptibility of critical electric infrastructure to electromagnetic pulse events and geomagnetic disturbances and determine whether and to what extent infrastructure affecting the transmission of electric power in interstate commerce should be hardened against such events and disturbances.
112	S 1546	DHS Authorization Act	Applies specified disclosure restrictions to critical electric infrastructure information submitted to FERC or DOE, or developed by a federal power marketing administration or the Tennessee Valley Authority, under this Act to the same extent as they apply to CI information voluntarily submitted to the DHS. Requires FERC and DOE to establish information sharing procedures on the release of CI information to entities subject to this act.

112	S 3414	Cybersecurity Act of 2012	Establishes a National Cybersecurity Council, to be chaired by the Secretary of the DHS, to identify categories of critical cyber infrastructure; establish a voluntary cybersecurity program for CI to encourage owners of CI to adopt such practices; develop procedures to inform owners and operators of CI of cyber threats, vulnerabilities, and consequences. Directs the Council to designate an agency to conduct top-level cybersecurity assessments of cyber risks to CI with voluntary participation from private sector entities. Directs the Council to identify Critical Cyber Infrastructure within each sector of CI and CI owners within each category and establish a procedure for owners of critical cyber infrastructure to challenge the identification. Directs the Council to identify CCI categories as a critical cyber infrastructure only if damage or unauthorized access could reasonably result in (1) the interruption of life-sustaining services sufficient to cause a mass casualty event or mass evacuations; (2) catastrophic economic damage to the U.S., including financial markets, transportation systems, or other systemic, long-term damage; or (3) severe degradation of national security. Requires the Council to establish procedures under which owners of critical cyber infrastructure shall report significant cyber incidents affecting critical cyber infrastructure. Provides for congressional review of critical cyber infrastructure determinations. Requires private sector coordinating councils within CI sectors established by the National Infrastructure Protection Plan to propose cybersecurity practices to the Council. Permits federal agencies with responsibilities for regulating the security of CI to adopt such practices as mandatory requirements. Directs the Council to establish the Voluntary Cybersecurity Program for Critical Infrastructure under which owners of CI who are certified to participate in the Program select and implement cybersecurity measures of their choosing that satisfy such cybersecurity practices in exchange for (1) liability protection from punitive damages; (2) expedited security clearances; and (3) prioritized technical assistance, real-time cyber threat information, and public recognition. Prohibits any of the above provisions relating to the CI public–private partnership from limiting the ability of a federal agency with responsibilities for regulating the security of CI from requiring that the cybersecurity practices adopted by the Council be met. Directs the Secretary to establish a Critical Infrastructure Cyber Security Tip Line. Requires the DHS and DOD to jointly establish academic and professional Centers of Excellence to protect CI in conjunction with international academic and professional partners.
113	HR 3032	Executive Cyberspace Coordination Act	Establishes the National Office for Cyberspace to serve as the principal office for coordinating issues relating to cyberspace. Requires the Director of the National Office for Cyberspace to review federal agency budgets relating to the protection of information infrastructures. Grants the Secretary of Homeland Security authority for the protection of the CI infrastructure, as defined by this Act.

(Continued)

(Continued)

113	HR 3410	Critical Infrastructure Protection Act	Amends the Homeland Security Act of 2002 to require the Secretary of Homeland Security to conduct outreach to educate owners and operators of CI of the threat of EMP events. Directs the Secretary to conduct research to mitigate the consequences of Electromagnetic Pulse Attack events, including an analysis of the risks to CI from a range of EMP events and available technology options to improve the resiliency of CI to such events.
113	HR 3696	National Cybersecurity and Critical Infrastructure Protection Act	Directs the Secretary of the DHS to coordinate with federal, state, and local governments, national laboratories, CI owners and operators, and other cross-sector coordinating entities to facilitate a national effort to strengthen and maintain CI from cyber threats; and ensure that the DHS policies enable CI owners and operators to receive appropriate and timely cyber threat information. In addition, the Secretary is to ensure that the allocation of federal resources is cost effective and reduces burdens on CI owners and operators. Directs the Secretary to oversee federal efforts to support the efforts of CI owners and operators to protect against cyber threats; to help share cyber threat information with owners and operators of CI; and to facilitate cyber incident response assistance by providing analysis and warnings related to threats to, and vulnerabilities of, CI information systems. Requires the Secretary to (1) designate CI sectors and (2) recognize, for each sector, a previously designated sector-specific agency (SSA), a Sector Coordinating Council (SCC), and at least one Information Sharing and Analysis Center (ISAC). Permits to be included as CI sectors: chemical; commercial facilities; communications; critical manufacturing; dams; Defense Industrial Base; emergency services; energy; financial services; food and agriculture; government facilities; healthcare and public health; information technology; nuclear reactors, materials, and waste; transportation systems; and water and wastewater systems. Directs the Secretary to implement procedures for continuous, collaborative, and effective interactions between the DHS and CI owners and operators. Directs the National Institute of Standards and Technology to facilitate the development of a voluntary, industry-led set of standards and processes to reduce cyber risks to CI.
113	HR 5712	DHS Private Sector Office Engagement Act of 2014	Amends the Homeland Security Act of 2002 with provisions establishing a Private Sector Office within the DHS. Gives the Assistant Secretary the responsibility to promote best practices regarding cyber security and CI protection to the private sector.
113	S2519	National Cybersecurity Protection Act	Amends the Homeland Security Act of 2002 to establish a national cybersecurity and communications integration center in the DHS that will be responsible for overseeing CI protection programs. Directs the Under Secretary to develop, maintain, and exercise adaptable cyber incident response plans to address cybersecurity risks to CI. Requires the Secretary to make the application process for security clearances relating to a classified national security information program available to sector coordinating councils, sector information sharing and analysis organizations, and owners and operators of CI. Prohibits this Act from being construed to grant the Secretary any authority to promulgate regulations or set standards relating to the cybersecurity of private sector CI that was not in effect on the day before the enactment of this act.

114	HR 85	Terrorism Prevention and Critical Infrastructure Protection Act	Directs the Secretary of Homeland Security to work with CI owners and operators and state, local, tribal, and territorial entities to take necessary steps to manage risk and strengthen the security and resilience of the nation's CI against terrorist attacks. The DHS should also establish terrorism prevention policies with international partners to strengthen the security and resilience of domestic CI and CI located outside of the U.S. The DHS is to establish a task force to conduct research into the best means to address the security and resilience of CI to reflect the interconnectedness and interdependency of CI. The security and resiliency of CI will be furthered by the establishment of the Strategic Research Imperatives Program. Finally, the DHS should make research findings available and provide guidance to federal civilian agencies for the identification, prioritization, assessment, and security of their CI. Authorizes the Secretary to consult with other federal agencies on how best to align federally funded research and development activities that seek to strengthen the security and resilience of the nation's CI. Requires the Secretary to develop a description of the functional relationships within the DHS and across the federal government related to CI security and resilience, and demonstrate a near real-time situational awareness, research-based pilot project for CI.
114	HR 1073	Critical Infrastructure Protection Act	Amends the Homeland Security Act of 2002 to require the DHS to conduct a campaign to educate owners and operators of CI, emergency planners, and emergency responders at all levels of government about EM threats. Directs the DHS to conduct research to mitigate the consequences of EM threats, including (1) a scientific analysis of the risks to CI from EM threats; (2) a determination of the CI that are at risk from EM threats; (3) an analysis of available technology options to improve the resiliency of CI to EM threats; and (4) the restoration and recovery capabilities of CI under differing levels of damage and disruption from various EM threats.
	HR 1560	Cyber Networks Act	Requires the national cybersecurity and communications integration center (NCCIC) to be the lead federal civilian agency for cross-sector sharing of information related to cyber threat indicators and cybersecurity risks for federal and nonfederal entities. Expands the NCCIC's functions. Directs the Under Secretary for Science and Technology to provide to Congress with an updated strategic plan to guide the overall direction of federal physical security and cybersecurity technology research and development efforts for protecting CI. Requires the plan to identify CI security risks and any associated security technology gaps. Requires the Under Secretary to produce a report on the feasibility of creating a risk-informed prioritization plan in the case that multiple CIs experience cyber incidents simultaneously.
114	HR 1731	National Cybersecurity Protection Advancement Act	Amends the Homeland Security Act of 2002 to require the DHS's NCCIC to oversee information sharing across CI sectors, with state and major urban area fusion centers and with small- and medium-sized businesses. Redesignates the DHS's National Protection and Programs Directorate as the Cybersecurity and Infrastructure Protection.

(Continued)

(Continued)

			Directs the Under Secretary for Science and Technology to provide an updated strategic plan to guide the overall direction of federal physical security and cybersecurity technology research and development efforts for protecting CI. The plan should identify CI security risks and any associated security technology gaps and identify programmatic initiatives for the advancement and deployment of security technologies for CI protection, including public–private partnerships, intergovernmental collaboration, university centers of excellence, and national laboratory technology transfers.
114	HR 2271	Critical Electric Infrastructure Protection Act	Amends the Federal Power Act to authorize the Department of Energy to issue orders for emergency measures to protect the reliability of either the bulk-power system or the defense critical electric infrastructure whenever the President issues a written directive identifying an imminent grid security emergency. Instructs DOE, before issuing an order for such emergency measures, to consult with governmental authorities in Canada and Mexico, regarding implementation of the emergency measures. Requires DOE to identify facilities in the U.S. and its territories that are critical to the defense of the U.S., and vulnerable to a disruption of the supply of electric energy provided by an external provider. Exempts critical electric infrastructure information from mandatory disclosure requirements under the Freedom of Information Act. Directs the Federal Energy Regulatory Commission to designate critical electric infrastructure information and prescribe regulations and orders prohibiting its unauthorized disclosure but also authorizes appropriate voluntary sharing with federal, state, local, and tribal authorities. Shields a person or entity in possession of critical electric infrastructure information from any cause of action for sharing or receiving information that was done in accordance with this Act.
114	HR 2402	Protecting Critical Infrastructure Act	Amends the Federal Power Act to exempt protected electric security information from mandatory public disclosure under the Freedom of Information Act. Prohibits any state, local, or tribal authority from disclosing such information pursuant to state, local, or tribal law.
114	HR 6381	DHS Reform and Improvement Act	Amends the Homeland Security Act of 2002 to prevent terrorism. Established an Office of Biometric Identity Management to share biometric data nationally and internationally; a National Computer Forensics Institute in the U.S. Secret Service; a Chemical, Biological, Radiological, Nuclear, and Explosives Office; and other agencies/offices. Requests that the DHS disseminate information as a way to prevent attacks by drones; asks the DHS to cooperate with foreign governments on border enforcement; Asks FEMA to award grants to help improve CI.
114	HR 5786	Community Protection and Preparedness Act	Asks the Department of Transportation to impose a fee for railroad tank cars used to carry certain flammable liquids.
114	HR 5843/HR 4860	US-Israel Cybersecurity Cooperation Enhancement Act	Requires the DHS to establish a grant program focused on cybersecurity research in cooperation with Israel.

114	HR 5537	Digital Global Access Policy Act (Digital GAP)	Asks the U.S. to cooperate with foreign governments and international organizations/businesses to promote first-time affordable Internet, removal of regulatory barriers to Internet access; supports Internet use to increase economic growth and cybersecurity and data protection
114	HR 5459	Cyber Preparedness Act	Amends the Homeland Security Act of 2002; requires the DHS to provide fusion centers with cybersecurity resources; the DHS must provide technical assistance and incident response capabilities and disseminate cybersecurity risk information to fusion centers
114	HR 5390	Cybersecurity and Infrastructure Protection Agency Act	Amends the Homeland Security Act of 2002 to designate the National Protection and Programs Directorate as the Cybersecurity and Infrastructure Protection Agency; to oversee efforts to protect the security of cyber and CI.
114	HR 5177	National Mitigation Investment Act	Authorizes the president to increase the maximum contributions for a major disaster. States must submit building codes that must meet nationally recognized building codes and standards in order to receive funds.
114	HR 5064	Improving Small Business Cyber Security Act	Makes grants for small businesses that have cybersecurity specialists.
114	HR 5050	Pipeline Safety Act	Directs the Department of Transportation to report on the status of pipeline safety requirements.
114	HR 4743	National Cybersecurity Preparedness Consortium Act	Allows the DHS to work with the National Cybersecurity Preparedness Consortium in an effort to increase cybersecurity and address threats of terrorism
114	S 2579	Drinking Water Safety and Infrastructure Act	Provides funds for public health emergencies related to public drinking water
114	HR 4350	Repeal of Cybersecurity Act	Repeals the Cybersecurity Act of 2015
114	S 2129	To Improve the Nation's Infrastructure	Provides funds to modernize and improve infrastructure
114	HR 3510	DHS Cybersecurity Strategy Act	Requires the DHS to develop a cybersecurity strategy
114	HR 3493	Security the Cities Act	Requires the Director for Domestic Nuclear Detection to create a program to enhance the ability of the U.S. to detect and prevent terrorist attacks that use nuclear or other radiological materials
114	HR 3490	Strengthening State and Local Cyber Crime Fighting Act	Creates a National Computer Forensics Institute to disseminate information related to cyber and electronic crimes and to educate and train personnel to investigate cyber crimes
114	HR 3313	Cyber Defense of Federal Networks Act	Requires the DHS to detect and remove intruders in the information systems of federal agencies; provide tools to agencies to improve and detect intrusions
114	HR 3305	EINSTEIN Act	Amends the Homeland Security Act to require the DHS to maintain systems to protect federal agency information systems.
114	S 1846	Critical Infrastructure Protection Act	Amends the Homeland Security Act to require the DHS to conduct a review of risk to CI and submit a strategy to protect the CI of the American Homeland.
114	HR 2534	SCAN Act	Scan Containers Absolutely Now Act for Every Pot.
114	HR 2271	Critical Electric Infrastructure Protection Act	Authorizes the Department of Energy to issue orders for emergency measures to protect the power system with a written notice from the president.
114	HR 1789	Tank Car Safety and Security Act	Directs the Department of Transportation to improve tank car standards if used to move flammable liquids.

(*Continued*)

(Continued)

114	HR 1753	Executive Cyberspace Coordination Act	Establishes the National Office for Cyberspace to serve as the principal office for coordinating issues pertaining to cyberspace.
114	HR 1731	National Cybersecurity Protection Advancement Act	Expands the composition of the National Cybersecurity and Communications Integration Center to include state and local governments on cybersecurity and others.
114	HR 1560	Protecting Cyber Networks Act	Requires the Director of National Intelligence to develop procedures to allow for the sharing of cyber threats.
114	S 754	To Improve Cybersecurity in the U.S.	Requires the Director of National Intelligence and the DHS to develop procedures to share cybersecurity threats with private organizations and nonfederal government agencies.
114	S 741	Water Infrastructure Resiliency and Sustainability Act	Requires the EPA to establish grants for projects that increase the resiliency of community water systems.
115	HR 945	Terrorism Prevention and CI Protection Act	Directs the DHS to work with CI owners and others to strengthen the security of the nation's CI against attacks' establish a terrorism prevention policy; to establish a task force to research best practices to address the security of CI.
115	S 3513	UAS CI Protection Act	Amends the FAA Extension, Safety, and Security Act to include RR facilities as CI.
115	HR 3958/S 79	Security Energy Infrastructure Act	Creates a pilot program to identify the security threats of entities in the energy sector.
115	S 1875	Flexible Grid Infrastructure Act	Requires the Federal Energy Regulatory Commission and the Department of Energy to address the security of the electric grid.
115	HR 1335	Cybersecurity Responsibility Act	Requires the Federal Communications Commission (FCC) to issue cybersecurity rules for communications networks, including wireless phones, Internet access services, and other services.
115	HR 2184/S 754	Cyber Scholarship Opportunities Act	Requires the Federal cyber scholarship-for-service program to include scholarship recipients who are seeking degrees in cybersecurity.
115	HR 1340	Interagency Cybersecurity Cooperation Act	Requires the FCC to establish an Interagency Communications Security Committee to review communications security reports from federal agencies.
115	S 2836/HR 6401	Preventing Emerging Threats Act	Authorizes the DHS and DOJ to authorize personnel to act to mitigate the threat that unmanned aircraft (drones) pose to the safety of facilities.
115	HR 1344/S 516	State Cyber Resiliency Act	Requires FEMA to administer a State Cyber Resilience Grant Program to assist states in preventing, preparing for, protecting against, and responding to cyber threats.
115	HR 1465	National Cybersecurity Preparedness Consortium Act	Authorizes the DHS to work with a consortium to support efforts to address cybersecurity risks and incidents, including threats or acts of terrorism.
115	H Res 334	Expressing the Sense of the House of Reps Regarding Grid Modernization	Urges the U.S. to promote the modernization of its energy delivery infrastructure.
115	HR 4668	Small Business Advanced Cybersecurity Enhancements Act	Directs the Small Business Administration to create a central small business cybersecurity assistance unit.
115	HR 935	Cyber Security Education and Federal Workforce Enhancement Act	Established within the DHS an Office of Cybersecurity Education and Awareness Branch to make recommendations regarding recruitment of cybersecurity professionals.

115	HR 4120	Grid Cybersecurity Research and Development Act	Requires the DOE to develop an initiative to mitigate the consequences of cyberattacks on the electric grid.
115	HR 59	Frank Lautenberg Memorial Secure Chemical Facilities Act	Sets provisions for regulations of security practices at chemical facilities.
115	HR 3776	Cyber Diplomacy Act	Sets forth U.S. international cyberspace policy.
115	HR 3010	Promoting Good Cyber Hygiene Act	Requires the National Institute of Standards and Technology to establish a list of best practices to defend against cybersecurity threats.
116	HR 7904/S 3728	Critical Infrastructure Employee Protection Act	Requires the DOT to support states to provide COVID testing for CI employees.
116	S 3668	Energy Infrastructure Protection Act	Authorizes FERC to provide assistance to owners of CI relating to methods used to defend against threats.
116	HR 7590	To establish in the Cybersecurity and Infrastructure Security Agency of the DHS a pilot program for the purpose of carrying out a talent exchange program between the private sector and the Cybersecurity and Infrastructure Security Agency	Establishes a pilot program for carrying out a talent exchange program between the private sector and the Cybersecurity and Infrastructure Security Agency to improve national security through public–private collaboration.
116	HR 680/S 174	Securing Energy Infrastructure Act	Establishes a program within DOE to identify the security vulnerabilities of some entities in the energy sector.
116	HR 7588	Strengthening the Cybersecurity and Infrastructure Security Agency Act	Directs CISA and GSA to carry out reviews related to the ability of CISA to carry out its mission.
116	S 4897	American Nuclear Infrastructure Act	Provides support for nuclear infrastructure in the U.S.
116	HR 4432	Protecting Critical Infrastructure Against Drones and Emerging Threats Act	Requires the DHS to address issues related to threats of drones and other emerging threats. The DHS must request information from other federal, state, local, and private sector agencies relating to threats associated with new technologies; develop and disseminate a security threat assessment regarding drones and other emerging threats associated with such new technologies; and establish a secure communications and information technology infrastructure, including data-mining and other advanced analytical tools.
116	S 4833	To authorize cybersecurity operations and missions to protect critical infrastructure by members of the National Guard in connection with training or other duty.	Authorizes members of the National Guard to perform, at the request of a state and in connection with training or other duty, cybersecurity operations or missions to protect critical infrastructure.

(*Continued*)

(Continued)

116	HR 2242	To amend the Robert T. Stafford Disaster Relief and Emergency Assistance Act to include certain services in the definition of critical services	Expands the definition of *critical services* in the Robert T. Stafford Disaster Relief and Emergency Assistance Act to include solid waste management, stormwater management, public housing, transportation infrastructure, and medical care.
116	HR 7590	To establish in CISA a pilot program to carry out a talent exchange program between the private sector and CISA	Establishes a pilot program for carrying out a talent exchange between the private sector and CISA to improve national security through public–private collaboration.
116	S 4157	National Infrastructure & Analysis Center Pandemic Modeling Act of 2020	Expands the authority of the National Infrastructure Simulation and Analysis Center to cover pandemics and other activities deemed appropriate by CISA. It also requires the center to provide modeling, simulation, and analyses to federal agencies upon request and not later than 15 days after completion.
116	S 3688	Energy Infrastructure Protection Act of 2020	Revises provisions related to protecting energy infrastructure (including critical electric infrastructure), including by authorizing the Federal Energy Regulatory Commission to provide assistance to certain entities relating to methods and tools that owners and operators of energy infrastructure may use to defend assets against threats.
116	S 4897	American Nuclear Infrastructure Act of 2020	Provides support and incentives for nuclear infrastructure and advanced nuclear technologies, requires licenses for exports of low-enriched uranium to be in compliance with nonproliferation standards and related requirements, allows the NRC to deny imports of certain enriched uranium supplies from Russia and China for national security purposes, creates a national strategic uranium reserve, and sets forth related requirements.
117	HR 2982	National Guard Cybersecurity Support Act	Authorizes the National Guard, at the request of a state and in connection with training, to protect CI.
117	HR 5491	Securing Systemically Important Critical Infrastructure Act	Sets out a process to designate elements of critical infrastructure as systemically important. *Critical infrastructure* refers to the machinery, facilities, and information that enable vital functions of governance, public health, and the economy.
117	HR 9337	Ensuring America's Critical Infrastructure Act	Expands the definition of *critical infrastructure* under the Critical Infrastructures Protection Act of 2001 in the USA PATRIOT Act. The term currently means systems and assets, whether physical or virtual, so vital to the U.S. that the incapacity or destruction of such systems and assets would have a debilitating impact on security, national economic security, or national public health or safety. The bill includes certain sectors as part of the existing definition, including communications, dams, emergency services, energy, financial services, food and agriculture, healthcare, information technology, transportation systems, and water and wastewater systems.
117	HR 3388	Protecting Critical Infrastructure Act of 2021	Increases federal criminal penalties for computer fraud and abuse offenses that involve critical infrastructure. The term *critical infrastructure* means systems and assets, physical or virtual, so vital to the U.S. that the incapacity or destruction of such systems and assets would have a debilitating impact on security, national economic security, national public health and safety, or any combination of those matters.

			Imposes a fine, a mandatory minimum prison term of 30 years, or both for a computer fraud or abuse offense that involves critical infrastructure. Directs the President to impose asset- and visa-blocking sanctions on foreign individuals and entities that access or attempt to access critical infrastructure.
117	HR 5440	Cyber Incident Reporting for Critical Infrastructure Act of 2021	Requires reporting and other actions to address cybersecurity incidents, including ransomware attacks. Entities that own or operate critical infrastructure must report cybersecurity incidents (e.g., ransomware attacks) within specified time frames while other entities may voluntarily report incidents. CISA must (1) carry out rulemaking to implement the reporting requirements, and (2) establish an office to receive and analyze such reports. To the extent practicable, CISA must align its rules with existing requirements related to the reporting of cybersecurity incidents.
117	HR 3848	Critical Supply Chains Commission Act	Establishes a National Commission on Critical Supply Chains to identify and investigate the dependencies, limitations, and risks associated with critical supply chains.
117	HR 3713	Space Infrastructure Act	Directs the DHS to designate space systems, services, and technology as a critical infrastructure sector. Under current law, *critical infrastructure* means systems and assets, whether physical or virtual, so vital to the U.S. that the incapacity or destruction of such systems and assets would have a debilitating impact on security, national economic security, national public health or safety, or any combination of those matters.
117	S 2377	Energy Infrastructure Act	Addresses energy infrastructure, clean energy supply chains, carbon capture and storage, ecosystem restoration, and western water infrastructure. Establishes new programs or expands existing programs to support energy infrastructure, including (1) the electric grid (e.g., infrastructure to prevent power outages, enhance grid cybersecurity, and integrate renewable energy); (2) nuclear energy (e.g., infrastructure for micro- or small modular reactors); (3) hydropower; (4) solar and other clean energy; (5) clean hydrogen; and (6) energy efficiency in buildings.
117	S 674	Public Health Infrastructure Saves Lives Act	Provides annual funding for the CDC to strengthen core public health infrastructure. Core public health infrastructure includes the elements and workforce capabilities that enable health departments to perform critical functions such as disease surveillance and emergency response.
117	HR 5135	GRID Act of 2021	Requires the DOD to implement a two-year pilot program to address vulnerabilities at critical defense facilities and their associated defense critical electric infrastructure. The DOD is authorized to make grants, enter into cooperative agreements, and supplement funds to support mitigating actions that address such vulnerabilities. The DOD must select at least three military installations that are designated as critical defense facilities to carry out the pilot program. The Government Accountability Office must conduct a review and report on the pilot program.
118	HR 3166		Requires CISA of the DHS to submit a report on the impact of the SolarWinds incident on information systems owned and operated by federal departments and agencies.

(*Continued*)

(Continued)

118	HR 1148/HR 1160	Critical Electric Infrastructure Cybersecurity Incident Reporting Act	Note: summary in progress.
118	HR 3059	To Amend Title 49 of the United States Code to define critical airport infrastructure	Note: summary in progress.
118	S 1050	Protect American Power Infrastructure Act	Note: summary in progress.
118	HR 1389	GRID Act	Note: summary in progress.
118	HR 763	Supply Chains Act	Note: summary in progress.
118	HR 891	Energy Resilient Communities Act	Note: summary in progress.
118	HR 278	Cyber Defense National Guard Act	Note: summary in progress.
118	HR 1166	Public Health Emergency Medical Supplies Enhancement Act	Note: summary in progress.
118	S 824	National Risk Management Act	Note: summary in progress.
118	S 939/HR 1724	Security Maritime Data from Communist China Act	Note: summary in progress.
118	HR 1219	Food & Agriculture Industry Cybersecurity Support Act	Note: summary in progress.
118	HR 280	Cyber Vulnerability Disclosure Reporting Act	Note: summary in progress.

8 National Perspective on Risk

DHS, FEMA, & CISA

Chapter Outline

Introduction

The attacks of September 11, 2001, were a pivotal moment in the nation's modern approach to critical infrastructure protection. Through the 9/11 Commission's recommendations, the federal government worked to develop a security framework that would not only protect our country from large-scale attacks but would also enhance federal, state, and local capabilities to prepare for, respond to, and recover from threats and disasters. Hence, the goal of managing risks became central to national security and a key responsibility of the newly formed Department of Homeland Security (DHS) in March 2003. Over the next 15 years, the DHS would lead the effort in building an integrated risk management (IRM) framework for the nation.

Under the USA PATRIOT Act, the Critical Infrastructure Protection Act of 2001 not only defined critical infrastructure but also mandated that the definition become U.S. policy. Furthermore, the National Preparedness Goal defined what it means for all communities to be prepared for the threats and hazards that pose the greatest risk to the security of the U.S. The goal describes this concept as:

> A secure and resilient nation with the capabilities required across the whole community to prevent, protect against, mitigate, respond to, and recover from the threats and hazards that pose the greatest risk.[1]

DOI: 10.4324/9781003434887-8

The myriad risks faced by the U.S. have increased exponentially, regardless of whether they are natural hazards, technological hazards, or terrorist events. The 2021 attacks on the U.S. Capitol; the East Palestine, Ohio, train derailment in 2022; and the Maui wildfire of 2023 are just a few recent examples demonstrating extreme loss of life, injuries, and economic activity. As these events illustrate, we live in a dynamic and uncertain world where previous events do not always shed light on the future.

What is National Risk Management?

The road to a national risk management policy came together in bits and pieces. Early on, the DHS assumed the lead role of risk management for the nation and did so through the Office of Risk Management and Analysis (RMA), which was situated within the National Protection and Programs Directorate (NPPD). The RMA, in coordination with the Risk Steering Committee (RSC), was responsible for developing and implementing a common framework for homeland security risk management. The efforts of the RMA produced three documents that were integral to developing this framework: the *DHS Risk Lexicon*, *DHS Risk Management Process*, and *Risk Management Fundamentals*. Collectively, these early documents created a starting point for understanding risk and were a critical first step toward using risk information and analysis to inform decision making.

DHS Risk Lexicon

To support the goal of building an IRM framework, the DHS RSC produced a document with definitions and terms that are fundamental to the practice of homeland security RMA. Initiated in 2008, the DHS Risk Lexicon provides a common language to improve the capability of the DHS to assess and manage homeland security risk. Initially, the DHS Risk Lexicon contained 23 terms developed by an RSC working group known as the Risk Lexicon Working Group. Validated against glossaries used by other countries and professional associations, the definitions serve as a tool to improve the capabilities of the DHS to assess and manage homeland security risk. In 2010, the second edition of the DHS Risk Lexicon was published with an additional 50 new terms and definitions.[1,2] These definitions are developed through a three-phase process and include:

- **Collection** Terms are collected from across the DHS and the risk community.
- **Harmonization** A single meaning for each term is produced by taking multiple, and often conflicting, definitions and synthesizing them.
- **Validation, Review, and Normalization** Non-DHS sources are used for validating harmonized definitions to guarantee that the definitions produced for use in the DHS are consistent with those used by the larger risk community. The entire Risk Lexicon Working Group is provided with proposed definitions for comment, which are debated and standardized for grammar and format.[2]

Among the key terms are definitions of risk, risk assessment, risk analysis, and risk-based decision making. The student of risk management policy would benefit from reviewing all the terms described in the document. The DHS Risk Lexicon provides a definition, sample usage of the term, and an annotation that describes in more detail how the term is used. For example, the definition of *risk-based decision making* is described as follows:

Definition: determination of a course of action predicated primarily on the assessment of risk and the expected impact of that course of action on that risk.

Sample Usage: After reading about threats and vulnerabilities associated with vehicle explosives, she practiced risk-based decision making by authorizing the installation of additional security measures.

Annotation: Risk-based decision making uses the assessment of risk as the primary decision driver, while risk-informed decision making may account for multiple sources of information not included in the assessment of risk as significant inputs to the decision process in addition to risk information. Risk-based decision making has often been used interchangeably but incorrectly with risk-informed decision making.[3]

The DHS Risk Lexicon is used by the DHS risk practitioners, decision makers, stakeholders, and state, local, tribal, and territorial government partners, as well as academia.[4] It is important to note that the DHS Risk Lexicon has helped in the development of institutional policy and technical guidelines, training and educational materials, as well as communications throughout the homeland security enterprise.[5] A challenge of the DHS Risk Lexicon is that it must be constantly updated and maintained. This is accomplished through a partnership between the RLWG and DHS Lexicon program. Overseeing the maintenance of existing terms and the addition of new terms is a major focus. In addition, consistency with related federal interagency efforts must be established. The Office of RMA continually collects information on risk-related lexicons and glossaries as they become available throughout the federal government.[6]

Risk Management Guidelines

These guidelines were published to serve as technically accurate primers for DHS risk analysis practitioners on key homeland security RMA processes and techniques. The first set of these guidelines was published in 2009 and included the following documents:

- Defining the Decision Context
- Developing Scenarios
- Designing Risk Assessments
- Analyzing Consequences
- Assessing Indirect Consequences in Risk Analysis
- Developing and Evaluating Alternative Risk Management Strategies
- Communicating Risk Analysis Results to Decision Makers

Risk Management Fundamentals

In 2011, an authoritative statement regarding the principles and process of homeland security risk management and what they mean to homeland security planning and execution was published. The release of this capstone document, *Risk Management Fundamentals*, was a major step forward in establishing a comprehensive homeland security risk doctrine.[7] The intent was to help develop a framework to make risk management an essential part of planning, preparing, and executing organizational missions. Homeland security leaders, supporting staffs, program managers, analysts, and operating personnel make use of this document in their efforts to promote risk management.[8] *Risk Fundamentals* is not a blueprint for homeland security action but rather a doctrine to support homeland security practitioners and their own experiences. Specifically, the doctrine offers these five areas of purpose:

1. Promote a common understanding of, and approach to, risk management;
2. Establish organizational practices that should be followed by DHS components;

3 Provide a foundation for conducting risk assessments and evaluating risk management options;
4 Set the doctrinal underpinnings for institutionalizing a risk management culture through consistent application and training on risk management principles and practices; and
5 Educate and inform homeland security stakeholders in risk management applications, including the assessment of capability, program, and operational performance, and the use of such assessments for resource and policy decisions.[9]

Policy for Integrated Risk Management

The Policy for IRM was established in May 2010, by then Secretary of Homeland Security, Janet Napolitano. Entitled "DHS Policy for Integrated Management," this document formalized many of the organizational aspects of the DHS risk effort. Specifically, it assigned lead responsibility to the NPPD and coordination authority to the Director of RMA. In addition, it established a number of key committees and processes to standardize risk across the DHS.[10] The policy supports the premise that security partners can most effectively manage risk by working together and that management capabilities must be built, sustained, and integrated with federal, state, local, tribal, territorial, nongovernmental, and private sector homeland security partners.[11]

Homeland Security Risk: Tenets and Principles

As suggested above, the guidelines presented in *Risk Management Fundamentals* are not designed to promote one way of doing risk management; rather, they offer broad guidelines that each organization may use to tailor to their own needs. The doctrine discourages a "one-size-fits-all" approach but does suggest that all DHS risk management programs be based on two key tenets:

- Risk management should enhance an organization's overall decision-making process and maximize its ability to achieve its objectives.
- Risk management is used to shape and control risk but cannot eliminate all risks.[12]

In addition to these two key tenets, DHS identifies five key principles for effective risk management which include:

- **Unity of effort**—*reiterates that homeland security risk management is an enterprise-wide process and should promote integration and synchronization with entities that share responsibility for managing risks.*
- **Transparency**—*establishes that effective homeland security risk management depends on open and direct communications.*
- **Adaptability**—*includes designing risk management actions, strategies, and processes to remain dynamic and responsive to change.*
- **Practicality**—*acknowledges that homeland security risk management cannot eliminate all uncertainty nor is it reasonable to expect to identify all risks and their likelihood and consequences.*
- **Customization**—*emphasizes that risk management programs should be tailored to match the needs and culture of the organization, while being balanced with the specific decision environment they support.*[13]

A Comprehensive Approach

Supporting the DHS Policy for IRM requires a comprehensive approach. According to the doctrine, *Risk Fundamentals*, a comprehensive approach improves decision making by allowing organizations to identify and balance internal and external sources of risk. Internal sources of risk

Table 8.1 DHS Organizational Risk Categories

	Strategic Risks	*Operational Risks*	*Institutional Risks*
Definition	Risk that affects an organization's vital interests or execution of a chosen strategy, whether imposed by external threats or arising from flawed or poorly implemented strategy.	Risk that has the potential to impede the successful execution of operations with existing resources, capabilities, and strategies.	Risk associated with an organization's ability to develop and maintain effective management practices, control systems, and flexibility and adaptability to meet organizational requirements.
Description	These risks threaten an organization's ability to achieve its strategy, as well as position itself to recognize, anticipate, and respond to future trends, conditions, and challenges. Strategic risks include those factors that may impact the organization's overall objectives and long-term goals.	Operational risks include those that impact personnel, time, materials, equipment, tactics, techniques, information, technology, and procedures that enable an organization to achieve its mission objectives.	These risks are less obvious and typically come from within an organization. Institutional risks include factors that can threaten an organization's ability to organize, recruit, train, support, and integrate the organization to meet all specified operational and administrative requirements.

Source: The Department of Homeland Security Risk Management Fundamentals (2011).

include financial stewardship, personnel reliability, and systems reliability. External sources of risk may be identified as those that are caused by external factors. Global, political, and societal trends, as well as hazards from natural disasters, terrorism, cybercrimes, pandemics, etc., are some examples.[14] Applying a comprehensive approach to risk management ensures that all risks are considered in a holistic manner. Risks should be managed as a system, while at the same time considering the underlying factors that directly impact organizational effectiveness and mission success.[15] The DHS identifies three organizational risk categories that demonstrate the holistic nature of a comprehensive approach: *strategic risks*, *operational risks*, and *institutional risks*. These categories are described in Table 8.1.

Key Practices

Included in this comprehensive approach are three key requirements. According to the DHS, effective management of risk is fostered and executed through:

1. A commitment and active participation by an organization's leadership;
2. A consistent approach across the organization; and
3. The ability to view risk on a comprehensive, enterprise-wide basis.[16]

DHS Risk Management Process

The DHS Policy for IRM directs DHS organizations to employ a standardized risk management process. It is the expressed purpose of this approach to encourage comparability and shared understanding of information and analysis in the decision-making process.[17] Comprised of seven planning and analysis efforts, the DHS Risk Management Process is as follows:

1. **Defining and framing the context** of decisions and related goals and objectives;
2. **Identifying the risks** associated with the goals and objectives;

Figure 8.1 Risk assessment.

Source: The Department of Homeland Security Risk Management Fundamentals (2011).

3 **Analyzing and assessing** the identified risks;
4 **Developing alternative actions** for managing the risks and creating opportunities and analyzing the costs and benefits of those alternatives;
5 **Making a decision** among alternatives and implementing that decision; and
6 **Monitoring** the implemented decision and comparing observed and expected effects to help influence subsequent risk management alternatives and decisions.
7 **Risk communications** underpin the entire risk management process (Figure 8.1).

Define and Frame the Context

This initial stage will inform and help shape the successive stages of the risk management cycle. An organization is likely to pull together a risk analysis and management team to tackle complex risk issues. These are often referred to as planning teams, a workforce, or a working group. To establish the context of risk, analysts must gain a strong understanding of the requirements and the environment in which the risks are to be managed. Among the considerations are policy concerns, goals and objectives, mission needs, decision makers and stakeholder interests, timeframe, resources, and risk tolerance.[18]

Identify Potential Risk

The homeland security enterprise covers a wide array of risks and as such makes identifying them complicated. As mentioned above, there are three organizational risk categories that can help sort out and identify potential risks: *strategic*, *operational*, and *institutional*. Some of the techniques the DHS suggests include making a list according to "unusual," "unlikely," and "emerging" risks.[19] Scenarios are also used as a tool in the identification of potential risks. These are hypothetical situations comprised of a hazard, an entity impacted by that hazard, and associated conditions including consequences when appropriate.[20] In later chapters you will use specific scenarios to identify potential risks for a sample risk assessment activity.

Assess and Analyze Risk

Assessing and analyzing risk includes the following components: *determining a methodology*, *gathering data*, *executing the methodology*, *validating and verifying the data*, and *analyzing*

Table 8.2 Homeland Security Risks: Likelihood and Consequences

Likelihood is the chance of something happening, whether defined, measured, or estimated in terms of general descriptors, frequencies, or probabilities.
Consequences (or impact) include the loss of life, injuries, economic impacts, psychological consequences, environmental degradation, and inability to execute essential missions.[21]

Source: The Department of Homeland Security Risk Management Fundamentals (2011).

the outputs. It is interesting to note that risk practitioners will often move back and forth between these various components. Rarely will they occur in a linear fashion. Specifically, these components include the following:

Methodology—When choosing a risk assessment methodology, it is important to stay within the organization's capabilities. The DHS defines "methodology" in its *Risk Fundamentals Doctrine* as "any logical process by which the inputs into an assessment are processed to produce outputs that inform the decision."[22] According to the DHS, the most important aspect to consider in selecting a methodology is the decision the assessment must inform. In other words, the methodology should be appropriate to inform the decision.[23] Complex methods should be avoided unless they are necessary to assess the risk. It is also a good strategy to look at similar assessments that have already been completed. Other considerations are data availability, time, financial and personnel constraints. Homeland security risks are also assessed in terms of *likelihood* and *consequences* (see Table 8.2).

It is important to note, however, that there is no single methodology that is appropriate for measuring the likelihood of consequences of every risk. Furthermore, each methodology requires independent judgment regarding its design, and in some cases likelihood and consequences may not be necessary for the assessment.

Methodologies also might include considering homeland security risks as a function of *threats, vulnerabilities, and consequences* (TVC). This can be especially useful when assessing critical infrastructure protection. The TVC framework will be discussed in the next chapter.

As stated above, simple methodologies are preferred, and most are sorted into qualitative and quantitative categories. Overall, the methodology that best meets the decision maker's needs is generally considered the best choice.[24]

Gathering Data—Sources for collecting data on risk information may come from a variety of places including historical records, models, simulations, and elicitations of experts in the field. It is necessary to consider all aspects of the decision, regardless of whether they can be quantified. For example, psychological impacts might be considered along with loss of life and financial losses. Sometimes there may be pieces of data that are unknown. These may be expressed as uncertainty in the outputs (i.e. a major earthquake in California might be estimated to fall within a range, with some values being more likely than others). Hence, it might be useful to consider the impact of the uncertainty along with the other pieces of data that have been collected.[25]

Validating and Presenting the Data—The DHS suggests that throughout the risk assessment process, the gathered data and evidence should be carefully studied and compared to previous work. The data and evidence should be analyzed to identify relevant and interesting features for the decision maker, who may have a specified area they wish to focus.[26]

Developing Alternative Actions

In the risk assessment process, the ultimate objective is to provide decision makers with a structured way to recognize and select risk management actions. Developing viable alternatives provides

Table 8.3 DHS Methods for Developing and Evaluating Alternatives

- Reviewing lessons learned from relevant past incidents;
- Consulting subject matter experts, best practices, and government guidelines;
- Brainstorming;
- Organizing risk management actions;
- Evaluating options for risk reduction and residual risk;
- Developing cost estimates for risk management actions;
- Comparing the benefit of each risk management action with its associated cost; and
- Eliminating potential options.[27]

Source: The Department of Homeland Security Risk Management Fundamentals (2011).

leaders with a clear image of the potential benefits and costs of specific risk assessment options. According to the DHS, the development of alternative risk management actions should:

- Be understanding to participants of the process, including decision makers and stakeholders;
- Match and comply with the organization's relevant doctrine, standards, and plans;
- Provide documentation with assumptions explicitly detailed;
- Allow for future refinements; and
- Include planning for assessment of progress toward achieving desired outcomes.[28]

Additionally, the DHS has outlined specific approaches that may be used to develop and evaluate alternatives (see Table 8.3).

Make Decision and Implement Risk Management Strategies

When a decision maker determines which alternatives are best for managing a specific risk, he or she can either decide to implement a new plan of action or maintain an existing policy. In this process, the decision maker must consider the feasibility of implementing the options and how various alternatives will affect and ultimately reduce the risk. Other considerations include sufficient resources, capabilities, time, policies, legal issues, and the potential impact on stakeholders. Additionally, the possibility of the action creating new risks for the organization must also be taken into account.[29] The strengths and weaknesses of the alternatives should be clearly articulated so the decision is made on sound judgment. Implementing the decision needs to include proper leadership and a comprehensive project management approach.

Evaluation and Monitoring

This phase focuses on evaluating and monitoring the performance of the risk option to determine whether it has achieved the state goals and objectives. The DHS warns that, in addition to assessing performance, organizations should guard against unintended adverse impacts such as creating additional risk or failing to recognize changes in risk characteristics.[30] While the implementation of the risk management program must be measured and improved, the action risk reduction measures must also be assessed. According to the DHS, the evaluation should be conducted in a way that is commensurate with both the level of risk and the scope of the mission.[31] Effectiveness criteria are often used in this phase. This consists of tracking and reporting on performance results with concrete, realistic measures. For example, in situations where the decision maker decided to do nothing, the continued appropriateness of accepting the risk may be the best possible measure. In other situations, the best measure is often the reduction of the likelihood or consequences associated with a risk.[32] Some methods used in this evaluation phase include red teaming (scenario role-playing), exercises, external reviews, and surveys.[33]

Risk Communications

Central to the DHS Risk Management Process are risk communications. For each element of the risk management process, effective communications with stakeholders, partners, and customers are critical. According to the DHS, there must be consistent, two-way communication throughout the process to ensure that decision makers, analysts, and officials in charge are able to implement any decision and share a common understanding of what the risk is and what factors may contribute to managing it.[34] It is important to note that communication requirements will differ according to the audience and timeframe. The DHS states that, typically, risk communication is divided between internal and external audiences and between incident and standard timeframes.[35] Table 8.4 illustrates these concepts.

According to the DHS, an incident does not represent a break in the risk management process but rather a temporary acceleration after which the process continues as normal.[36] Furthermore, risk communications will be most effective if guided by the following:

- Plan for communications;
- Maintain trust;
- Use language appropriate to the audience;
- Be both clear and transparent;
- Respect the audience's concerns;
- Maintain the integrity of information.[37]

The FEMA Perspective

The FEMA is an operational component of the DHS. FEMA's perspective on risk encompasses a number of resources that help communities plan for, respond to, and recover from all types of hazards. FEMA takes its lead from Presidential Policy Directive 8 (PPD-8). This national preparedness directive recognizes that preparedness is a shared responsibility. Commonly referred to as

Table 8.4 DHS Risk Communication Audiences

Internal Risk Communications—Some risk communications are internal to an organization, for example, between analysts and decision makers.

External Risk Communications—Occur when the public and cross-agency nature of homeland security risk necessitates that the DHS communicate with external stakeholders, partners, and the public.

Incident Communications (also referred to as 'crisis communications')—This takes place under different conditions than standard communications. During a crisis the need to explain the situation clearly becomes a priority. Public officials must operate under extraordinary time constraints to get factual and potentially life-saving information.

Standard Communications—After a crisis event, standard communications should resume so that all stakeholders build a common understanding of what has happened, why certain decisions were made, and how to move forward.[37]

the "*whole community* approach," this perspective calls on federal agencies and communities to work together are six specific elements: Goal, Approach, Frameworks, Annual Report, and Federal Plans. Specifically, these elements include the following:

1. **The National Preparedness Goal**, as described in the Introduction to this chapter, details the necessary characteristics of a community that is truly prepared to handle various risks. Examples of such risks include natural disasters, man-made hazards, pandemics, terrorist attacks, and cyberattacks. The five key mission areas defined by the National Preparedness Goal are Prevention, Protection, Mitigation, Response, and Recovery. These mission areas are used to establish and evaluate 32 "core capabilities" that are needed to assess risk.[38] The following are four examples of core capabilities and the mission areas they fall under:

   a. Search and Rescue Operations (Response);
   b. Intelligence and Information (Prevention, Protection);
   c. Threats and Hazards Identification (Mitigation);
   d. Housing (Recovery).

By identifying the greatest risks to a community, leaders may use these core capabilities and mission areas to address preparedness weaknesses.

2. **The National Preparedness System (or Approach)** organizes a community so that it progresses toward the National Preparedness Goal. FEMA identifies six main ingredients of the National Preparedness System.[39] In short, communities must:

   a. **Identify and assess risk,** which is often done through data collection and analysis on historical and present-day risks;
   b. **Estimate capability requirements,** which can be achieved by assessing a community's core capabilities via the five key mission areas;
   c. **Build and sustain capabilities:** How must the community allocate its resources to bolster its core capabilities?
   d. **Plan to deliver capabilities,** where leaders communicate and coordinate action plans with the proper organizations;
   e. **Validate capabilities** through evaluations, simulations, and other exercises to determine strengths and weaknesses in a community's core capabilities;
   f. **Review and update** core capabilities, action plans, and allocation of resources with a frequency that is on par with the ever-evolving risks to the community.

3. **The National Planning Frameworks** detail how the whole community works toward the National Preparedness Goal. There exists a framework for each of the five key mission areas:

   a. **National Prevention Framework:** What are the capabilities necessary for the whole community to avoid a credible, imminent threat?
   b. **National Protection Framework:** What actions does the whole community take to secure its members from such a threat?
   c. **National Mitigation Framework:** How does the whole community lessen the impact of disasters, including loss of life and property, when they occur?
   d. **National Response Framework:** What are the capabilities needed to save lives, property, and the environment after a disaster?
   e. **National Recovery Framework:** How does the whole community ensure its effective recovery in the aftermath of a disaster?

The common thread linking each of these frameworks together is cohesive cooperation. All members of the whole community must work together to answer these vital questions and achieve the goals of the five key mission areas.

The National Preparedness Report (or Annual Report) synthesizes the status of the core capabilities of the whole community, including its progress and continued weak points. After thorough data collection and analysis of major disasters, FEMA releases a yearly National Preparedness Report that addresses the current risk landscape of the Nation.[40]

Federal Interagency Operational Plans detail the federal government's support of state and local efforts to address the five key mission areas and identify the capabilities of the entire nation.[41]

Another responsibility of FEMA is to provide a number of risk management resources to communities. These resources involve Hazard Mitigation Planning; Climate Resilience; Building Science (i.e. the study of natural hazards and their impact on structural architecture and engineering); Earthquake Risk; Dam Safety; Hurricane, Wind, & Water Surge Hazards; and Hazardous Response Capabilities (i.e. chemical, biological, radiological, and nuclear hazards).

Finally, FEMA offers planning guides to individuals and communities to assess the risks posed to a community and implement different types of mitigation measures. These comprehensive guides provide a carefully laid-out method to engage the whole community in thinking through potential crises and the required capabilities and resources needed in the event of a catastrophic event. One of these is the *Comprehensive Preparedness Guide (CPG) 101: Developing and Maintaining Emergency Operation Plans*.[42] The other guide, which is covered in the next chapter, is an update to CPG 101 entitled *CPG 201: The Threat and Hazard Identification and Risk Assessment (THIRA) and the Stakeholder Preparedness Review Guide (SPR)*.[43]

The CISA Perspective

The Cybersecurity and Infrastructure Security Agency (CISA) was established in 2018 under President Trump by Executive Order 13800: The Cybersecurity and Infrastructure Security Act of 2018. This amended the Homeland Security Act of 2002 to redesignate DHS's National Protections and Programs Directorate as CISA, officially transferring the directorate's resources and responsibilities over to the new agency.

Since 2018, CISA has taken the lead on our national risk management initiatives. The agency supports both cybersecurity and critical infrastructure risk management and assessment. Like FEMA, CISA is an operational component of the DHS. CISA works to understand, manage, and mitigate risk to our nation's cyber and physical infrastructure and encourages collaboration between government agencies, the private sector, and key stakeholders. Initiatives within CISA involve the advancement of, for example:

a Space system security, which focuses on the risks posed to near-Earth satellites;
b Pipelines – a physical infrastructure that often transports flammable or toxic materials;
c Positioning, navigation, and timing services, such as the Global Positioning Navigation (GPS);
d Security to fifth-generation (5G) wireless technology for telecommunication;
e Protection against electromagnetic pulse (EMP) attacks, which are man-made and targeted, and a geomagnetic disturbance (GMD) caused by space weather, which arises from the interaction between plasma from the Sun and the Earth's magnetic field. EMPs and GMDs cause immense damage to the nation's electrical grid, communication technology, water and wastewater systems, and transportation.[44]

To ensure the nation's resilience against targeted and un-targeted events, CISA directs assessments alongside government agencies, interested stakeholders, and the private industry. These assessments identify the capabilities and weaknesses of infrastructure and offer insight into what would happen in the event that the infrastructure is damaged and strategies and plans to protect critical infrastructure against such catastrophic incidents.

Protecting election infrastructure is another area of interest for CISA. Election security includes voter registration databases; IT systems used to count, audit, and report results; voting systems; and polling places. CISA offers assistance and resources to state and local governments, election officials, those in the private sector, and federal partners in regard to election security.

Finally, in September 2022, CISA released its inaugural comprehensive strategic plan to help progress the agency toward its goals over the next three years. The 2023–2025 Strategic Plan identifies four key areas to guide future risk assessments on critical infrastructure:

1. **Cyberspace**—CISA aims to mitigate risks to the nation's cyberspace by adopting a proactive mindset to prevent incidents and minimize their impacts should they occur.
2. **Critical infrastructure**—CISA plays a key role in coordinating the private sector, government agencies, and stakeholders to ensure the resilience of the nation's critical infrastructure and rapid and effective response if incidents were to arise. Assessments identify vulnerabilities and capabilities of numerous infrastructure sectors and follow an "all-hazards" approach, working to minimize risks posed by cyberattacks, natural hazards, and other threats.
3. **Collaboration and information sharing**—A core belief of CISA, collaboration, and proper communication between relevant parties drives forth the stability of the nation against all hazards. Partnerships between government, industry, academic, and international collaborators allow for CISA to better prevent, prepare for, and respond to risks against critical infrastructure.
4. **One CISA**—An internal goal, CISA, is built on concrete principles such as innovation, teamwork, clear communication, inclusion, and trust. CISA identifies itself as a team that operates as "One CISA," which, in their own words, means they "will work smart to operate in an efficient and cost-effective manner."[45]

Conclusion

An understanding of the DHS, FEMA, and CISA perspectives on risk management is an important objective for the student of risk policy. While the DHS initially took the lead on risk management in the creation of March 2003, since that time, FEMA and CISA have added responsibilities to support our nation's risk management initiatives. All three departments work in tandem to support risk resilience.

Key Terms

Adaptability
All Hazards
CISA
Consequences
DHS Risk Lexicon
DHS Risk Management Process
FEMA
Institutional Risks
IRM
Likelihood
Operational Risks
Risk
Risk-Based Decision Making
Risk Communications
Risk Fundamentals
Strategic Risks
Unity of Effort
Whole Community

Review Questions

1. How does the DHS define risk?
2. Explain the documents *Risk Management Fundamentals*, *The DHS Risk Lexicon*, and *The DHS Risk Management Process*.
3. Describe the "whole community" approach to preparedness. For each of the national planning frameworks, how might preparedness responsibilities be shared between individual families and the state, local, and federal governments?
4. Define the concept of "all hazards." How might CISA use the all-hazards approach to ensure election security?

Notes

1 https://www.fema.gov/emergency-managers/national-preparedness/goal#:~:text=The%20National%20Preparedness%20Goal%20describes,greatest%20risks%20to%20the%20nation.
2 US DHS, "DHS Risk Lexicon," 2010, p. 2
3 US DHS, "DHS Risk Lexicon," 2010, p. 33.
4 US DHS, "DHS Risk Lexicon," 2010, p. 43.
5 US DHS, "DHS Risk Lexicon," 2010, p. 43.
6 US DHS, "DHS Risk Lexicon," 2010, p. 42.
7 US DHS, "Strategies and Methods for Informing Risk Management: An Alternative Perspective," A White Paper, 2011, p. 10.
8 US DHS, "Risk Management Fundamentals," 2011, p. 5.
9 US DHS, "Risk Management Fundamentals," 2011, p. 5–6.
10 Napolitano, Janet (2010). "DHS Policy for Integrated Risk Management," memorandum, May 27.
11 US DHS, "Risk Management Fundamentals," 2011, p. 1.
12 US DHS, "Risk Management Fundamentals," 2011, p. 11.
13 US DHS, "Risk Management Fundamentals," 2011, p. 11-12.
14 US DHS, "Risk Management Fundamentals," 2011, p. 13.
15 US DHS, "Risk Management Fundamentals," 2011, p. 13.
16 US DHS, "Risk Management Fundamentals," 2011, p. 14.
17 US DHS, "Risk Management Fundamentals," 2011, p. 15.
18 US DHS, "Risk Management Fundamentals," 2011, p. 16.
19 US DHS, "Risk Management Fundamentals," 2011, p. 18.
20 US DHS, "Risk Management Fundamentals," 2011, p. 19.
21 US DHS, "Risk Management Fundamentals," 2011, p. 20.
22 US DHS, "Risk Management Fundamentals," 2011, p. 20.
23 US DHS, "Risk Management Fundamentals," 2011, p. 20.
24 US DHS, "Risk Management Fundamentals," 2011, p. 21.
25 US DHS, "Risk Management Fundamentals," 2011, p. 21.
26 US DHS, "Risk Management Fundamentals," 2011, p. 22.
27 US DHS, "Risk Management Fundamentals," 2011, p. 24.
28 US DHS, "Risk Management Fundamentals," 2011, p. 22.
29 US DHS, "Risk Management Fundamentals," 2011, p. 24.
30 US DHS, "Risk Management Fundamentals," 2011, p. 25.
31 US DHS, "Risk Management Fundamentals," 2011, p. 25.
32 US DHS, "Risk Management Fundamentals," 2011, p. 26.
33 US DHS, "Risk Management Fundamentals," 2011, p. 26.
34 US DHS, "Risk Management Fundamentals," 2011, p. 15.
35 US DHS, "Risk Management Fundamentals," 2011, p. 26.
36 US DHS, "Risk Management Fundamentals," 2011, p. 26.
37 US DHS, "Risk Management Fundamentals," 2011, p. 27.
38 https://www.fema.gov/emergency-managers/national-preparedness/goal
39 https://www.fema.gov/emergency-managers/national-preparedness/system
40 https://www.fema.gov/emergency-managers/national-preparedness#reports
41 https://www.fema.gov/emergency-managers/national-preparedness/frameworks/federal-interagency-operational-plans
42 https://www.fema.gov/sites/default/files/2020-05/CPG_101_V2_30NOV2010_FINAL_508.pdf
43 https://www.fema.gov/sites/default/files/2020-04/CPG201Final20180525.pdf
44 https://www.cisa.gov/topics/risk-management
45 https://www.cisa.gov/strategic-plan

9 Methods of Risk Assessment

Chapter Outline

Introduction

As presented in Chapter 8, the Department of Homeland Security (DHS) and its components, the Federal Emergency Management Agency (FEMA) and the Cybersecurity and Infrastructure Security Agency (CISA), have made strides to standardize risk and vulnerability assessments through various approaches. Over the years, a considerable amount of money has been spent on various methods, including software programs, to manage risks. Programs such as Risk Analysis and Management for Critical Asset Protection (RAMCAP), CARVER, the Target Analysis and Vulnerability Assessment, and the Partnership for Safe and Secure Communities (PASCOM) have been used at the federal, state, and local levels.

Risk assessment under the DHS, FEMA, and CISA has expanded on these earlier methods to incorporate an all-hazards, whole-community approach. In this chapter, you will explore different methods of risk assessment with particular emphasis on the Threat and Hazard Identification and Risk Assessment (THIRA). THIRA is a community-based, whole-community risk assessment process supported by FEMA and outlined by the 2013 National Infrastructure Protection Plan as a method to integrate human, physical, and cyber elements of critical infrastructure risk.

A Brief Discussion of Earlier Risk Assessment Methods

Enhancing the security of our nation's critical infrastructure has advanced tremendously since 911. Along the way, there have been numerous approaches to assessing and managing risk. Many of these methods were borrowed from the private sector and have their roots in the fields of engineering, science, and the military. However, the transfer of these techniques to the public sector has

DOI: 10.4324/9781003434887-9

not been seamless. Risk management can mean something different to everyone, and the quest to standardize a common risk approach has been a major goal of the DHS. The current approach favored by the DHS and the FEMA is the Integrated Risk Management (IRM) approach. The goal of the IRM is to identify, evaluate, prioritize, counter, and monitor the likelihood, vulnerability, and consequences of threats, natural hazards, and natural disasters to local people, property, infrastructure, and environment.[1] This approach serves as the basis of the THIRA, as described in *Comprehensive Preparedness Guide (CPG) 201: Threat and Hazard Identification and Risk Assessment Guide.*[2]

RAMCAMP, CARVER, and PASCOM

Before THIRA, other methodologies were used by state and local governments in conjunction with the DHS in their efforts to standardize risk assessment. One of these, RAMCAP is a framework for analyzing and managing risks associated with terrorist attacks against critical infrastructures. The purpose is to provide government decision makers with essential information about consequences and vulnerabilities in the private sector, which owns 85% of the nation's critical infrastructure.[3] The program is a seven-step process for asset analysis (asset characterization, threat characterization, consequence analysis, vulnerability analysis, threat assessment, risk assessment, and risk management). It is unique in that it facilitates the comparison of risks within a sector and across multiple sectors by employing common terminology and standardized measurement metrics.[4] This methodology is sector-specific in application, meaning the process is tailored to specific aspects of the 16 critical infrastructure sectors. These are compiled in documents called sector-specific guidance documents (SSGs). These SSGs assist companies in identifying and reporting on the vulnerabilities and potential consequences of terrorism by providing guidance on how to complete both preliminary and in-depth assessments.

CARVER was originally developed by the U.S. military to identify areas within critical or military infrastructures that may be vulnerable to an attack by U.S. Special Forces. It is a six-step approach to conducting security vulnerability assessments on critical infrastructure. It identifies the critical component of an asset that meets that requirement:

- **C**ritically;
- **A**ccessibility;
- **R**ecuperability;
- **V**ulnerability;
- **E**ffect;
- **R**ecognizability.

Later, it was adopted by the FDA and the U.S. Department of Agriculture for the food and agriculture critical infrastructure sectors. The approach allows food companies to analyze and identify critical areas that are most likely targets of an attack.[5] CARVER and Shock will be discussed in greater detail in Chapter 10.

As part of the Partnerships for Safe and Secure Communities, PASCOM was developed as a process for communities to systematically identify critical assets, conduct community vulnerability assessments, and develop executive preparedness programs in a manner that is tailored to the individual community's profile. PASCOM is an assessment tool that communities may use to identify critical assets, create threat scenarios, assess vulnerabilities, and analyze risk.[6]

Federal Guidelines for Risk Assessment

Current methods of risk assessment must begin with a discussion of federal guidelines. On March 30, 2011, President Barack Obama signed Presidential Policy Directive 8 (PPD-8): National Preparedness. This directive was a result of the realization that first responders cannot do it all by themselves and began a new chapter with the intent and scope of national preparedness.[7] The new policy theme that emerged was that capabilities required a whole-community approach—from the federal government to individual citizens. In this approach, leaders at all levels of government, private industry, nonprofit organizations, and the public must work together in a systematic effort to keep the nation safe and resilient when struck by catastrophic events such as natural disasters, acts of terrorism, cyber threats, technological incidents, and pandemics.[8] The goal of the policy directive is to ensure that federal departments and agencies work with the whole community to develop a National Preparedness Goal and a system to guide and track activities toward that preparedness goal. Within this framework there are five mission areas of national preparedness: *prevention*, *protection*, *mitigation*, *response*, and *recovery*. By far, the most significant of these missions is prevention. Methods of risk assessment are embodied in the idea that prevention is the key to thwarting future catastrophic events. If you prevent an event from happening—it cannot occur.

Within the National Preparedness Goal are 31 core capabilities categorized by the five mission areas (see Table 9.1). The core capabilities are essential elements that are needed to achieve the National Preparedness Goal.

To meet the National Preparedness Goal, the National Preparedness System (NPS) was developed to provide an integrated set of guidance, programs, and processes. The NPS provides an all-of-nation approach for building and sustaining a cycle of preparedness activities over time. There are six essential parts to this system:

- Identifying and Assessing Risk;
- Estimating Capability Requirements;
- Building and Sustaining Capabilities;
- Planning to Deliver Capabilities;
- Validating Capabilities;
- Reviewing and Updating.[9]

These six components of the NPS include specific resources and tools to assist communities in building strong preparedness programs. In support of these components, National Planning Frameworks were established to set the strategy and doctrine for building, sustaining, and delivering the 31 core capabilities identified in the National Preparedness Goal. These frameworks cover all five mission areas and are built to be scalable, flexible, and adaptable. They also provide a common terminology and overall approach. The strength of the National Planning Frameworks is that they are movable—and easily adapted to various jurisdictions. Things tend to change quickly and the frameworks have the ability to be expanded or contracted based on need. The frameworks address the roles of individuals, nonprofit entities and governments, nongovernmental organizations, the private sector, communities, critical infrastructure, governments, and the nation as a whole.[10] In terms of developing a risk assessment plan, the National Planning Frameworks are crucial—especially when creating a THIRA. The frameworks contain detailed information about the 31 core capabilities, which are needed to properly define desired outcomes, set capability targets, and specify appropriate resources.[11]

Table 9.1 Core Capabilities by Mission Area

Prevention	*Protection*	*Mitigation*	*Response*	*Recovery*
Planning Public Information and Warning Operational Coordination				
Forensics and attribution	Access control and identity verification	Community resilience	Critical transportation	Economic recovery
Intelligence and information sharing	Cybersecurity	Long-term vulnerability reduction	Environmental response/health and safety	Health and social services
Interdiction and disruption	Intelligence and information sharing	Risk and disaster resilience assessment	Fatality management services	Housing
Screening, search, and detection	Interdiction and disruption	Threats and hazard identification	Infrastructure systems	Infrastructure systems
	Physical protective measures		Mass care services	Natural and cultural resources
	Risk management for protection programs and activities		• Mass search and rescue operations	
	Screening, search, and detection		On-scene security and protection	
	Supply chain integrity and security		Operational communications	
			Public and private services and resources	
			Public health and medical services	
			Situational assessment	

Source: FEMA, "Jurisdictional Threat and Hazard Identification and Risk Assessment Training Support Package," October 2014, pp. 1–8.

Threat and Hazard Identification and Risk Assessment Process

THIRA process is a framework that provides a comprehensive approach for identifying risks and associated impacts. This process helps a community to understand which risks it faces by providing a common assessment in which it can identify threats and hazards of greatest concern.[12] Such threats and hazards are those that would most stress the core capabilities of the community. This three-step risk assessment process helps communities understand their risks and what they need to do to address those risks by answering the following questions (see Figure 9.1):

1 What threats and hazards can affect our community?
2 If they occurred, what impacts would those threats and hazards have on our community?
3 Based on those impacts, what capabilities should our community have?

Step 1: Identify Threats and Hazards of Concern

In Step 1 of the THIRA process, communities develop a list of community-specific threats and hazards. These will be unique to your own jurisdiction; therefore, community leaders and first

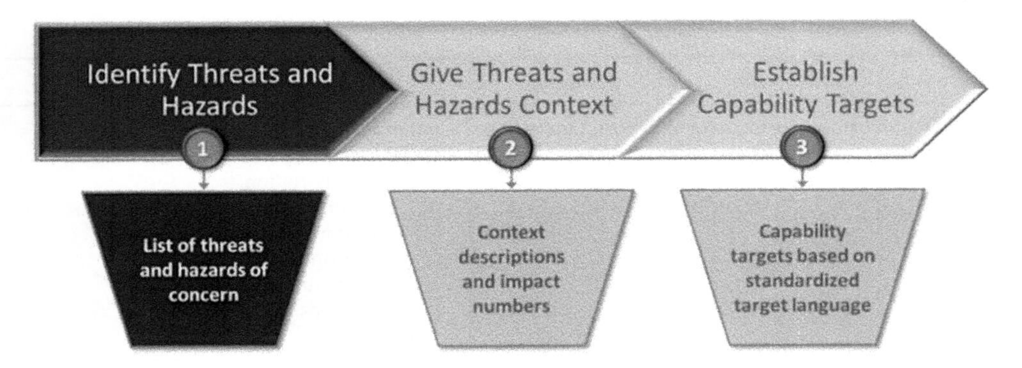

Figure 9.1 The three-step THIRA process.

responders will need to collaborate on identifying the specifics of these threats and hazards of concern. Three categories must be taken into consideration (see Table 9.2):

- Natural hazards
- Technological hazards
- Human-caused incidents

Additionally, there are two factors that must be considered when selecting threats and hazards for inclusion in the THIRA:

1 The likelihood of the incident, and
2 The significance of the threat/hazard effects.[13]

Table 9.3 shows an example of what Step 1 looks like in the THIRA process.

Table 9.2 Threat and Hazard Examples

Natural	*Technological*	*Human-Caused*
• Avalanche	• Airplane crash	• Biological attack
• Animal disease outbreak	• Dam failure	• Chemical attack
• Drought	• Levee failure	• Cyber incident
• Earthquake	• Mine accident	• Explosives attack
• Epidemic	• Hazardous materials release	• Radiological attack
• Flood	• Power failure	• Sabotage
• Hurricane	• Radiological failure	• School and workplace violence
• Landslide	• Train derailment	
• Pandemic	• Urban conflagration	
• Tornado		
• Tsunami		
• Volcanic eruption		
• Wildfire		
• Winter storm		

Source: FEMA, "Jurisdictional Threat and Hazard Identification and Risk Assessment Training Support Package," October 2014, pp. 1–22.

Table 9.3 Step 1 Examples

Natural	*Technological*	*Human-Caused*
Resulting from acts of nature	Involves accidents or the failures of systems and structures	Caused by the intentional actions of an adversary
Earthquake	Accidental chemical releases	Improvised explosive device (IED)

Source: FEMA, "Jurisdictional Threat and Hazard Identification and Risk Assessment Training Support Package," October 2014, pp. 1–21.

Step 2: Give Threats and Hazards Context

In Step 2 of the THIRA process, context descriptions are given, which outline the conditions, including time and location, under which a threat or hazard might occur. Essentially the question asked is, "How will the threat or hazard affect your community?" It is suggested that during this step communities seek out experts or analyze statistics to better inform their descriptions.[14] It is important that communities consider only those threats that would affect them. For example, if you live in Cleveland, Ohio, and are developing a THIRA, you would not add context to the natural hazard of an avalanche. However, you would add context to the threat of a winter storm or tornado.

Step 3: Establish Capability Targets

Once you have established your context descriptions in Step 2, you will use them in Step 3 to assess each threat and hazard in context and develop a specific capability target for each of the 31 core capabilities identified in the National Preparedness Goal (see Table 9.1).[15] This is an important step as the capability targets define what it would take for the community to successfully meet the challenge of the threat.[16] In this step, communities can also begin to identify preparedness activities. This might include establishing a list of resources and venues in the community where shelters could be established in the event the threat or hazard does occur.

The Benefits of Conducting a THIRA

There are a number of benefits to using the THIRA method of risk assessment in your community. These include:

- Long-term strategy and risk-based decision making;
- Gap analysis and shortfall planning;
- Standardized process/risk management aid;
- Tie to National Preparedness Report (NPR) findings; and
- Compliance with grant requirements.[17]

Long-term Strategy and Risk-Based Decision Making

A major goal of the THIRA is to get communities thinking in terms of strategies that are long term and specific. It is an economic reality that most jurisdictions are not fully equipped with the resources necessary to meet the challenges that a major disaster would present. The THIRA process empowers communities to plan for the long term by encouraging the whole community to focus on those likely threats and hazards that would have a significant impact on them.[18] The ability to identify future trends and challenges and then meet them with the appropriate resources is a critical aspect of this long-term strategy. Decisions must be risk-based to avoid implementing an unsound strategy.

Gap Analysis and Shortfall Planning

The THIRA does not include a process for measuring gaps—it only identifies capability targets and resource needs. However, by performing additional analyses and comparing them to previous ones, planners can determine if gaps exist. In their shortfall planning, communities may have to rely on mutual aid agreements, memorandums of understanding with private and nonprofit sector partners, or other formalized processes to meet the capabilities they lack.[19] Strategies to build the capabilities they lack may be part of the long-term planning process and may include such things as planning, personnel, equipment, training, exercise needs, and ways to resource these needs.[20]

Standardized Process/Risk Management Aid

The THIRA process provides a common framework for identifying community-specific threats and hazards. It is also a means by which jurisdictions can learn from each other as they consider threats and hazards and develop preparedness activities. This standardized process also serves as an excellent risk management tool for all stakeholders. By participating in the THIRA process, communities begin to understand their particular risks. Once the THIRA is completed, they are able to make smart, information-based decisions on how to manage those risks and develop needed capabilities.[21]

Tie to National Preparedness Report Findings

In addition to informing individual communities of specific threats and hazards and the capabilities and resources needed to address them, the THIRA helps to inform and develop the NPR. If you recall, the NPR is required annually by PPD-8, and it summarizes progress in building, sustaining, and delivering the 31 core capabilities. It provides a national perspective on critical preparedness trends and is used to inform program priorities, allocate resources, and communicate with stakeholders on common issues.[22]

Compliance with Grant Requirements

To receive some types of federal assistance through the various DHS grant programs, many local, tribal, and state entities have been required to develop and maintain a THIRA to support their State Preparedness Report. Some of the programs that require THIRA are the Homeland Security Grant Program, the Tribal Homeland Security Grant Program, and the Emergency Management Performance Grants. The requirements usually mandate that the THIRA be updated annually and that they include a capability estimation process for a subset of the core capabilities.

To understand how the THIRA process is implemented, it is useful to take a deeper look at these three steps and examine the specific factors that must be included.

Step 1: Identify Threats and Hazards of Concern

In this first step, communities must identify jurisdictional threats and hazards by considering threat and hazard groups, sources of information, and essential key factors. Here, communities will develop a list of specific threats and hazards that would overwhelm the community's core capabilities (Table 9.1). Throughout this step, communities should consider all of the different types of threats and hazards and the likelihood and significance of possible threats before including them in the actual THIRA document.[23] Remember, the THIRA process considers three different types of threats and hazards: *natural*, *technological*, and *human-caused*. Developing an understanding of which threats and hazards pose the greatest risk is the goal. As such, it would be helpful to review what significant threats and hazards have occurred in the past. Considering whether the previous incident could happen again and have the same impact on the community should also be debated. For example, a city in the Midwest might consider the effect of flash flooding after a significant amount of rainfall in a short period of time. The history of this natural threat should be researched to determine the likelihood of it occurring again. Has anything been done to mitigate the effects of a similar incident occurring again? If so, the community should explore the state of their current flood control measures, i.e. dams and levees.[24] Additionally, local leaders should be included in this first step to help build community support and to appropriate and or request necessary resources.[25] It may also be beneficial to consider the threats and hazards of similar communities and how they have responded. The following is a list of some of the sources communities can access to help identify threats and hazards in this first step:

- State and local homeland security and emergency management laws, plans, policies, and procedures;
- Existing hazard and risk assessments;
- Local regional and neighboring community THIRAs;
- After-action reports from community exercises;
- Analysis of critical infrastructure interdependencies;
- Fusion Centers—bulletins and assessments;
- Whole-community partners such as
  - Emergency management/homeland security agencies;
  - Local and state hazard mitigation offices;
  - Local or regional national weather service offices;
  - Tribal governments;
  - FEMA regional offices;
  - Private sector partners;
  - Local/state fire, police emergency medical services, and health departments;
  - DHS Protective Services Advisors;
  - DHS Regional Cyber Security Analysts;
  - Volunteer Organizations Active in Disasters;
  - Colleges/universities and other research organizations.[26]

While completing the first step of the THIRA process, communities should consider only those threats and hazards that are deemed to be of greatest concern. The DHS recommends that two key factors be used to assist in identifying those threats and hazards of major concern: the likelihood of the incident and the significance of the threat/hazard effects on the community.[27] Likelihood is explained as the chance of something happening, whether defined, measured, or estimated objectively or subjectively. It is best determined by focusing on only those threats and hazards that could plausibly occur. While historical information is useful in this analysis, future likelihood should also be considered. A good example is the notion that a terrorist attack will never occur in a particular jurisdiction. Recent events have shown that a terrorist event can occur almost anywhere. Communities should not assume that since a terrorist attack has not occurred in their jurisdiction it will not happen in the future. Rather, they need to consult with local law enforcement and other intelligence gathering agencies (such as the Joint Terrorism Task Force and fusion centers) to determine the likelihood of an attack happening. Similarly, natural disasters may be evaluated based on past occurrences, but changing weather patterns should also be studied. The unpredictability of earthquakes, tornados, hurricanes, and flooding, and the devastation they bring, should be considered along with the historical data of the jurisdiction.[28]

The second factor to consider is what effects will the threats or hazards have on the community. The DHS advises that communities only consider those threats and hazards that would pose a serious strain on the community's ability to respond, that would cause operational coordination problems, or that would cause great economic or social challenges.[29] A good strategy is for communities to focus on threats beyond which they are currently prepared to manage. They should also consider the effects of other factors that may make a given threat or hazard worse.[30] These might include shifting demographics toward coastal areas prone to natural disasters, systemic failures of critical infrastructure (such as power grid failure and loss of potable water system), and reliance on technology. In some cases these may adversely impact the disaster management services the community can activate, making a minor threat much worse.[31]

Step 2: Give Threats and Hazards Context

Adding context to the local threats and hazards identified in Step 1 allows jurisdictional leaders and stakeholders to define the circumstances under which the threats or hazards might occur. This allows a more robust description of specific situations and gives planners a sense of scale for the impacts of the threats and hazards. The key is to develop context descriptions that are both detailed and relevant. The DHS describes seven threat and hazard description factors: *time*, *place*, *adverse conditions*, *demographics*, *climate*, *built environment*, and *community infrastructure* (see Table 9.4).[32] It is critical to bring these potential events into reality for your jurisdiction. You want this stage of the planning process to be outside of your current capabilities.

Step 3: Establish Capability Targets

In this third step of the THIRA process, jurisdictions will focus on the core capabilities, estimate the consequences of a threat or hazard, determine the desired outcomes, and finally develop capability target statements. Capability targets are best described as what the community wants to achieve in a particular area of preparedness.[33] These should be both measurable and specific in content. The impacts of the threats and hazards identified in Step 2 of the THIRA process should be used here to develop the capability targets. Essentially, communities must identify the core capabilities that will be needed (Table 9.1) and consider both the *impacts* of threats and the *desired*

Table 9.4 Example of a Partial THIRA

Threat/Hazard	*Earthquake*
Context description	A magnitude 7.8 earthquake along the Mainline Fault occurring at approximately 2:00 pm on a weekday with ground shaking and damage expected in 19 counties, extending from Alpha County in the south to Tau County in the north, and into the Zeta Valley
Core Capability: Mass Search and Rescue Operations	
Capability target	Within 72 hours, rescue: • 5,000 people in 1,000 completely collapsed buildings • 1,000 people in 2,000 non-collapsed buildings • 20,000 people in 5,000 buildings • 1,000 people from collapsed light structures
Resource Requirements	
Resources	Number Required
Type I US&R task forces	10
Type II US&R task forces	38
• Collapse search and rescue (S&R) Type III teams	100
Collapse S&R Type IV teams	20
Canine S&R Type I teams	20

Source: FEMA, "Jurisdictional Threat and Hazard Identification and Risk Assessment Training Support Package," October 2014, pp. 1–26.

outcomes of community threat management. Impacts describe how a threat or hazard will affect core capability and are linked to the size and complexity of the incident.[34] It is essential that community planners be able to interpret the impacts an incident could have on a jurisdiction. These impacts can be expressed in a variety of ways, such as:

- Size;
- Complexity;
- Casualties;
- Disruption of critical infrastructure;
- Communications;
- Economic impacts.

Estimating impacts should involve whole-community partners and local experts. Modeling software can be used to allow planners to better estimate the breadth and seriousness of the impacts of a given threat. One of these is HAZUS-MH—a geographic information system for natural hazards used by the FEMA. It is a multi-hazard risk assessment tool that can model four types of hazards: flooding, hurricanes, coastal surges, and earthquakes. Another notable tool is the Johns Hopkins University Hospital's National Center for the Study of Preparedness and Catastrophe Event Response suite. This includes three interactive modeling tools to help communities estimate human impacts: Emergency Mass Casualty Planning Scenarios (EMCAPS) 2.0, Surge, and Flucast. EMCAPS 2.0 allows local disaster management planners, hospitals, and medical planners to model different disaster scenarios to better estimate the human impacts of incidents. Surge supports medical planners to assess current hospital surge capacity, and Flucast is a tool for hospitals to determine weekly flu cases based on historical data.[35]

The *desired outcomes* are those standards to which a community will have successfully managed the incident. These may be expressed as the level of effort in percentages and timeframe of service delivery.[36] Measurable percentages are most often used for the prevention, protection, and mitigation mission areas. For example, if the desired outcome is screening for threats, the outcome description might be to screen 100% of baggage, mail, and targeted cargo.[37] The timeframe is most often used for response and recovery mission areas. Success is often measured by communities delivering capabilities within a stated timeframe. For example, recovering all fatalities within 24 hours of the incident, setting up warming shelters for displaced populations within 48 hours, and/or completing search and rescue operations within 72 hours. Tables 9.5 and 9.6 illustrate some examples of desired outcomes expressed as measurable percentages and as timeframes.

The next step is to create capability target statements. These are concise definitions of success that include both detailed, capacity-specific impacts of an incident, and the measurable desired outcomes to which consequences must be managed. Capability target statements are starting points for communities to estimate the resources they will need. This is essentially why communities do the THIRA process! Capability target statements may be simple or complex depending on how the community is assessing the threat. A simple example may come from combining the largest impacts with the corresponding desired outcomes. A complex example would go deeper and look at how different threats and hazards affect the same core capability in different ways.[39] The advantage of developing a complex capability target is that it will account for the desired outcomes for delivering service during the worst impacts and allow for a more accurate view of given incidents and the resources needed to successfully manage them.[40]

Jurisdictional leaders must learn how to calculate and develop written resource requirements for any situation. Calculating the resource requirements for a community may be accomplished by looking at their existing and required capabilities and the threat or hazard of concern. By using the identified capability targets (Step 3), the operational capacities of resources, and available time,

Table 9.5 Threat and Hazard Description Factors

Time

Jurisdictional leaders and stakeholders should look at the impact an incident would have at a particular time of day and whether the jurisdiction would be able to manage it. Some examples are a shopping mall on Black Friday, day versus weekday, holidays, or annual festivals. The season of the year should also be considered as to which would have the greatest impact on the jurisdiction's ability to respond.

Place

Where is the location the incident may occur? Is it a heavily populated area, a commercially zoned location, or a business district of the community? Do adequate entrance and exit routes exist to allow first responders to quickly contain the incident and to provide for a well-ordered evacuation?

Adverse Conditions

Are there other circumstances that could influence the jurisdiction's ability to manage the incident? Has the impact of atmospheric conditions been considered, such as wind direction and speed or relative humidity, which might accelerate the harmful impact of a threat? What if there are multiple incidents occurring at the same time? Is the jurisdiction able to manage more than one event?

Demographics

What are the demographics of the jurisdiction? Is there a large retirement population living in special housing? Will they require assistance in the event of an evacuation? What percentage of the community is below the age of 18 years of age? Do you have a population consisting of non-English speaking citizens? How will you help them understand the nature of the incident and instructions for their safety?

Climate

Climate is important to consider, especially during a response. Does the jurisdiction exist in an area where severe summer heat or harsh winter cold is the norm? Is the jurisdiction located in an area of climate change where, for example, coastal areas will be at risk due to rising sea levels and the potential for more intense storms?

Built Environment

All of the man-made structures within a geographical space that humans use for work, residences, or leisure activities make up the built environment. Leaders must consider how the built environment impacts the jurisdiction's ability to manage the incident. Is there a need for more emergency medical services due to a large retirement community? Is the community rural with a limited built environment to support an adequate response process? Are there volunteer services (police, fire, and EMS)? Does the community need to rely on support from outside the community for these services, thus lengthening the time for response?

Community Infrastructure

A community's infrastructure includes all the assets and organizations in the public and private sectors that are part of normal life and economic activity in the area. A jurisdiction may begin defining the community infrastructure in the 16 different critical sectors:

- Chemical
- Commercial facilities
- Communications
- Critical manufacturing
- Dams
- Defense industrial base
- Emergency services
- Energy
- Financial services
- Food and agriculture
- Government facilities
- Healthcare and public health
- Information technology
- Nuclear reactors, materials, and waste
- Transportation systems
- Water and wastewater systems

Community infrastructure consists of not only local critical infrastructure but also secondary local assets that might be relied on in the event that the critical assets are affected by a threat or hazard. Such assets include interconnecting links, such as roads, rail, bridges, phone lines, power lines, and broadcast towers; natural resources (i.e. water); places of local cultural significance; and any other unique assets or facilities.[38]

Source: FEMA, "Jurisdictional Threat and Hazard Identification and Risk Assessment Training Support Package," October 2014.

Table 9.6 Example of Percentage-based Desired Outcomes

Outcome Type	*Example Outcome Description*
Screening for threats	Screen 100% of targeted cargo, conveyances, mail, baggage, and people associated with an imminent terrorist threat or act using technical, non-technical, intrusive, or non-intrusive means.
Verifying identity	Ensure 100% verification of identity to authorize, grant, or deny physical and cyber access to specific locations, information, and networks.

Source: FEMA, "Jurisdictional Threat and Hazard Identification and Risk Assessment Training Support Package," October 2014, pp. 4–19.

communities can best determine what resources they will need to manage an incident. A simple calculation example is as follows:

Simple Calculation Example

- Mission-critical activity: Search buildings for survivors;
- Identified Resource: Urban Search and Rescue (US&R) Task Forces.[41]

The calculation process should be approached carefully with emphasis placed on finding the most appropriate resources for achieving the capability targets. In addition, the typical performance characteristics of those resources and how they might fare in local conditions should be considered. The level of detail should be kept to a minimum. The DHS recommends using Tier I resource types, when possible. Table 9.7 illustrates a full calculation example.

Example of a Completed THIRA

Table 9.9 shows a completed THIRA from the State of New Mexico. Notice the progression from capability target to resource requirement for the Critical Transportation Core Capability. Four different threats/hazards are identified: earthquake, terrorism, wildfire, and mass migration. In this example, notice the use of detail and the inclusion of key points from each of the four steps in the THIRA process (Tables 9.8–9.10).

Applying THIRA Results to Policy Decisions

Once the final calculations have been completed, communities may use the results of the THIRA to make critical decisions about how the community can achieve its capability targets through the allocation of resources. The results may indicate that the community must either sustain current capabilities or it may expose capability shortfalls and gaps.[42] Conducting the THIRA is also beneficial in that community planners can use it to achieve buy-in from government leaders and the whole community to support making the necessary investments to build new or expand current capabilities.[43] The THIRA is scalable, meaning it can be adjusted to fit the needs and profile of a particular community or jurisdiction. In that sense, the results can inform policy decisions at the smallest of levels (i.e. local, rural, tribal). THIRA is strategic planning that may assist communities in developing better preparedness and mitigation activities and ultimately reduce the amount of resources needed in the future. It also fosters communications between policymakers, state and local authorities, emergency managers, and other stakeholders, ultimately contributing to policy decisions for emergency preparedness and disaster plans.

Table 9.7 Example Time-Based Desired Outcomes

Outcome Type	*Example Outcome Description*
Completing operations	Complete evacuation of a neighborhood within four hours
Establishing services	Establish feeding and sheltering operations for displaced populations within 24 hours
Service duration	Maintain behavioral screening checks for the affected population for one month
Combination	Establish feeding and sheltering operations within 24 hours and maintain services for a period of two weeks

Source: FEMA, "Jurisdictional Threat and Hazard Identification and Risk Assessment Training Support Package," October 2014, pp. 4–20.

Table 9.8 Tier I and Tier II Type Resources

Tier I
- National Incident Management System (NIMS)

The complete list of typed resources may be found at: https://rtlt.preptoolkit.org/Public
Tier II
- State, tribal, and local-typed standardized, deployable, resources
- No FEMA-involved in defining or inventorying these resources
- Includes Mutual aid resources and Emergency Management Assistance Compact resources

Source: FEMA, "Jurisdictional Threat and Hazard Identification and Risk Assessment Training Support Package," October 2014, pp. 5–8, 5–9.

Table 9.9 Calculation Example

Capability Target	*Search 42 Type 1 Structures in 72 Hours*
Resource	Type 1 US&R task force
Operational capacity	Type 1 US&R task force can search two (2) Type 1 structures a day
Time available	Three days (72 hours)
Calculated resource requirement	Seven (7) Type 1 US&R task forces

Source: FEMA, "Jurisdictional Threat and Hazard Identification and Risk Assessment Training Support Package," October 2014, pp. 5–12.

Conclusion

Significant strides to standardize risk and vulnerability assessment have been made through an IRM approach. FEMA encourages the THIRA process and supports its use by communities to identify capability targets and resource requirements necessary to address anticipated and unanticipated risks. Consistent application of the THIRA provides an important tool for integrating the whole community and for policymakers to make informed decisions on critical infrastructure protection.

Key Terms

Capability Target
Context
Core Capability
Desired Outcome
Human-Caused Hazard
Impact
Likelihood
Natural Hazard
Resource Requirement
Technological Hazard
Threat/Hazard Effect
Whole Community

Table 9.10 Example of a Completed THIRA: State of New Mexico

Critical Transportation Core Capability Desired Outcomes and Impacts

Threat/Hazard	*Desired Outcomes*	*Impacts*
Earthquake	During the first four hours, DHSEM will coordinate with local Emergency Management to establish physical access by air for search and rescue and emergency lifesaving equipment and personnel. Within the first four hours, DHSEM in coordination with ESF #1 and ESF # 13 will assist local Emergency Management programs in identifying traffic control packages necessary to secure the affected area. During the first 24 hours, DHSEM will assist local Emergency Management to establish full and unrestricted physical access by land to the affected area(s) for the delivery of required resources to save lives; evacuate citizens, at risk and special needs population and animals from the affected area(s). During the first 24 hours, coordinate activities to acquire resources for local Emergency Management programs to meet the critical sustainment needs of citizens and emergency response personnel in the impacted area(s) for two weeks.	Extensive rockslides and debris are blocking major transportation and evacuation routes on Highways 502, 501, and 4. There is damage to two bridges on major roadways that has made them impassable. The need to maximize available ground transportation for movement of resources, personnel, and non-life threating medical issues and evacuees is essential.
Terrorism		Downtown area will be impassable. Debris removal will have to be completed to regain use of roadway. Roadways are within the perimeter of the incident. A large number of road blocks will be needed to control access. Busses will be needed to transport people from the area.
Wildfire		Debris from burned vegetation, trees, and homes could be blocking road way hindering evacuation from the area. Security concerns exist throughout incident area, so the need for security road blocks exist to keep criminal elements from entering area, and ensure evacuees do not return to area before it is rendered safe.
Mass Migration		Six buses will be needed to move 2,000 evacuees from approximately five collection points. Anticipate need for wreckers to move stranded cars blocking roadways, ESF 1 will coordinate this.

Capability Targets

During the first four hours of the incident, DHSEM will coordinate with local Emergency Management to establish physical access by air for search and rescue and emergency lifesaving equipment and personnel. Within the first four hours, DHSEM in coordination with ESF # 1 and ESF # 13 will assist local Emergency Management programs in identifying traffic control equipment necessary to secure the affected area. Needed resources are 5–10 programmable directional signs, 100 road barriers and over 1,000 traffic cones. During the first 24 hours, DHSEM will assist Local Emergency Management to establish full and unrestricted physical access by land to the affected area(s) for the delivery of required resources to save lives; evacuate citizens, at risk and special needs population and animals from the affected area(s). Support is needed for 3,000 injured citizens and 6,250 displaced citizens, 625 at risk and special needs population, and 250 animals in the affected area (s). During the first 24 hours, coordinate activities to acquire resources for local Emergency Management programs to meet the critical sustainment needs of citizens and emergency response personnel in the impacted area(s) for two weeks.

(*Continued*)

Resource Requirements: NIMS Tier I Resources (Continued)

Group	*Resource*	*Type*	*Number Required*
Public Works	Buses	Type I	12
Animal Health Emergency	Large Animal Transport Team	Type I	2
Animal Health Emergency	Small Animal Transport Team	Type I	10
Emergency Medical Services	Air Ambulance (Rotary-Wing)	Type II	2
Emergency Medical Services	Air Ambulance (Fixed-Wing)	Type II	2
Incident Management	Evacuation Liaison Team (ELT)	Type I	1
Incident Management	Airborne Transport Team (Fixed-Wing)	Type II	2
Law Enforcement	Mobile Field Force Law Enforcement (Crowd Control Teams)	Type II	150
Emergency Medical Services	Ambulance Strike Team	Type II	20
Emergency Medical Services	Ambulance Strike Team	Type IV	20
Public Works	Civil/Field Engineer	Type I	10
Fire and Hazardous Materials	HazMat Entry Team	Type II	5

Resource Requirements: Other Resources

Resource	*POETE*	*Number Required*
National Guard Search and Rescue Helicopters (Medivac)	Equipment	2
National Guard Soldiers for traffic control along evacuation/reentry routes	Team	100
NMDOT Directional Road Signs	Equipment	10
NMDOT Jersey Barriers	Equipment	100

Source: New Mexico Department of Homeland Security and Emergency Management, “2013 State of New Mexico THIRA.” http://www.nmdhsem.org/local-thira-guidance.aspx

Notes

1 DHS, "Risk Management Fundamentals," US Department of Homeland Security May 2018, 2011.
2 FEMA, "Comprehensive Preparedness Guide (CPG) 201: Threat and Hazard Identification and Risk Assessment Guide." US Department of Homeland Security May 2018, 2013.
3 "Aiding the Fight Against Terrorism, ASME-ITI Gets Contract to Develop RAMCAMP™ Guidelines to Protect Critical Infrastructure." *Business Wire*, 2006.
4 "Aiding the Fight Against Terrorism, ASME-ITI Gets Contract to Develop RAMCAMP™ Guidelines to Protect Critical Infrastructure." *Business Wire*, 2006.
5 Government Training, Inc., "CARVER+Shock Vulnerability Assessment Tool." 2006.
6 Eastern Kentucky University Justice and Safety Center Website. http://jsc.eku.edu/pascom
7 Caudle, S. (2012). "Homeland Security: Advancing the National Strategic Position." *Homeland Security Affairs*, Vol. xiii. pp. 3–4.
8 FEMA, "Jurisdictional Threat and Hazard Identification and Risk Assessment Participant Guide," October 2014.
9 FEMA, "Jurisdictional Threat and Hazard Identification and Risk Assessment Participant Guide," October 2014, pp. 1–9.
10 FEMA, "Jurisdictional Threat and Hazard Identification and Risk Assessment Participant Guide," October 2014, pp. 1–11.
11 FEMA, "Jurisdictional Threat and Hazard Identification and Risk Assessment Participant Guide," October 2014, pp. 1–11.
12 FEMA, "Jurisdictional Threat and Hazard Identification and Risk Assessment Participant Guide," October 2014, pp. 1–20.
13 FEMA, "Jurisdictional Threat and Hazard Identification and Risk Assessment Participant Guide," October 2014, pp. 1–22.
14 FEMA, "Jurisdictional Threat and Hazard Identification and Risk Assessment Participant Guide," October 2014, pp. 1–23.
15 FEMA, "Jurisdictional Threat and Hazard Identification and Risk Assessment Participant Guide," October 2014, pp. 1–24.
16 FEMA, "Jurisdictional Threat and Hazard Identification and Risk Assessment Participant Guide," October 2014, pp. 1–24.
17 FEMA, "Jurisdictional Threat and Hazard Identification and Risk Assessment Participant Guide," October 2014, pp. 1–27.
18 FEMA, "Jurisdictional Threat and Hazard Identification and Risk Assessment Participant Guide," October 2014, pp. 1–27.
19 FEMA, "Jurisdictional Threat and Hazard Identification and Risk Assessment Participant Guide," October 2014, pp. 1–27, 28.
20 FEMA, "Jurisdictional Threat and Hazard Identification and Risk Assessment Participant Guide," October 2014, pp. 1–28.
21 FEMA, "Jurisdictional Threat and Hazard Identification and Risk Assessment Participant Guide," October 2014, pp. 1–28.
22 FEMA, "Jurisdictional Threat and Hazard Identification and Risk Assessment Participant Guide," October 2014, pp. 1–28.
23 FEMA, "Jurisdictional Threat and Hazard Identification and Risk Assessment Participant Guide,"October 2014, pp. 2–4.
24 FEMA, "Jurisdictional Threat and Hazard Identification and Risk Assessment Participant Guide," October 2014, pp. 2–8.
25 FEMA, "Jurisdictional Threat and Hazard Identification and Risk Assessment Participant Guide," October 2014, pp. 2–7.
26 FEMA, "Jurisdictional Threat and Hazard Identification and Risk Assessment Participant Guide," October 2014, pp. 2–8.
27 FEMA, "Jurisdictional Threat and Hazard Identification and Risk Assessment Participant Guide," October 2014, pp. 2–9.
28 FEMA, "Jurisdictional Threat and Hazard Identification and Risk Assessment Participant Guide," October 2014, pp. 2–8.
29 FEMA, "Jurisdictional Threat and Hazard Identification and Risk Assessment Participant Guide," October 2014, pp. 2–10.

30 FEMA, "Jurisdictional Threat and Hazard Identification and Risk Assessment Participant Guide," October 2014, pp. 2–10.
31 FEMA, "Jurisdictional Threat and Hazard Identification and Risk Assessment Participant Guide," October 2014, pp. 2–10.
32 FEMA, "Jurisdictional Threat and Hazard Identification and Risk Assessment Participant Guide," October 2014, pp. 3–6.
33 FEMA, "Jurisdictional Threat and Hazard Identification and Risk Assessment Participant Guide," October 2014, pp. 4–8.
34 FEMA, "Jurisdictional Threat and Hazard Identification and Risk Assessment Participant Guide," October 2014, pp. 4–9.
35 FEMA, "Jurisdictional Threat and Hazard Identification and Risk Assessment Participant Guide," October 2014, pp. 4–14, 4–15.
36 FEMA, "Jurisdictional Threat and Hazard Identification and Risk Assessment Participant Guide," October 2014, pp. 4–18.
37 FEMA, "Jurisdictional Threat and Hazard Identification and Risk Assessment Participant Guide," October 2014, pp. 4–19.
38 FEMA, "Jurisdictional Threat and Hazard Identification and Risk Assessment Participant Guide,"October 2014, pp. 3–8, 3–9, and 3–10.
39 FEMA, "Jurisdictional Threat and Hazard Identification and Risk Assessment Participant Guide," October 2014, pp. 4–21, 4–22.
40 FEMA, "Jurisdictional Threat and Hazard Identification and Risk Assessment Participant Guide," October 2014, pp. 4–23.
41 FEMA, "Jurisdictional Threat and Hazard Identification and Risk Assessment Participant Guide," October 2014, pp. 5–11.
42 FEMA, "Jurisdictional Threat and Hazard Identification and Risk Assessment Participant Guide," October 2014, pp. 5–16.
43 FEMA, "Jurisdictional Threat and Hazard Identification and Risk Assessment Participant Guide," October 2014, pp. 5–16, 5–17.

Appendix

Class Activities: Develop a Sample THIRA

Activity 1: Identify Threats and Hazards

Break into teams and create a list of jurisdictional threats and hazards. Each team will be assigned a local jurisdiction and should choose three threats and three hazards, one from each category (natural, technological, and human-caused). Present your findings to the class.

Activity 2: Contextualize Threats and Hazards

Identify Community Infrastructure: Each team will be assigned a critical infrastructure sector from the list of 16 categories created by the DHS. Within that sector, each team should identify as many specific examples of CI as they can that exist within their jurisdiction. Next, prioritize your identified specific examples (by which ones are most critical to the jurisdiction) and add up to two other characteristics, elements, events, or other factors important to the jurisdiction that might be useful in evaluating the significance of threats or hazards. Present your findings to the class.

Develop Context Descriptions for Jurisdictional Threats and Hazards: Each team will then choose *one* incident from the list of three threats and hazards of significant concern identified in activity one and create a context description. This should include the conditions (at a minimum time and location) most relevant to the community. Use Table 9.5 to develop your context descriptions. Present your findings to the class.

Activity 3: Establish Capability Targets

Estimate Impacts of a Threat of Hazard: Using the context descriptions developed in Activity 2, choose five core capabilities, one from each of the five mission areas (prevent, protect, mitigation, response, and recovery) that would pertain to managing the threat or hazard (see Table 9.1). Next, estimate the impacts of the threat or hazard on the chosen capabilities. Present your findings to the class.

Develop Desired Outcomes and Capability Target Statements: Using all of the information created in the previous activities, each team will develop one desired outcome for each of the five core capabilities identified in the *Estimate Impacts of a Threat of Hazard* activity (above). Use the estimated impacts and the desired outcomes to develop one capability target statement for each of the core capabilities. Present your findings to the class.

10 Sector-Specific Agencies' Approaches to Risk

Food and Agriculture Sector, Water and Wastewater Sector, and Energy Sector

Chapter Outline

Introduction

In early 2013, President Obama announced Presidential Policy Directive 21 (PPD-21): *Critical Infrastructure Security and Resilience* thus setting the stage for the next era in critical infrastructure protection policies. With this new plan, the number of critical sectors was reduced from 18 to 16, and a reorganization of a number of the sector-specific agencies (SSAs) and their sectors took place. Each SSA develops a sector-specific plan (SSP) through a coordinated effort involving its public and private sector partners. These plans detail how the National Infrastructure Protection Plan (NIPP) risk management framework is implemented within the context of the unique characteristics and risk environment of each sector.[1] SSPs are important in that they guide each sector to meet evolving threats.

SSPs were established in 2015 and address the increasing nexus between cyber and physical security, as well as the continued interdependence between various sectors risks associated with climate change and an aging and outdated infrastructure.[2] These newer risks present continued challenges to critical infrastructure protection and the management of risk. The following chapters will explore the roles and responsibilities of these SSAs, each sector's own approach to risk management, and the policy implications. The sectors covered will serve as an example of the varied

DOI: 10.4324/9781003434887-10

approaches and how each SSP tailors national strategic guidance to the unique operating conditions and risk landscape of its respective sectors.

Food and Agriculture Sector Profile

Protecting the nation's food and agriculture infrastructure is a complex undertaking with responsibilities that extend beyond the boundaries of the U.S. The Food and Agriculture (FA) Sector comprises systems that are almost entirely owned by private interests, which operate in highly competitive global markets.[3] The FA sector covers more than 935,000 restaurants and institutional food service establishments and an estimated 114,000 supermarkets, grocery stores, and other food retailers. With an estimated 2.1 million farms, and more than 200,000 registered food manufacturing, processing, and storage facilities, this sector accounts for roughly one-fifth of the nation's economic activity.[4] As with all of the sectors, the FA sector has critical interdependencies with many sectors but specifically

- Water and Wastewater sector—for clean irrigation and processed water;
- Transportation sector—for movement of products and livestock;
- Energy sector—to power the equipment needed for agriculture production and food processing; and
- Chemical sector—for fertilizers and pesticides used in the production of crops.[5]

Some of the sector risks include food contamination and disruption (accidental or intentional), disease and pests, severe weather (droughts, floods, and climate variability), and cybersecurity. Under PPD-21, the U.S. Department of Agriculture (USDA) and the Department of Health and Human Services (HHS) were designated as the SSAs for the FA sector. The HHS has delegated this responsibility to the Food and Drug Administration (FDA).[6]

Goals and Priorities of the FA Sector

The goals outlined in the FA SSP for the years 2015–2019 include the following:

- **Goal 1:** Continue to promote the combined federal, SLTT, and private sector capabilities to prevent, protect against, mitigate, respond to, and recover from manmade and natural disasters that threaten the national FA infrastructure.
- **Goal 2:** Improve sector situational awareness through enhanced intelligence communities and information sharing among all sector partners.
- **Goal 3**: Assess all-hazards risks, including cybersecurity, to the FA sector.
- **Goal 4:** Support response and recovery at the sector level.
- **Goal 5:** Improve analytical methods to bolster prevention and response efforts, as well as increase resilience in the FA sector.[7]

The FA sector outlines five priorities to support the furtherance of these goals:

- **Priority 1:** Improve the ability to prevent, detect, and respond to animal and plant disease outbreaks and food contamination, whether naturally occurring or intentional, through the expansion of laboratory systems and qualified personnel.

- **Priority 2:** Enhance and integrate existing information-sharing approaches.
- **Priority 3:** Raise awareness of and evaluate potential cyber risks, and encourage FA sector members to use the National Institute of Standards and Technology (NIST) Cybersecurity framework.
- **Priority 4:** Continue to resolve decontamination and waste management-related issues.
- **Priority 5:** Engage all levels of the FA sector in national planning efforts and goals.[8]

FA Sector: Assessing Risk

As discussed in earlier chapters, the NIPP risk management framework calls for critical infrastructure partners to assess risk from any scenario as a function of consequence, vulnerability, and threat: $R = f(C,V,T)$. When conducting a risk assessment, the FA sector generally focuses on systems and networks as well as individual assets.[9] Remember, the purpose of conducting a risk assessment is to decide where to put limited resources while having the greatest impact. In the FA sector, risk assessments of food safety are conducted to determine the quantitative or qualitative value of risk attributed to exposure to food contamination by either a biological or chemical hazard.[10] The FA sector uses a number of different methodologies to identify and determine each component of risk (consequence, vulnerability, and threat). The following discussion outlines the methods used for each of these three components.

Reportable Data (Consequence)

This first component of risk is assessed through the accumulation of reportable data (i.e. illness and death and economic impact). The USDA and FDA both have structures in place to monitor adverse events. Data is then collected to produce a clear picture of the consequences of each type of disaster.[11]

CARVER Plus Shock Method (Vulnerabilities)

Vulnerability assessments are useful to assist SSAs in identifying the products of highest concern, threat agents likely to be used, points in the production process where intentional contamination is most likely to occur, laboratory testing and research needs, and potential countermeasures that may be taken.[12] The CARVER Plus Shock methodology is used by the FA sector to determine the vulnerabilities in its assets, systems, and networks. This is accomplished by encompassing the consequences and threats.

It is important to note the relationship between the opportunity (vulnerability) and outcome (consequence) of an attack in the FA sector. The definition of vulnerability, as stated in the NIPP, presents a challenge for application to the FA sector. The NIPP defines vulnerability as a physical feature or operational attribute that renders an entity open to exploitation or susceptible to a given hazard. However, many of the FA sector's interdependent systems are not physical structures like buildings, bridges, or dams. Rather, they are open areas such as farms, ranches, or livestock transport areas.[13]

These systems are susceptible to natural threats like disease and foodborne pathogens, and, as a result, it may not be feasible to prevent the introduction of threat agents. In response, the FA sector supports timely awareness by veterinarians, agriculture producers, and nationally coordinated disease surveillance programs that have the ability to target different threat agents in its systems.[14] Moreover, because the interdependent relationships within and among other sectors present additional vulnerabilities for the FA sector, there is a need to clearly identify these points of dependence on critical partner sectors. Coordination with those SSAs must be established to address, mitigate, and strengthen these vulnerabilities.

As stated above, vulnerability and consequence assessments must be conducted together to determine risk in the FA sector. The CARVER plus Shock method has been adapted from the military version (CARVER) for use in the food industry. It is an offensive targeting prioritization tool that allows the user to think like an attacker to identify the most attractive targets for an attack. A food production facility or process may be assessed using this method to determine the most vulnerable points and focus resources on protecting those weaknesses in their infrastructure.[15] CARVER is an acronym for six attributes used to evaluate the attractiveness of a target for an attack:

- **C**ritically—measure of public health and impacts of an attack
- **A**ccessibility—ability to physically access and egress from the target
- **R**ecuperability—ability of the system to recover from an attack
- **V**ulnerability—ease of accomplishing an attack
- **E**ffect—amount of direct loss from an attack as measured by loss in production
- **R**ecognizability—ease of identifying target

CARVER Plus Shock is a modified version that evaluates a seventh attribute, the combined health, economic, and psychological impacts of an attack, or the Shock attributes of a target.[16] Scales have been developed for each of the seven attributes that may be used to rank the attractiveness of a target. While these scales were developed with the mindset that mass mortality is a goal of terrorist organizations, it is important to remember that any intentional food contamination could also have a major psychological and economic impact on the affected industry.[17] Tables 10.1–10.6 illustrate these attributes and the scales used by agencies for scoring each one.

There are five essential steps to the CARVER Plus Shock process. These are:

Step 1—Establishing Parameters
Step 2—Assembling Experts
Step 3—Detailing Food Chain Supply
Step 4—Assigning Scores
Step 5—Applying What Has Been Learned.[18]

Table 10.1 CARVER Plus Shock Attributes and Scale for Criticality

Criticality: A target is critical when the introduction of threat agents into food at this location would have a significant health or economic impact. Example metrics are:

Critically Criteria	*Scale*
Loss of over 10,000 lives **OR** loss of more than $100 billion. (Note: if looking on a company level, loss of > 90% of the total economic value for which you are concerned.[a])	9–10
Loss of life is between 1,000 and 10,000 **OR** loss of between $10 billion and $100 billion. (Note: if looking at a company level, a loss of between 61% and 90% of the total economic value for which you are concerned.[a])	7–8
Loss of life between 100 and 1000 **OR** loss of between $1 and $10 billion. (Note: if looking at a company level, a loss of between 10% and 30% of the total economic value for which you are concerned.[a])	5–6
Loss of life less than 100 **OR** loss of between $100 million and $1 billion. (Note: if looking on a company level, loss of between 10% and 30% of the total economic value for which you are concerned.[a])	3–4
No loss of life **OR** loss of less than $100 million. (Note: if looking at a company level, loss of <10% of the total economic value for which you are concerned.[a])	1–2

[a] The total economic value for which you are concerned depends upon your perspective. For example, for a company this could be the percent of a single facility's gross revenues, or percentage of a single facility's gross revenues, or percentage of a company's gross revenues lost from the effect on a single product line. Likewise, a state could evaluate the effect of the economic loss caused by an attack on a facility or farm by the proportion of the state's economy contributed by that commodity.

Source: FDA, Protecting and Promoting Your Health, "Carver+Shock Primer: An Overview of the Carver Plus Shock Method for Food Vulnerability Assessments," 2009.

Table 10.2 CARVER Plus Shock Attributes and Scale for Accessibility

Accessibility: A target is accessible when an attacker can reach the target to conduct the attack and egress the target undetected. Accessibility is the openness of the target to the threat. This measure is independent of the probability of successful introduction of threat agents. Example metrics are:

Accessibility Criteria	*Scale*
Easily Accessible (e.g. target is outside the building and no perimeter fence). Limited physical or human barriers or observation. Attacker has relatively unlimited access to the target. Attack can be carried out using medium or large volumes of containment without undue concern detection. Multiple sources of information concerning the facility and the target are easily available.	9–10
Accessible (e.g. target is inside the building but in an unsecured part of the facility). Human observation and physical barriers are limited. Attacker has access to the target for an hour or less. Attack can be carried out with moderate to large volumes of containment, but requires the use of stealth. Only limited specific information is available on the facility and the target.	7–8
Partially Accessible (e.g. inside the building, but in a relatively unsecured, but busy, part of the facility). Under constant possible human observation. Some physical barriers may be present. Contaminants must be disguised, and time limitations are significant. Only general, non-specific information is available on the facility and the target.	5–6
Hardly Accessible (e.g., inside the building in a secured part of the facility). Human observation and physical barriers with an established means of detection. Access generally restricted to operators or authorized persons. Containment must be disguised and time limitations are extreme. Limited general information available on the facility and the target.	3–4
Not Accessible. Physical barriers, alarms, and human observation. Defined means of intervention in place. Attacker can access the target for less than 5 minutes with all equipment carried in pockets. No useful policy available information concerning the target.	1–2

Source: FDA, Protecting and Promoting Your Health, "Carver+Shock Primer: An Overview of the Carver Plus Shock Method for Food Vulnerability Assessments," 2009.

Table 10.3 CARVER Plus Shock Attributes and Scale for Recuperability

Recuperabilty: A target's recuperability is measured in the time it will take for the specific system to recover productivity

Recuperabilty Criteria	*Scale*
>1 year	9–10
6 months to 1 year	7–8
3–6 months	5–6
1–3 months	3–4
<1 month	1–2

Source: FDA, Protecting and Promoting Your Health, "Carver+Shock Primer: An Overview of the Carver Plus Shock Method for Food Vulnerability Assessments," 2009.

Step 1—Establishing Parameters

Prior to scoring, the scenarios and assumptions you plan to use in the analysis must be determined. Initially, you need to decide what you are trying to protect and what you are trying to protect it from. This should include:

- What food supply chain you are going to assess (e.g. hot dog production versus deli meat production versus chicken nugget production and the overall assessment based on a generic process from farm to table versus post-slaughter processing in a specific facility);

Table 10.4 CARVER Plus Shock Attributes and Scale for Effect

Effect: Effect is a measure of the percentage of system productivity damaged by an attack at a single facility. Thus, effect is inversely related to the total number of facilities producing the same product. Example metrics are:

Effect Criteria	*Scale*
Greater than 50% of the system's production impacted	9–10
25–50% of the system's production impacted	7–8
10–25% of the system's production impacted	5–6
1–10% of the system's production impacted	3–4
Less than 1% of the system's production impacted	1–2

Source: FDA, Protecting and Promoting Your Health, "Carver+Shock Primer: An Overview of the Carver Plus Shock Method for Food Vulnerability Assessments," 2009.

Table 10.5 CARVER Plus Shock Attributes and Scale for Recognizability

Recognizabilty: A target's recognizability is the degree to which it can be identified by an attacker without confusion with other targets or components. Example metrics are:

Recognizabilty Criteria	*Scale*
The target is clearly recognizable and requires little or no training for recognition	9–10
The target is easily recognizable and requires only a small amount of training for recognition	7–8
The target is difficult to recognize or might be confused with other targets or target companies and requires some training for recognition	5–6
The target is difficult to recognize. It is easily confused with other targets or components and requires extensive training for recognition	3–4
The target cannot be recognized under any conditions, except by experts	1–2

Source: FDA, Protecting and Promoting Your Health, "Carver+Shock Primer: An Overview of the Carver Plus Shock Method for Food Vulnerability Assessments," 2009.

- What the endpoint of concern is (e.g., foodborne illness and death versus economic impacts);
- What type of attacker and attack you are trying to protect against. Attackers could range from disgruntled employees to international terrorist organizations. Different attackers have different capabilities and different goals. One example is a major assumption used by the Food Safety and Inspection Service (FSIS) and FDA in their vulnerability assessments is that one of the goals is to terrorist organizations is to cause mass mortality by adding acutely toxic agents to food products.[19] That assumption has a major impact on the scoring of the various parts of the supply chain;
- What agent(s) might be used. The agent used in your scenario will impact the outcome of the assessment. Potential agents include biological, chemical, or radiological agents.[20]

Step 2—Assembling Experts

The assessment team should consist of subject matter experts from the following industries: food production, food science, toxicology, epidemiology, microbiology, medicine (human and veterinarian), radiology, and risk assessment. Once the team has been assembled, the CARVER–Shock method should be applied to each element of the food system infrastructure. Using the scenario and assumptions detailed in Step 1, values from one to ten for each attribute should be assigned.[21]

Table 10.6 CARVER Plus Shock Attributes and Scale for Shock

Shock: Shock is the final attribute considered in the methodology. Shock is the combined measure of health, psychological, and collateral national economic impacts of a successful attack on the target system. Shock is considered on a national level. The psychological impact will be increased if there are a large number of deaths or the target has historical, cultural, religious, or other symbolic significance. Mass casualties are not required to achieve widespread economic loss or psychological damage. Collateral economic damage includes such items as decreased national economic activity, increased unemployment in collateral industries, etc. Psychological impact will be increased if victims are members of sensitive subpopulations such as children or the elderly. The metrics for this criterion are:

Shock Criteria	*Scale*
Target has major historical, cultural, religious, or other symbolic importance. Loss of over 10,000 lives. Major impact on sensitive subpopulations, e.g. children or elderly. The national economic impact is more than $100 billion.	9–10
Target has high historical, cultural, religious, or other symbolic importance. Loss between 1,000 and 10,000 lives. Significant impact on sensitive subpopulations, e.g. children or elderly. National economic impact between $10 and $100 billion.	7–8
Target has moderate historical, cultural, religious, or other symbolic importance. Loss of life between 100 and 1,000. Moderate impact on sensitive subpopulations, e.g. children or elderly. National economic impact between $1 and $10 billion.	5–6
Target has little historical, cultural, religious, or other symbolic importance. Loss of life less than 100. Small impact on sensitive subpopulations, e.g. children or elderly. National economic impact between $100 million and $1 billion.	3–4
Target has no historical, cultural, religious, or other symbolic importance. Loss of life less than 10. No impact on sensitive subpopulations, e.g. children or elderly. National economic impact less than $100 million.	1–2

Source: FDA, Protecting and Promoting Your Health, "Carver+Shock Primer: An Overview of the Carver Plus Shock Method for Food Vulnerability Assessments," 2009.

Step 3—Detailing Food Supply Chain

In this step, the system under evaluation should first be described. The analysis should include a graphical representation (flow chart) of the system and its subsystem, complexes, components, and nodes (its smaller structural parts). For example, if you are evaluating hot dog production, the food system is hot dog production, which can be broken down into subsystems (production of live animals subsystem, slaughter/processing subsystem, distribution subsystem). Those subsystems can be further broken down into complexes (e.g. slaughterhouse facility and processing facility), which can then be broken down into components and include the raw materials receiving area, processing area, storage area, and shipping area) and the smallest possible nodes (e.g. individual pieces of equipment).[22]

Step 4—Assigning Scores

Once the infrastructure has been broken down into its smallest parts (i.e. components and nodes), these can be ranked or scored for each of the seven CARVER–Shock attributes to calculate an overall score for that node. The nodes with the higher overall scores are those that are potentially the most vulnerable nodes (i.e. most attractive targets for an attacker). The rationale for a particular consensus score should be captured.[23]

Step 5—Applying What Has Been Learned

In this final step, you have already identified the critical nodes of the system and are now ready to develop a plan to establish countermeasures that minimize the attractiveness of the nodes as targets. Countermeasures might include enhancements to physical security, personnel security, and operational security that help to reduce an attacker's access to the product or process.[24]

Final Calculations and Interpretation

Once the attribute scales have been completed for a given node within the food supply system, the ranking on all of the scales can be totaled to give an overall value for that node. This process should be repeated for each node within a food supply system. A comparison of the overall values for all the nodes can be made to rank the vulnerability of the different nodes relative to each other. Nodes with the highest total rating have the highest potential vulnerability and should be the focus of countermeasure efforts.[25]

Federal Policy on Vulnerability Assessments

Federal guidelines (HSPD-9) require the USDA and FDA to conduct vulnerability assessments of the FA sector and to update them every two years. This includes efforts by the Strategic Partnership for Agro-terrorism (SPPA) Initiative, which, between the years 2005 and 2008, conducted over 50 vulnerability assessments on a variety of food and agricultural products, processes, or commodities under the regulatory authority of the FDA and USDA.[26] Additionally, the USDA has conducted more than 30 vulnerability assessments and updates including products and factors such as deli meats, establishment size, ground beef, hot dogs, imported food products, liquid eggs, ready-to-eat meals, National School Lunch Program, ready-to-eat chicken, threat agents, transportation, and water used in food. The FDA also conducted 18 vulnerability assessments and updated 16 of its original assessments conducted under the SPPA Initiative.[27]

The National Counterterrorism Center and THIRA (Threat)

The final component of a risk assessment is to determine the threat. The National Counterterrorism is the lead entity in determining this final component for the FA sector. All FA sector threats deemed credible by law enforcement agencies are investigated further with the help of FA sector partners. The Department of Homeland Security (DHS) and the Intelligence Community provide critical information to the FA sector, which helps in determining the criticality of known risks.

As presented in Chapter 9, the Threat and Hazard Identification and Risk Assessment (THIRA) also assists the FA sector in risk assessment by helping government and private sector partners (i.e. any entity receiving federal grants for preparedness activities) understand the risks within their community and estimate capability requirements. Those critical infrastructures within the FA sector are largely privately owned and operated, which requires a community approach to risk assessments as outlined by THIRA. Efforts such as those outlined in the SPPA, and in FSIS's cybersecurity vulnerability initiative, are central to securing the vast and open network of systems that comprise the FA sector.[28]

Water and Wastewater Systems Sector Profile

With more than 153,000 public drinking water systems and approximately 16,500 publicly owned treatment works in the U.S., the Water and Wastewater sector infrastructure is one of the most critical to protect. Safe drinking water and properly treated wastewater are vital for human consumption and activity, the prevention of disease, and the health of the environment. The Water and Wastewater sector infrastructure consists of drinking water and wastewater systems, most of which are found in local municipalities, and has a long history of implementing programs that provide clean and safe water.[29] Under the guidance of the Safe Drinking Water Act and Clean Water Act, drinking water and wastewater utilities have been conducting routine, daily, weekly, and monthly water quality monitoring for over 30 years.[30]

Today, research continues to advance ways to improve the quality of water and the safety of these systems. Under Homeland Security Presidential Directive 7 (HSPD-7), the U.S. Environmental Protection Agency (EPA) is designated as the sector-specific agency for this critical infrastructure. The sector is susceptible to risks associated with malicious acts, natural disasters, and denial of service attacks. These could result in a large number of illnesses or casualties and negative economic impacts. Additionally, critical services such as firefighting and healthcare (hospitals), and other dependent and interdependent sectors such as energy, transportation, and FA, would suffer damaging effects from a denial of potable water or properly treated wastewater.[31]

Drinking Water and Wastewater

Safe drinking water is essential to the life of an individual and society as a whole. The effects of drinking water contamination would be devastating and far-reaching with significant consequences to public health, the economy, and the environment. Likewise, a disruption of a wastewater treatment utility or service can cause loss of life, economic impacts, and severe public health and environmental impacts. Both drinking water and wastewater systems can be divided into three elements: physical, cyber, and human.[32] Table 10.7 shows the elements contained in both water and wastewater components.

Goals and Priorities of the Water and Wastewater Sector

Table 10.8 illustrates the vision and mission statements of both the Water and Wastewater sector and the EPA. These statements guide the sector and are the foundation for its goals.

The Water and Wastewater SSP includes four goals, each with a set of objectives, which guide the strategic planning process for sector protection and resilience. Table 10.9 shows these specific goals and objectives.

Table 10.7 Drinking Water and Wastewater Elements

	Physical Elements	*Cyber Elements*	*Human Elements*
Drinking Water	Water source Conveyance Raw water storage Treatment Finished water storage Distribution system Monitoring system	Supervisory Control and Data Acquisition (SCADA) system	Employees and contractors
Wastewater	Collection Raw influent storage Treatment Treated wastewater storage Effluent/discharge Monitoring system	Supervisory Control and Data Acquisition (SCADA) system	Employees and contractors

Source: DHS, Water and Wastewater Sector-Specific Plan, 2010, pp. 8–11.

Table 10.8 Vision and Mission Statements of the Water and Wastewater Sector and EPA

Water and Wastewater Sector Vision Statement:
A secure and resilient drinking water and wastewater infrastructure that provides clean and safe water as an integral part of daily life, ensuring the economic vitality of and public confidence in the nation's drinking water and wastewater service through a layered defense of effective preparedness and security practices in the sector.
EPA's Water Security Mission Statement:
To provide national leadership in developing and promoting programs that enhance the sector's ability to prevent, detect, respond to, and recover from all hazards.

Source: DHS, "Water and Wastewater Sector-Specific Plan," 2010, p. 15.

Water and Wastewater Systems Sector: Assessing Risk

There is a diversity of assets in the Water and Wastewater sector such as size, treatment complexity, disinfection practices, and geographic locations. As such, a multitude of all-hazard risk assessment methodologies have been developed. These are used by sector owners and operators and address a full range of utility components, including the physical plant (physical); employees (human); information technology/Supervisory Control and Data Acquisition and communications (cyber); and customers.[33] As with all sectors, the Water and Wastewater sector conducts full risk assessments using the formula: Risk (R) = f {Consequence (C), Threat (T), Vulnerability (V)}.[34] Risk assessments are conducted at the asset level by owners and operators and are based on local conditions, threats, and other factors; approaches to and the results of an assessment may vary for each type of asset.[35]

The EPA partners with the DHS and Water sector partners to come up with risk assessment documents and tools to assist owners and operators in conducting local assessments. These initiatives are ongoing to improve the security of the sector. There are three risk assessment tools which have been widely used across the Water and Wastewater sector:

1 Risk Assessment Methodology—Water (RAM-W);
2 Security and Environmental Management Systems (SEMS) emergency response checklist;
3 Vulnerability Self-Assessment Tool (VSAT).[36]

In 2010, the ASME Innovative Technologies Institute and the American Water Works Association developed a risk and resilience standard for the Water and Wastewater sector based on Risk Analysis and Management for Critical Asset Protection (RAMCAP) Plus. RAMCAP Plus establishes a common terminology and measurement metrics for analyzing risk and resilience. This process originally consisted of a seven-step methodology that enables asset owners to perform analyses of their risks and risk reduction options[37] RAMCAP was first identified by the DHS as a model for sector-specific guidance in 2005 and used for various national assets including nuclear power plants, petroleum refineries, and water and wastewater systems.[38] When the Water and Wastewater sector developed its first version of its SSP, the goals included improving the identification of vulnerabilities, threats, and consequences to utility owners and operators that could implement risk-based approaches to enhance the security and resilience of their assets. ASME ITI, under the sponsorship of the DHS, initiated a discussion with the Water and Wastewater sector to consider the development of sector-level guidance based on RAMCAP Plus, and hence it was put in place.[39]

RAMCAP Plus is a process for analyzing and managing the risks and resilience associated with malicious attacks and natural disasters against critical infrastructures. The standard proposes specific threat scenarios and comprises seven interrelated analytic steps. Table 10.10 lists these steps and a brief overview of what they each entail. These steps provide a foundation for owners and operators to collect, interpret and analyze data, and use those results in their decision-making processes for managing risk and resilience.

Table 10.9 Water and Wastewater Sector Goals and Objectives

Goal 1	*Sustain protection of public health and the environment*
The nation relies on the sustained availability of safe drinking water and on the treatment of wastewater to maintain public health and environmental protection. To better protect public and environmental health, the Water and Wastewater Sector works to ensure the continuity of both drinking and wastewater services.	
Objective 1	Encourage integration of security concepts into daily business operations at utilities to foster a security culture.
Objective 2	Evaluate and develop surveillance, monitoring, warning, and response capabilities to recognize and address all-hazards risks at Water sector systems that affect public health and economic viability.
Objective 3	Develop a nationwide laboratory network for water quality protection that integrates federal and state laboratory resources and uses standardized diagnostic protocols and procedures or develop a supporting laboratory network capable of analyzing threats to water quality.
Goal 2	*Recognize and reduce risk*
With an improved understanding of the vulnerabilities, threats, and consequences, owners and operators of utilities can continue to thoroughly examine and implement risk-based approaches to better protect, detect, respond to, and recover from all hazards.	
Objective 1	Improve identification of vulnerabilities based on knowledge and best available information, with the intent of increasing the sector's overall protection posture.
Objective 2	Improve identification of potential threats through knowledge base and communications—with the intent of increasing the overall protection posture of the sector.
Objective 3	Identify and refine public health and economic impact consequences of manmade or natural incidents to improve utility risk assessments and enhance the sector's overall protection posture.
Goal 3	*Maintain a resilient infrastructure.*
The Water and Wastewater sector will investigate how to optimize the continuity of operations to ensure the economic vitality of communities and the utilities that serve them. Response and recovery from an incident in the sector will be crucial to maintaining public health and confidence.	
Objective 1	Emphasize continuity of drinking water and wastewater services as it pertains to utility emergency preparedness, response, and recovery planning.
Objective 2	Explore and expand the implementation of mutual aid agreements/compacts in the Water and Wastewater sector. The sector has significantly enhanced its resilience through agreements among utilities and States; increasing the number and scope of these will further enhance resilience.
Objective 3	Identify and implement key response and recovery strategies. Response and recovery from an incident in the sector will be crucial to maintaining public health and confidence.
Objective 4	Increase understanding of how the sector is interdependent with other critical infrastructure sectors. Sectors such as Healthcare and Public Health and Emergency Services are largely dependent on the Water and Wastewater s for their continuity of operations, while the Water and Wastewater sector is dependent on sectors such as Chemical and Energy for continuity of its operations.
Goal 4	*Increase communication, outreach, and public confidence*
Safe drinking water and water quality are fundamental to everyday life. An incident in the Water sector could have significant impacts on public confidence. Fostering and enhancing the relationships between utilities, government, and the public can mitigate negative perceptions in the face of the incident.	
Objective 1	Communicate with the public about the level of protection and resilience in the Water and Wastewater sector and provide outreach to ensure the public's ability to be prepared for and respond to a natural disaster or manmade incident.
Objective 2	Enhance communication and coordination among utilities and Federal, State, and local officials and agencies to provide information about threats.
Objective 3	Improve relationships among all Water and Wastewater sector partners through a strong public–private partnership characterized by trusted relationships.

Source: DHS, Water and Wastewater Sector-Specific Plan, 2010, pp. 16–17.

Table 10.10 Seven Analytic Steps of RAMCAP Plus

Step 1: Asset Characterization: What are the critical assets?
Step 2: Threat Characterization: What reasonably possible event can harm or disrupt them?
Step 3: Consequence Analysis: What would the event cost in terms of human suffering or economic loss?
Step 4: Vulnerability Analysis: Where are the assets most open to harm from an event?
Step 5: Threat Assessment: What is the likelihood that an event will occur?
Step 6: Risk/Resilience Assessment: Risk = Consequences × (Vulnerability × Threat). Resilience = Service Outage × (Vulnerability × Threat).
Step 7: Risk/Resilience Management: What are the options to reduce risks and increase resilience? What is the benefit/cost ratio of the options?

Source: Morely, K., and Brashear, J. (2010). "Protecting the Water Supply." *Mechanical Engineering*, Vol. 132, Issue 1, pp. 34–36.

Water and Wastewater Sector-Specific Initiatives/Policies

The Water and Wastewater sector has made significant progress in addressing the all-hazards risk environment of today. Some of the sector's accomplishments include:

- The formation of specialized working groups and the development of quality products such as the *Roadmap to a Secure & Resilient Water Sector* and the *Roadmap to Secure Control Systems in the Water Sector* (the latter addressed activities to mitigate cybersecurity risk over the next ten years);
- Water security initiative pilots that were implemented in six U.S. cities (Cincinnati, Dallas, New York City, Philadelphia, and San Francisco). Information gathered from the initial pilot was used to publish three interim guidance documents to advise utilities regarding the design, development, and use of contaminant monitoring and warning systems; and
- Establishment of intrastate mutual aid and assistance agreements, such as Water and Wastewater Agency Response Networks within 47 states to foster a utilities-helping-utilities approach to response and recovery efforts following incidents or events.[40]

Finally, in compliance with HSPD-7, all SSAs provide the DHS with annual reports that serve as a primary tool for assessing performance and reporting on progress in the sector. The Water and Wastewater sector must also provide data to support this report. The Sector Annual Report accomplishes the following purposes:

1 Provides a common vehicle across all sectors to communicate critical infrastructure and key resource (CIKR) protection performance and progress to CIKR partners and other government entities;
2 Establishes a baseline of existing sector-specific CIKR protection programs and initiatives;
3 Identifies plans for SSA resource requirements and budget;
4 Determines and explains how sector efforts support the national effort;
5 Provides an overall progress report for the sector;
6 Provides feedback to the DHS, sectors, and other government entities to illustrate the continuous improvement of CIKR protection activities; and
7 Helps identify and share beneficial practices from successful programs.[41]

Energy Sector Profile

PPD-21 identifies the Energy sector as uniquely critical because it provides an "enabling function" across all critical infrastructure sectors. With over 80% of the country's energy infrastructure owned by the private sector, protecting the energy sector requires a strong partnership between government and industry partners. The energy infrastructure is divided into three interrelated segments: electricity, oil, and natural gas. These segments are widely diverse, geographically dispersed, and are often interdependent of one another.[42]

According to the DHS, the U.S. electricity segment alone contains more than 6,413 power plants and approximately 1,075 gigawatts of installed generation. Additionally, about 48% of electricity is produced by combusting coal (primarily transported by rail), 20% is found in nuclear power plants, and 22% by combusting natural gas. Hydroelectric plants (6%), oil (1%), and renewable sources (solar, wind, and geothermal) (3%) make up the remaining generation. Furthermore, the heavy reliance on pipelines to distribute products across the nation emphasizes the interdependencies between the Energy and Transportation Systems sectors.[43] The Energy sector is subject to regulation in various forms as they are often overseen under numerous jurisdictions. The complexities of the operating structure, along with the evolving threat, make this sector especially challenging to protect. The Department of Energy (DOE) is the SSA for this sector.[44]

The Energy Sector Goals and Priorities

As with other critical infrastructure sectors, the Energy sector has developed a sector vision, goals, and priorities that are aligned and in support of the NIPP critical infrastructure security and resilience goals. Table 10.11 provides an overview of the national vision, goals, and priorities for the Energy sector.

The Energy Sector: Assessing Risk

The Energy sector consists of highly diverse assets, systems, functions, and networks. It also has a considerable amount of experience in risk management. Over the years, the industry has responded to the increased need for enterprise-level security efforts and business continuity plans and continues to assess the security vulnerabilities of single-point assets such as refineries, storage terminals, and power plants, as well as networked features such as pipeline, transmission lines, and cyber systems.[45] It is interesting to note that the types of threats faced by the electricity, oil, and natural gas industries vary widely, as does the meaning of "risk" as perceived by each organization. A discussion of these is important before we delve into the overall sector efforts.

Electricity Subsector Risks and Threats

Risk assessments of the Electricity subsector are conducted by a wide variety of organizations. For example, the North American Reliability Corporation (NERC) assesses risks in terms of the potential impact on the reliability of the bulk power system. The question asked here might be, "Did an event result in the loss or interruption of service to customers?" On the other hand, private companies and utilities examine risks and threats as they relate to the operational and financial security of each company. The question here might be, "Could a threat negatively impact the company's financial health?"[46] While there are differences across the subsector as to what constitutes risk, the Electricity subsector has identified several issues as the key risks and threats to its infrastructure. These include:

- Cyber and physical security threats;
- Natural disasters and extreme weather conditions;
- Workforce capability ("aging workforce") and human errors;
- Equipment failure and aging infrastructure;
- Evolving environmental, economic, and reliability regulatory requirements; and
- Changes in the technical and operational environment, including changes in fuel supply.[47]

Table 10.11 National and Energy Sector Critical Infrastructure Vision, Goals, and Priorities

Vision Statement

A nation in which physical and cyber critical infrastructure remain secure and resilient, with vulnerabilities reduced, consequences minimized, threats identified and disrupted, and response and recovery hastened.

National and Energy Sector Critical Infrastructure Goals

- Assess and analyze threats to, vulnerabilities of, and consequences to critical infrastructure to inform risk management activities.
- Secure critical infrastructure against human, physical, and cyber threats through sustainable efforts to reduce risk, while accounting for the costs and benefits of security investments.
- Enhance critical infrastructure resilience by minimizing the adverse consequences of incidents through advance planning and mitigation efforts, as well as effective responses to save lives and ensure the rapid recovery of essential services.
- Share actionable and relevant information across the critical infrastructure community to build awareness and enable risk-informed decision making.
- Promote learning and adaptation during and after exercises and incidents.

Electricity Subsector Priorities	*Oil and Natural Gas Subsector Priorities*
Tools and Technology—Deploying tools and technologies to enhance situational awareness and security of critical infrastructure. • Deploying the propriety government technologies on utility systems that enable machine-to-machine information sharing and improved situational awareness of threats to the grid. • Implementing the National Institute of Standards and Technology (NIST) Cybersecurity framework. **Information Flow**—Making sure actionable intelligence and threat indicators are communicated between the government and industry in a time-sensitive manner. • Improving the bidirectional flow of threat information; • Coordinating with interdependent sectors. **Incident Response**—Planning and exercising coordinated responses to an attack. • Developing playbooks and capabilities to coordinate industry-government response and recovery efforts. • Ongoing assessments of equipment-sharing programs.	The Oil and Natural Gas Subsector Coordinating Council strives to provide a venue for industry owners and operators to mutually plan, implement, and execute sufficient and necessary sector-wide security programs; procedures and processes; information exchange; accomplishment assessment; and progress to strengthen the security and resilience of its critical infrastructure. • Priorities are placed in the following: • Partnership coordination; • Implementation and communication; • Identification of sector needs/gaps and/or best practices; • Information sharing; and • Business continuity.

Source: DHS, "Energy Sector-Specific Plan," 2015, pp. 3–4.

Oil and Natural Gas Subsector Risk and Threats

The Oil and Natural Gas subsector has a worldwide geographic presence and, as such, faces a diverse risk environment. Key risks in the oil and natural gas industry include:

- Natural disasters and extreme weather conditions;
- Regulatory and legislative changes—including environmental and health—as well as increased cost of compliance;

- Volatile oil and gas prices and demands;
- Operational hazards including blowouts, spills, and personal injury;
- Disruption due to political instability, civil unrest, or terrorist activities;
- Transportation infrastructure constraints impacting the movement of energy resources;
- Inadequate or unavailable insurance coverage;
- Aging infrastructure and workforce; and
- Cybersecurity risks, including insider threats.[48]

Cybersecurity

A growing and evolving security challenge for the Energy sector is that of cybersecurity. A common framework is needed to guide the public–private partnerships that the Energy sector encompasses. As discussed in previous chapters, policies to address evolving cyber threats were outlined by President Obama in the 2013 Executive Order (EO) 13636, "Improving Critical Infrastructure Cybersecurity," one of which directed the NIST to work with stakeholders to develop a cybersecurity framework.[49] To comply with this standard, the DOE collaborated with industry partners to develop the Energy Sector Cybersecurity Framework Implementation Guidance. As mentioned above, the Energy sector organizations have a strong track record of working together to develop cybersecurity standards, tools, and processes that ensure uninterrupted services. The DOE worked with the Electricity subsector and Oil and Natural Gas subsector Coordinating Councils along with other SSAs to develop this Framework Implementation Guidance specifically for energy sector owners and operators.[50] This was designed to assist energy sector organizations in four specific areas to:

- Characterize their current and target cybersecurity posture;
- Identify gaps in their existing cybersecurity risk management programs, using the framework as a guide, and identify areas where current practices may exceed the Framework Implementation;
- Effectively demonstrate and communicate their risk management approach and use of the framework to both internal and external stakeholders.

There are a myriad of cybersecurity risk management tools, processes, standards, and guidelines widely used by energy sector organizations. An example of one of these tools is the Cybersecurity Capability Maturity Model (C2M2). Developed in 2011, this program is a public–private partnership effort to improve Energy sector cybersecurity capabilities and to understand the cybersecurity posture of the industry.[51] In 2014, two distinct C2M2s were developed—one for the Electricity subsector and another for the Oil and Natural Gas subsector. Other notable tools include the Cyber Resilience Review and the Cyber Security Evaluation Tool (CSET). Table 10.12 describes an example of cybersecurity risk management approaches used across the energy sector.

In addition to approaches used cross-sector, the Electricity subsector and the Oil and Natural Gas subsector each have tailored standards or cybersecurity approaches that many organizations may use either voluntarily or by requirement.[52] Tables 10.13 and 10.14 illustrate examples of risk management approaches that are applicable only to specific subsectors.

Finally, energy sector organizations can map their current cybersecurity approach to the framework elements, using specific mappings as a guide. The Framework Implementation approach consists of seven steps that outline the activities that should be taken in a cybersecurity risk assessment. These are:

- Step 1: Prioritize and Scope—organization decides how and where it wants to use the framework

Table 10.12 Example Cybersecurity Risk Management Approaches Used in Energy Sector

Name	*Summary*	*Additional Information*
Cybersecurity Capability Maturity Model (C2M2), both Electricity subsector and Oil and Natural Gas subsector-specific versions	Used to assess an organization's cybersecurity capabilities and prioritize their actions and investments to improve cybersecurity	http://energy.gov/oe/cybersecurity-capability-maturity-model-c2m2
Cyber Resilience Review	Evaluates an organization's operational resilience and cybersecurity practices across ten domains	https://www.us-cert.gov/ccubedvp/self-service-crr
Cyber Security Evaluation Tool (CSET)	Guides users through a step-by-step process to assess their control system and information technology network security practices against recognized industry standards	http://ics-cert.us-cert.gov/Assessments
Electricity Subsector Cybersecurity Risk Management Process Guideline	Enables organizations to apply effective and efficient risk management processes and tailor them to meet their organizational requirements	http://energy.gov/oe/downloads / cybersecurity-risk-management-process-rmp-guideline-final-may-2012

Source: U.S. Department of Energy, "Energy Sector Cybersecurity Framework Implementation Guidance," 2015, p. 5.

Table 10.13 Examples of Electricity Subsector Cybersecurity Risk Management Approaches

Name	*Summary*	*Additional Information*
Critical Infrastructure Protection (CIP) Standards	The North American Electric Reliability Corporation (NERC) CIP Standards provide a set of regulatory cybersecurity requirements to assist in securing the energy system assets that operate and maintain the bulk electric grid.	http://www.nerc.com/pa/Stand/Pages/CIPStandards.aspx
Interagency Report (IR) 7628, Guidelines for Smart Grid Cyber Security	The National Institute of Standards and Technology (NIST) guidelines present an analytical framework to develop effective cybersecurity strategies tailored to their particular smart grid-related characteristics, risks, and vulnerabilities.	http://csrc.nist.gov/publications/PubsNISTIRs.html#NIST-IR-7628

Source: U.S. Department of Energy, "Energy Sector Cybersecurity Framework Implementation Guidance," 2015, p. 6.

- Step 2: Orient—*organization identifies in-scope systems and assets (i.e. people, information, technology, and facilities)*
- Step 3: Create a Current Profile—*organization identifies its current cybersecurity and risk management state*
- Step 4: Conduct a Risk Assessment—*perform a risk assessment for the in-scope portion of the organization*

Table 10.14 Examples of Oil and Natural Gas Cybersecurity Risk Management Approaches

Name	*Summary*	*Additional Information*
Control Systems Cyber Security Guidelines for the Natural Gas Pipeline Industry	The Interstate Natural Gas Association of America guideline assists operators of natural gas pipelines in managing their control systems' cyber security requirements. It sets forth and details the unique risk and impact-based differences between the natural gas pipeline industry and the hazardous liquid pipeline and liquefied natural gas operators.	http://www.ingaa.org/
RP 780 Risk Assessment Methodology	The American Petroleum Institute (API) document provides guidance on risk assessment for oil and natural gas operations.	http://www.api.org/publiccations-standards-and-statistics
Chemical Facilities Anti-Terrorism Standards	The risk-based performance standards (RBPS) from the Department of Homeland Security (DHS) provide guidance on physical and cybersecurity for organizations handling chemicals of interest. RBPS 8 specifically requires facilities regulated by the Chemical Facilities Antiterrorism Standards (CFATS) to address cybersecurity in their facility security plan.	http://www.dhs.gov/chemical-facility-anti-terrorism-standards

Source: U.S. Department of Energy, "Energy Sector Cybersecurity Framework Implementation Guidance," 2015, p. 6.

- Step 5: Create a Target Profile—*organization identifies goals that will mitigate risk commensurate with the risk to organizational and critical infrastructure objectives*
- Step 6: Determine, Analyze, and Prioritize Gaps—*gaps should be identified in both the target profile and target tier*
- Step 7: Implement Action Plan—*the organization implements the action plan and tracks its progress over time, ensuring that gaps are closed and risks are monitored.*[53]

The Framework Implementation should be included in an organization's risk management process. Intended as a continuous activity, it should be repeated according to organization criteria. Additionally, a plan to communicate progress to the appropriate stakeholders should be implemented. Validation and feedback at each step are encouraged for process improvement and overall efficiency and effectiveness.[54]

Conclusion

Each critical infrastructure sector has distinctive characteristics, operating models, and risk profiles. Protecting our nation's food supply, water systems, and energy networks requires strong policy proposals and methods that reflect an understanding of the current risk landscape. The DHS and sector partners have worked together to address these security challenges and build resilience within unique risk management perspectives. The FA, Water and Wastewater, and Energy sectors illustrate not only the vastness of critical infrastructures but the variety of approaches and policies used to managing risk. CARVER, THIRA, and RAMCAP Plus are just some of the methods favored by these sectors. While some of the sectors share risk approaches, the extent to which risk can be estimated also varies by hazard. Therefore, adopting the appropriate policies and methods is critical for successful risk management.

Key Terms

CARVER Plus Shock Method
RAM-W
Sector-Specific Plan (SSP)
SEMS
VSAT

Review Questions

1. What is a sector-specific agency (SSA)? Explain the sector-specific plan (SSP), and how each is used by the individual sectors.
2. Describe the missions of the critical infrastructure sectors presented in this chapter.
3. Identify and explain the various methods each sector takes to manage risk. Is there an overlap in methods?
4. How do different hazards impact the various approaches taken by each sector to manage risk?

Notes

1 DHS, "Food and Agriculture Sector." https://www.dhs.gov/food-and-agriculture-sector
2 DHS, "2015 Sector-Specific Plans." http://www.dhs.gov/2015-sector-specific-plans.
3 DHS, "Food and Agriculture Sector-Specific Plan," 2015.
4 DHS, "Food and Agriculture Sector-Specific Plan," 2015.
5 DHS, "Food and Agriculture Sector." https://www.dhs.gov/food-and-agriculture-sector
6 DHS, "Food and Agriculture Sector-Specific Plan," 2015, pp. 5–7.
7 DHS, "Food and Agriculture Sector-Specific Plan," 2015, p. 14.
8 DHS, "Food and Agriculture Sector-Specific Plan," 2015, p. 14.
9 DHS, "Food and Agriculture Sector-Specific Plan," 2015, p. 23.
10 DHS, "Food and Agriculture Sector-Specific Plan," 2015, p. 23.
11 DHS, "Food and Agriculture Sector-Specific Plan," 2015, p. 24.
12 DHS, "Food and Agriculture Sector-Specific Plan," 2015, p. 24.
13 DHS, "Food and Agriculture Sector-Specific Plan," 2015, p. 25.
14 DHS, "Food and Agriculture Sector-Specific Plan," 2015, p. 25.
15 DHS, "Food and Agriculture Sector-Specific Plan," 2015, p. 26.
16 FDA, Protecting and Promoting Your Health, "CARVER + Shock Primer: An Overview of the Carver Plus Shock Method for Food Sector Vulnerability Assessments," 2009, p. 1.
17 FDA, Protecting and Promoting Your Health, "CARVER+Shock Primer: An Overview of the Carver Plus Shock Method for Food Sector Vulnerability Assessments," 2009, p. 2.
18 FDA, Protecting and Promoting Your Health, "CARVER+Shock Primer: An Overview of the Carver Plus Shock Method for Food Sector Vulnerability Assessments," 2009, pp. 1–2.
19 FDA, Protecting and Promoting Your Health, "CARVER+Shock Primer: An Overview of the Carver Plus Shock Method for Food Sector Vulnerability Assessments," 2009, p. 2.
20 FDA, Protecting and Promoting Your Health, "CARVER+Shock Primer: An Overview of the Carver Plus Shock Method for Food Sector Vulnerability Assessments," 2009, pp. 1–2.

21 FDA, Protecting and Promoting Your Health, "CARVER+Shock Primer: An Overview of the Carver Plus Shock Method for Food Sector Vulnerability Assessments," 2009, pp. 1–2.
22 FDA, Protecting and Promoting Your Health, "CARVER+Shock Primer: An Overview of the Carver Plus Shock Method for Food Sector Vulnerability Assessments," 2009, p. 2.
23 FDA, Protecting and Promoting Your Health, "CARVER+Shock Primer: An Overview of the Carver Plus Shock Method for Food Sector Vulnerability Assessments," 2009, p. 2.
24 FDA, Protecting and Promoting Your Health, "CARVER+Shock Primer: An Overview of the Carver Plus Shock Method for Food Sector Vulnerability Assessments," 2009, p. 2.
25 FDA, Protecting and Promoting Your Health, "CARVER+Shock Primer: An Overview of the Carver Plus Shock Method for Food Sector Vulnerability Assessments," 2009, p. 5.
26 DHS, "Food and Agriculture Sector-Specific Plan," 2015, p. 27.
27 DHS, "Food and Agriculture Sector-Specific Plan," 2015, p. 27.
28 DHS, "Food and Agriculture Sector-Specific Plan," 2015, p. 24.
29 DHS, "Water Sector-Specific Plan," 2010, pp. 1–5.
30 DHS, "Water Sector-Specific Plan," 2010, pp. 1–5.
31 DHS, "Water Sector-Specific Plan," 2010, p. 1.
32 DHS, "Water Sector-Specific Plan," 2010, pp. 8–11.
33 DHS, "Water Sector-Specific Plan," 2010, p. 23.
34 DHS, "Water Sector-Specific Plan," 2010, p. 24.
35 DHS, "Water Sector-Specific Plan," 2010, p. 24.
36 DHS, Water Sector-Specific Plan, 2010, p. 27.
37 Morely, K., and Brashear, J. (2010). "Protecting the Water Supply." *Mechanical Engineering*, Vol. 132, Issue 1, pp. 34–36.
38 Morely, K., and Brashear, J. (2010). "Protecting the Water Supply." *Mechanical Engineering*, Vol. 132, Issue 1, pp. 34–36.
39 Morely, K., and Brashear, J. (2010). "Protecting the Water Supply." *Mechanical Engineering*, Vol. 132, Issue 1, pp. 34–36.
40 DHS, "Water Sector-Specific Plan," 2010, preface.
41 DHS, "Water Sector-Specific Plan," 2010, p. 38.
42 DHS, "Energy Sector-Specific Plan," 2015, p. 3.
43 DHS, "Energy Sector." https://www.dhs.gov/energy-sector.
44 DHS, "Energy Sector-Specific Plan," 2015, p. 3.
45 DHS, "Energy Sector-Specific Plan," 2015, p. 13.
46 DHS, "Energy Sector-Specific Plan," 2015, p. 5.
47 DHS, "Energy Sector-Specific Plan," 2015, p. 5.
48 DHS, "Energy Sector-Specific Plan," 2015, p. 6.
49 DHS, "Energy Sector-Specific Plan," 2015, p. 14.
50 US Dept. of Energy, "Energy Sector Cybersecurity Framework Implementation Guidance," January 2015, p. 1.
51 US Dept. of Energy, "Energy Sector Cybersecurity Framework Implementation Guidance," January 2015, p. 1.
52 US Dept. of Energy, "Energy Sector Cybersecurity Framework Implementation Guidance," January 2015, p. 6.
53 US Dept. of Energy, "Energy Sector Cybersecurity Framework Implementation Guidance," January 2015, pp. 8–18.
54 US Dept. of Energy, "Energy Sector Cybersecurity Framework Implementation Guidance," January 2015, p. 18.

11 Sector-Specific Agencies' Approaches to Risk

Healthcare and Public Health Sector, Transportation Systems Sector, and Emergency Services Sector

Chapter Outline

Introduction

Many of the 16 critical infrastructure sectors are highly dependent on fellow sectors, meaning they rely on the services, systems, and processes to operate. An increased reliance upon information technology and telecommunications has increased the dependence of one sector upon another. The Healthcare and Public Health (HPH) sector, Transportation Systems sector, and Emergency Services sector (ESS) are good examples of these interdependencies and the risks of cascading failures from a single disaster. A classic example is the September 11, 2001 terrorist attacks which shut down our transportation networks, impaired emergency services, and challenged area hospitals. The sectors covered in this chapter each offer a unique policy framework and approach to risk management. It is important to note that a heavy reliance on cyber technologies permeate these sectors, creating newer risks and interdependencies.

Healthcare and Public Health (HPH) Sector Profile

The HPH is a vast and diverse sector employing about 13 million personnel and representing an estimated 16.2% ($2.2 trillion) of our nation's gross domestic product.[1] Included in this sector are

DOI: 10.4324/9781003434887-11

acute care hospitals, ambulatory healthcare, and a sizeable and multifaceted public–private healthcare system. A large portion of private sector enterprises are also included such as those manufacturing, distributing, and selling drugs, vaccines, medical supplies, and equipment.[2] The HPH protects all sectors of the economy from hazards such as terrorism, infectious disease outbreaks, and natural disasters. As with other critical sectors, the vast majority of the HPH sector is privately owned and operated and therefore information sharing and collaboration with other sectors is essential.[3]

It is important to note that, while healthcare tends to be delivered and managed locally, the public health component of the sector, focused primarily on population health, is managed across all levels of government.[4] The Department of Health and Human Services (HHS) is designated as the sector-specific agency (SSA) for this sector. Threats to the HPH sector come not only from natural disasters and terrorism, but high costs, limited resources, and excessive demands on personnel, equipment, and systems. The COVID-19 pandemic of 2020 highlighted the healthcare and public health sector and many of its vulnerabilities. Since then, the sector has worked to improve its risk profile shoring up its weak and fragmented infrastructure. We will return to the impact of COVID-19 on critical infrastructure in Chapter 13.

Goals and Priorities of the HPH Sector

The HPH sector vision promotes a resilient national healthcare and public health infrastructure by striving to protect its workforce and ensure its ability to respond to and recover from routine and emergency events.[5] This vision, along with the HPH sector mission and goals can be found in Table 11.1.

The HPH Sector: Assessing Risk

Conducting risk assessment in the HPH sector is customary to achieve compliance with safety, physical security, and information security regulations. Meeting these regulatory requirements

Table 11.1 Healthcare and Public Health Sector Vision, Mission, and Goals

HPH Sector Vision

The HPH sector will achieve overall resilience against all hazards. It will prevent or minimize damage to, or destruction of, the nation's healthcare and public health infrastructure. It will strive to protect its workforce and preserve its ability to mount timely and effective responses (without disruption to services in unaffected areas) and to recover from both routine and emergency situations.

Mission Statement for the HPH Sector

To sustain the essential functions of the nation's healthcare and public health delivery system and to support effective emergency preparedness and response to nationally significant hazards by implementing strategies, evaluating risks, coordinating plans and policy advice, and providing guidance to prepare, protect, prevent, and when necessary, respond to attacks on the nation's infrastructure, and support the necessary resilience in infrastructure to recover and reconstitute healthcare and public health services.

Goals for the Healthcare and Public Health Sector

Goal 1: Service Continuity—Maintain the ability to provide essential health services during and after disasters or disruptions in the availability of supplies or supporting services such as water and power.

Goal 2: Workforce Protection—Protect the sector's workforce from the harmful consequences of all hazards that may compromise their health and safety and limit their ability to carry out their responsibilities.

Goal 3: Physical Asset Protection—Mitigate the risks posed by all hazards to the sector's physical assets.

Goal 4: Cybersecurity—Mitigate risks to the sector's cyber assets that may result in disruption to or denial of health services.

Source: DHS, "Healthcare and Public Health Sector-Specific Plan," 2010.

includes hospitals needing to achieve certification requirements for reimbursement by the Federal Medicare program; pharmaceutical companies meeting regulations to safeguard the effectiveness of their products; and those aspects of the healthcare system that require data to comply with security and privacy rules as found in the Health Insurance Portability and Accountability Act of 1996.[6] In addition to meeting regulatory requirements, the HPH sector supports risk assessment to expose underlying vulnerabilities and potential points of failure in the system. The HPH sector is vast and complex, and it is difficult to apply one risk analysis method to the entire sector. Organizations in the sector use a variety of tools and methods to assess vulnerabilities and consequences. The following discussion outlines some of these.

Strategic Homeland Security Infrastructure Risk Analysis

SHIRA provides a common framework that sectors can use to assess the economic, loss of life, and psychological consequences resulting from terrorist incidents and natural hazards and domestic threats.[7] It is a threat-based approach and is the result of an integrated "fusion" effort between the infrastructure protection and intelligence communities. SHIRA is similar to the military decision-making process where intelligence initiates the planning and all functional areas participate in the entire process.

The SHIRA model begins with the SSA working with the intelligence community to rank the threat and define current themes. In this case, it is the HHS agency. Vulnerabilities and consequences are next assessed. Finally, the HITRAC calculates the risk ratings. The risk ratings are used to compare the risks faced by a sector or associated with an attack method or threat theme. The simple formula is Risk = Threat × Vulnerability × Consequence.

On behalf of the HPH sector, the Risk Assessment Working Group uses the SHIRA methodology to develop scenarios of real-world events that could impact critical sector services, including the delivery of care and the medical supply chain.[8] Some of the scenarios that have been developed include those relating to biological, cyber, vehicle-born explosive devices, and insider threats.[9] The SHIRA analysis favors those events that have high consequences and have a possibility of occurring. Sample targets include laboratories handling biological select agents and toxins, manufacturing facilities, medical supply storage facilities, and cyber infrastructure.[10]

HPH Sector and Cyber Security

As previously mentioned, under Presidential Policy Directive 21 (PPD-21), the Department of HHS has SSA responsibility for the HPH sector. HHS implements its SSA role for the HPH Sector through the CIP Program within the Office of the Assistant Secretary for Preparedness and Response (ASPR). Through this program, the HHS works in voluntary partnership with public and private sector entities to enhance their security and resilience with respect to all hazards, including cyber threats.[11]

Cybersecurity in the HPH Sector is becoming more of a concern as medical devices have become increasingly linked to the Internet, hospital networks, and other medical devices. This reliance on technology for medical purposes has opened up a new set of vulnerabilities. Devices can malfunction, treatments can be interrupted, and there may be breaches of patient information. In light of PPD-21, the federal government has worked with the HPH sector, including private entities, to manage risk and strengthen the security and resilience of this critical infrastructure against cyber threats.

For example, in 2014 the Center for Devices and Radiological Health finalized its guidance for the industry entitled, "Content of Premarket Submissions for Management of Cybersecurity in Medical Devices."[12] Additionally, the National Institute of Standards and Technology (NIST) published a voluntary risk-based framework focusing on enhanced cybersecurity. The Cybersecurity

Framework is for organizations of any size in any of the 16 critical infrastructure sectors that either already have a mature cyber risk management and cybersecurity program or that do not have such programs. Many of the sectors have adopted this framework in their sector-specific plans. The HPH sector has also utilized this framework to help manage and limit cybersecurity risks. The framework comprises three primary components: *core*, *implementation tiers*, and *profile* and provides an assessment mechanism that enables organizations to determine their current cybersecurity capabilities. Additionally, organizations can set individual goals for a target state and establish a plan for improving and maintaining cybersecurity programs.[13] Table 11.2 illustrates the framework components.

The four elements of the Framework Core are considered essential to an effective cybersecurity program. These five core elements are further described in Table 11.3.

The Framework Implementation Tiers provide context on how an organization views cybersecurity risk and the processes in place to manage that risk. Ranging from Partial (Tier 1) to Adaptive (Tier 4), each tier describes an increasing degree of rigor and sophistication in cybersecurity risk management practices. Further, each tier provides information as to the extent to which risk management is informed by business needs and is integrated into an organization's overall risk management practices.[14] Table 11.4 illustrates these four tiers and specific components. Considered within the tier selection process are an organization's current risk management practices, threat environment, legal and regulatory requirements, business/mission objectives, and organizational constraints.[15] Furthermore, organizations should consider external guidance from federal government departments and agencies, Information Sharing and Analysis Centers (ISACs), existing maturity models, or other sources to determine which tier they should select. It is important to note that these tiers do not represent maturity levels, meaning the successful implementation of the framework is not based upon tier determination but rather on the achievement of the outcomes described in the organization's target profile.[16]

Finally, the Framework Profile enables an organization to establish a roadmap for reducing cybersecurity risk that is aligned with organizational and sector goals. Since many organizations are complex, there may be a need to create multiple profiles that are aligned with particular components and specific needs.[17] Framework profiles can be used to describe the current state or target state of cybersecurity activities. The Current Profile indicates the cybersecurity outcomes that are currently being achieved, and the target profile indicates outcomes that are needed to achieve the desired cybersecurity risk management goals. A comparison of the two profiles may be useful to reveal gaps in risk management actions that need to be addressed. Furthermore, organizations that

Table 11.2 Cybersecurity Framework Components

Framework Core	*Framework Implementation Tiers*	*Framework Profile*
Cybersecurity activities and informative references, organized around particular outcomes. Enables communication of cyber risks across an organization. Comprises four elements: 1 Identity 2 Protect 3 Detect 4 Respond 5 Recover	Describes how cybersecurity risk is managed by an organization and the degree the risk management practices exhibit key characteristics. Tiers are identified as: • Tier 1: Partial • Tier 2: Risk Informed • Tier 3: Repeatable • Tier 4: Adaptive	Aligns industry standards and best practices to the Framework Core in a particular implementation scenario. Supports prioritization and measurement while factoring in business needs

Source: NIST, "Framework for Improving Critical Infrastructure Cybersecurity," February 12, 2014.

Table 11.3 Cybersecurity Framework Core Elements

Functions	*Definitions*	*Categories*
Identity	An understanding of how to manage cybersecurity risks to systems, assets, data, and capabilities	Asset management, business environment, governance, risk assessment, and risk management strategy
Protect	The controls and safeguards necessary to protect or deter cybersecurity threats	Access control, awareness and training, data security, data protection processes, maintenance, and protective technologies
Detect	Continuous monitoring to provide proactive and real-time alerts of cybersecurity-related events	Abnormalities and events, continuous monitoring, and detection processes
Respond	Incident response activities	Response planning, communications, analysis, mitigation, and improvements
Recover	Develop and implement the appropriate activities to maintain plans for resilience and to restore any capabilities or services that were impaired due to a cybersecurity event	Recovery planning, improvements, and communications

Source: NIST, "Framework for Improving Critical Infrastructure Cybersecurity," February 12, 2014.

Table 11.4 Cybersecurity Framework Implementation Tiers

Tier 1: Partial	*Tier 2: Risk Informed*	*Tier 3: Repeatable*	*Tier 4: Adaptive*
Organizational cybersecurity risk management practices are not formalized, and risk is managed in an ad hoc manner with limited awareness of risk and no collaboration with other entities.	Risk management processes and programs are in place but may not be established by organizational-wide policy; collaboration is understood but the organization lacks formal capabilities. Information is shared informally.	Formal policies for risk management processes and programs are established and there is an organization-wide approach to managing cyber risk. The organization understands its dependencies and partners. External participation with partners enables collaboration and risk-based management decisions in response to events.	Risk management processes and programs are based on lessons learned and are embedded in the organizational culture. The organization manages risk and actively shares information with partners to ensure accurate, current information is being shared.

Source: NIST, "Framework for Improving Critical Infrastructure Cybersecurity," February 12, 2014.

make this comparison a priority are better prepared to gauge resource estimates, which are needed for such items as staffing and funding. Overall, this risk-based approach enables an organization to achieve cybersecurity goals in a cost-effective and prioritized manner.[18]

Efforts to mitigate cybersecurity in the HPH Sector have been ongoing with sector members increasing their own cybersecurity activities. One example is the American Hospital Association (AHA), which has been working to increase awareness of cyber threats among hospital executives by providing them with risk resources. The AHA has provided a series of threat briefings and teleconferences on cyber threats to their members, which have been supported by presenters from the HHS CIP Branch, the DHS, National Security Council Staff, and others.[19]

In response to cybersecurity efforts for medical devices, the FDA has published a number of guidelines including Guidance for Industry: Cybersecurity for Networked Medical Devices Containing Off-the-Shelf Software (issued January 14, 2005), FDA Safety Communication: Cybersecurity for Medical Devices and Hospital Networks (issued June 13, 2013), and Draft Guidance for Industry and FDA Staff: Content of Premarket Submissions for Management of Cybersecurity in Medical Devices (issued June 14, 2013).[20] Another example is an internal agency program, the HHS Cybersecurity Program. This information security and privacy program helps protect HHS against potential information technology threats and vulnerabilities and is dedicated to compliance with federal mandates and legislation. HHS also participates in CyberRx, a series of industry-wide exercises to stimulate cyberattacks on healthcare organizations.[21]

HPH Sector: Policy Initiatives

The HHS/ASPR's CIP program supports key activities and policies for risk assessment including infrastructure risk analysis and prioritization, cybersecurity initiative coordination, emergency operation liaison with private sector partners during emergencies, and sector lead for developing, evaluating, and implementing all hazards CIP measures.[22] Included among these initiatives are:

1 The CIP can provide subject matter expert (SME) support for partners requesting guidance on best practices for both physical and cyber critical infrastructure protection (CIP). The HHS/ASPR's CIP Program can tap into its private sector partners in the following healthcare and public health subsectors:

   - Direct Healthcare
   - Plans and Payers
   - Mass Fatality Management
   - Pharmaceuticals
   - Public Health
   - Labs
   - Blood
   - Medical Materials
   - Health IT and Medical Technology.[23]

2 Another key initiative includes opportunities for two-way sharing of information with national-level private sector partners. This includes a wide representation of experts in all aspects of healthcare and public health CIP. The CIP team partners in the exchange of critical expertise in areas such as cybersecurity, physical security, workforce protection, and supply chain management.[24]

3 The Homeland Security Information Network for the Healthcare and Public Health community (HSIN-HPH) is an infrastructure protection tool for the HPH Sector. It is the nation's primary web portal for public and private collaboration to protect critical infrastructure and key resources. Through HSIN-HPH, sensitive but unclassified information can be shared with trusted partners. Specifically, users have access to:

   - Timely, relevant, and actionable information about threats, vulnerabilities, security, policy, cybersecurity, and incident response and recovery activities affecting the healthcare and public health community;
   - Alerts and notifications of credible threats;
   - Best practices for protection and preparedness measures for HPH stakeholders;

- CIKR preparedness and resilience analysis and research products;
- Communication and collaboration with other SMEs.[25]

In addition to these initiatives, the Office of Policy and Planning advises HHS and ASPR leaders through policy options and strategic planning to support domestic and international public health emergency preparedness and response activities. Some of these include:

1 Leads an integrated approach to policy development and analysis within ASPR;
2 Analyzes proposed policies, presidential directives, and regulations, and develops short and long-term policy objectives for ASPR;
3 Serves as ASPR's focal point for the National Security Council policy coordination activities and represents the ASPR, as appropriate, in interagency policy coordination meetings and related activities;
4 Studies public health preparedness and response issues, identifies gaps in policy, and initiates policy planning and formulation to fill identified gaps; and
5 Leads the implementation of the Pandemic and All Hazards Preparedness Act and is responsible for developing the quadrennial National Health Security Strategy (NHSS) and the NHSS Biennial Implementation Plan for public health emergency preparedness and response.[26]

Transportation Systems Sector Profile

The U.S. transportation network is expansive, open, and easily accessible by land, sea, and air. This complex set of interconnected systems includes a wide variety of airways, roads, tracks, terminals, and passages that provide essential services. The DHS identifies seven interconnected subsectors: *aviation*, *highway infrastructure and motor carrier*, *maritime*, *mass transit and passenger rail*, *pipeline systems*, *freight rail*, and *postal and shipping*. These systems transport not only people but also food, water, fuel, and other supplies vital to public health, safety, security, and the economic well-being of our country.[27] The Department of Homeland Security (DHS) Transportation Security Administration (TSA), the U.S. Coast Guard (USCG), and the Department of Transportation (DOT) are the Co-SSAs for the Transportation Systems Sector. Table 11.5 presents an overview of the seven key subsectors of the Transportation System Sector.

The Transportation Systems' sector-specific plan details how the NIPP risk management framework is implemented within the context of the distinct characteristics and environment of the sector. Under the 2013 PPD-21, the Postal and Shipping sector was consolidated within the Transportation Systems sector.

Similar to other sectors, the vast majority of the transportation infrastructure is owned by the private sector and is composed of physical, human, and cyber components. All of the remaining 15 CIKR sectors are highly dependent on transportation services, which in turn are dependent on the Energy, Communications, Information Technology, Chemical, and Critical Manufacturing sectors.[28]

Transportation System Sector Mission and Goals

The mission of the Transportation System sector is to continuously improve the risk posture of the transportation system. This mission and stated goals are illustrated in Table 11.6.

Table 11.5 7 Key Subsectors of the Transportation System Sector

Aviation: includes aircraft, air traffic control systems, approximately 450 commercial airports and 19,000 additional airports, heliports, and landing strips. This mode includes civil and joint-use military airports, heliports, short takeoff and landing ports, and seaplane bases.
Highway Infrastructure and Motor Carrier: encompasses nearly four million miles of roadway, almost 600,000 bridges, and some 400 tunnels in 35 states. Vehicles include automobiles, motorcycles, and trucks carrying hazardous materials, other commercial freight vehicles, motorcoaches, and school buses.
Maritime Transportation System: consists of about 95,000 miles of coastline, 361 ports, 25,000 miles of waterways, 3.4 million square miles of Exclusive Economic Zone, and intermodal landside connections, which allow the various modes of transportation to move people and goods, to, from, and on the water.
Mass Transit and Passenger Rail: includes service by buses, rail transit (commuter rail, heavy rail—also known as subways or metros—and light rail, including trolleys and streetcars), long distance rail—namely Amtrak and Alaska Railroad—and other, less common types of service (cable cars, inclined planes, funiculars, and automated guideway systems).
Pipeline Systems: consist of vast networks of pipelines that traverse hundreds of thousands of miles throughout the country, carrying nearly all of the nation's natural gas and about 65% of hazardous liquids, as well as various chemicals. These include approximately 2.2 million miles of natural gas distribution pipelines, about 168,900 miles of hazardous liquid pipelines, and more than 109 liquefied natural gas processing and storage facilities.
Freight Rail: consists of seven major carriers, hundreds of smaller railroads, over 140,000 miles of active railroad, over 1.3 million freight cars, and roughly 20,000 locomotives. Further, over 12,000 trains operate daily. The Department of Defense has designated 30,000 miles of track and structure as critical to mobilizing and resupplying U.S. forces.
Postal and Shipping: moves over 574 million messages, products, and financial transactions each day. Postal and shipping activity is differentiated from general cargo operations by its focus on letter or flat mail, publications, or small- and medium-size packages and by service from millions of senders to nearly 152 million destinations.

Source: DHS, "Transportation Systems Sector," https://www.dhs.gov/transportation-systems-sector

Table 11.6 Transportation Systems Sector Mission and Goals

Transportation Systems Sector Mission
To achieve a sustained reduction in the impact of incidents on the sector's critical infrastructure
Goals for the Transportation Systems Sector
Goal 1: Prevent and deter acts of terrorism using, or against, the transportation system.
Goal 2: Enhance the all-hazards preparedness and resilience of the global transportation system to safeguard U.S. national interests.
Goal 3: Improve the effective use of resources for transportation security.
Goal 4: Improve sector situational awareness, understanding, and collaboration.

Source: DHS, "Transportation Systems Sector-Specific Plan," 2010.

Transportation System Sector: Assessing Risk

Risk assessments in the Transportation System sector use the basic risk equation of Risk= *f* {Threat, Vulnerability, Consequence}. Because risk assessments of the transportation system examine the probability and the consequence of an undesirable event affecting, or resulting from, sector assets, systems, or networks, transportation risk is characterized in two fundamentally and non-mutually exclusive ways: (1) Risk to the Transportation System, and (2) Risk from the Transportation System.[29] Information considered in an assessment of transportation assets and systems might include cargo or passenger volume, proximity to population centers, and system dependence on a particular asset.

There are three classes of risk assessments that may be used in the sector. These assessments may vary in methodology depending on their scope and purpose and are depicted as Mission, Asset, and System-Specific Risk Assessments, model risk assessments, and sector cross-model risk assessments. Table 11.7 details these three classes of risk assessments used in the Transportation System sector.

The primary risk assessment methods used by the transportation modes along the three class levels (as listed in Table 11.7) are the Transportation Sector Security Risk Assessment (TSSRA), the Baseline Assessment for Security Enhancement (BASE), and the Maritime Security Risk Analysis Model (MSRAM). The following is a brief discussion of each of these primary methods.

Transportation Sector Risk Assessment

The TSSRA is an analytical technique that ranks the risks associated with multiple attack scenarios in each mode and compares these risks across the sector. It is an example of both a cross-modal (Class 3) and a modal risk assessment (Class 2). The TSSRA process allows the sector to evaluate those scenarios that present the highest relative risk. The focus is on a wide-ranging set of possible scenarios, including cyber incidents, for different arrangements of transportation assets, attack types, and targets. Fault tree analysis is used to determine the risk. Costs, benefits, and perceived effectiveness of current and proposed measures are then calculated. The final risk scores are presented to decision makers and stakeholders, who in turn make informed decisions on Transportation Systems sector priorities.[30]

Table 11.7 Three Classes of Risk Assessments in the Transportation Systems Sector

Class 1: Mission, Asset, and System-Specific Risk Assessments (MASSRA)	*Class 2: Modal Risk Assessments*	*Class 3: Cross-Model Comparative Analysis*
MASSRA focus on one or more of the risk elements or on scenario-specific assessments. An example is a blast effect analysis on a certain type of conveyance. These assessments generally do not cross jurisdictional lines and have a narrow, specific focus. They provide a detailed analysis of infrastructure vulnerabilities and can be used to determine which countermeasures should be used to mitigate risk.	These are used to identify how best to determine or validate high-risk focus areas within a mode of transportation. These modal risk assessments may also help to establish the sector's priorities for a specific mode. TSA's Transportation Sector Security Risk Assessment (TSSRA) tool is used to conduct modal security risk assessments for each of the primary transportation modes, as well as sub-modal groups such as the school bus transportation system. Further, the USCG uses the Maritime Security Risk Analysis Model (MSRAM) and other inputs to provide maritime risk information to TSSRA.	Class 3 assessments are cross-modal risk assessments focusing on two or more modes, or on the entire sector. TSSRA is also an example of a cross-modal comparative analysis method. These analyses help identify strategic planning priorities and define long-term visions. Additionally, these inform key leadership decisions, including investments in countermeasures. A good example is a sector-wide assessment that could identify an IED attack on underwater tunnels as a top threat.
Risk Assessment Classes: Summary		
These three risk assessment types may be conducted concurrently and/or independently by various sector partners. Once the assessments take place and the results are analyzed and disseminated, they are sent to the sector's leadership as tools to aid in decision-making processes.[31]		

Source: DHS, "Transportation Systems Sector-Specific Plan," 2010, pp. 34–35.

Baseline Assessment for Security Enhancement

The BASE is a comprehensive security assessment program designed to evaluate posture in 17 Security and Emergency Management Action Items foundational to an effective security program. Some of these action items are the agency's security plan, background investigation of employees, security training, drills and exercises, public awareness, facility security and access control, and cybersecurity.[32] Conducting a BASE informs security priorities, the development of security enhancement programs, allows for the allocation of resources and assists in the compilation of smart security practices for mass transit and passenger rail agencies.[33] Furthermore, BASE is an example of a mission-specific assessment that focuses on vulnerability and effective security implementation. In the BASE program, TSA takes responsibility for reviewing the implementation of security actions jointly developed by TSA, DOT's Federal Transit Administration, and sector partners from mass transit and passenger rail systems.[34]

Maritime Security Risk Analysis Model

The USCG uses MSRAM specifically for assessing maritime security and risks. MSRAM is an example of a scenario-based risk assessment that falls into both the modal risk assessment (Class 2) and mission-specific risk assessment categories (Class 1).[35] The MSRAM methodology uses a combination of target and attack mode scenarios and assesses risk in terms of threat vulnerability and consequences. It is a powerful tool that allows the Federal Maritime Security Coordinators and Area Maritime Security Committees to perform detailed scenario risk assessments on all of the maritime sectors.[36] Furthermore, MSRAM is used by all levels of government—federal, state, and local.

Transportation System Sector Policies and Priorities

Information gathered from an assessment in the Transportation System sector is then analyzed in combination with other factors in the decision environment. Collectively, this enables stakeholders and decision makers to set policy priorities and strategies for the sector. By prioritizing specific areas of risk, the sector can also determine resource allocation and budget needs. Table 11.8 illustrates examples of some factors that the sector considers when developing policy priorities and strategies based on risk assessment.

Table 11.8 Factors Impacting Transportation Systems Sector Priorities and Policies

Intelligence and Risk Assessment	**Legislative and Executive Requirements**
• Modal Assessments • Intelligence Reports • Unknown Risk Hedges • Comprehensive Reviews • Sector Comparative Analysis • Mission and Asset-Specific Assessments	• DHS Priorities • Presidential Directives • Congressional Mandates • GAO Recommendations
Safety, Privacy, and Stakeholder Concerns	**Budget and Implementation Constraints**
• Transportation Flow • Safety/Security Conflicts • Privacy/Security Conflicts • Unfunded Mandate Issues	• Sector Capabilities • Budget & Acquisition Cycles • Federal Regulation Timelines • Competing National Budget Priorities

Source: DHS, "Transportation Systems Sector-Specific Plan," 2010, p. 6.

Emergency Services Sector Profile

The ESS includes a diverse array of disciplines and capabilities that support a broad range of services in the areas of prevention, preparedness, response, and recovery. The ESS serves as the primary protector of the other 15 critical infrastructures, operating as the first line of defense. The ESS sector is so critical, that a failure or disruption in the sector could result in substantial damage, loss of life, major public health problems, long-term economic loss, and cascading disruptions to the other critical sectors.[37] Potential risks to the ESS include natural disasters and extreme weather patterns; cyberattacks or disruptions, violent extremist and terrorist attacks; as well as chemical, biological, radiological, and nuclear events.

As with other critical sectors, the risk of cyberattacks continues to evolve, especially with the reliance on cyber systems for emergency operations. Communications, data management, biometric activities, telecommunications, and electronic security systems are just some of the ways in which technological advancements in the sector also present risk. Additionally, increased expectations from the public for emergency response, along with limited financial resources, create an environment where risk to the sector is a serious challenge.[38] The DHS is designated as the SSA for the ESS.

ESS Key Operating Characteristics

Millions of trained personnel, in both paid and volunteer capacities, make up the ESS community. The majority of these come from state, local, tribal, and territorial (SLTT) levels of government. Some examples include city police departments, fire and rescue services, county sheriff's offices, public works departments, and at the federal level the Department of Defense police and fire departments.[39] With a focus on the protection of other sectors and the public, the ESS consists of key operating characteristics. The DHS 2015 ESS-specific plan identifies these characteristics as follows:

- The ESS is the most geographically distributed sector with more than 2.5 million personnel serving every location in all 50 States, five territories, and the District of Columbia;
- *First response can greatly affect the resulting severity and duration of emergency events*;
- *Adaptability and flexibility are hallmarks of ESS operations*;
- *Sector operations are personnel-driven, but highly dependent on communications, information technology, and transportation systems*;
- *ESS personal are operating in a limited resource environment.*[40]

The ESS is made up of five disciplines: law enforcement, emergency management, fire and rescue services, emergency medical services, and public works. The sector consists of systems and networks composed of three components: human, physical, and cyber. Table 11.9 shows these components in detail.

As noted above, the majority of ESS personnel are public employees in the SLTT levels of government. The private sector, however, also contributes to the ESS with private industrial fire departments, private security officers, and private EMS providers. While the sector as a whole is not regulated, there are numerous regulations that govern many emergency response functions at the SLTT level including hazardous materials, fire and rescue, and public utilities. Finally, the ESS has critical sector interdependencies with the energy, communications, transportation systems, water, healthcare and public health, and information sectors.[41] Specialized emergency services are also provided through individual personnel and teams. These specialized capabilities are listed below:

Table 11.9 Emergency Services Sector Components and Disciplines

Three Components		*Five Distinct Disciplines*		
Human	**More than 2,500,000** career and volunteer ESS personnel in five disciplines	**Law Enforcement** **Emergency Management**	**Fire and Rescue Services** **Emergency Medical Services** **Equipment** Specialized for discipline and capability (e.g. personal protective, communications, surveillance). **Internet** The Internet is widely used by the sector to provide information and distribute alerts, warnings, and threats relevant to the sector.	**Public Works** **Vehicles** Specialized for disciplines and capability (e.g. ambulances, HazMat, aircraft, and watercraft) **Information Networks** Computer-aided dispatch and watch and warning systems, information-sharing portals, and social networking are leveraged to keep the ESS informed and connected.
Physical	**Facilities** for daily operations, support, training, or storage			
Cyber	**Virtual Operations** Emergency operations communications, database management, biometric activities, and security systems are frequently operated in cyberspace.			

Source: DHS, "Emergency Services Sector-Specific Plan," 2015.

- Tactical Teams;
- Hazardous Devices Team/Public Safety Bomb Disposal;
- Public Safety Dive Teams/Maritime Units;
- Canine Units;
- Aviation Units;
- Hazardous Materials;
- Search and Rescue (SAR);
- Public Safety Answering Points;
- Fusion Centers;
- Private Security Guard Forces;
- National Guard Civil Support.[42]

ESS Current Risks

It can be argued that the ESS has one of the most critical missions of all sectors. Risks to its operations and functions could have devastating impacts on public safety, the protection of other infrastructures, and first responders and others working in the ESS. The DHS 2015 ESS-specific plan identifies four of the most significant sector threats, and their associated risks: cyber infrastructure attacks or disruptions, natural disasters and extreme weather, violent extremist and terrorist attacks,

Table 11.10 Significant Emergency Services Sector Risks

Cyber Infrastructure Attacks or Disruptions
Advancements in technology have increased the dependency of the Emergency Services Sector on cyber-based infrastructures and operations. These include emergency operations communications, data management, biometric activities, telecommunications (i.e. computer-aided dispatch), and electronic security systems. Reliance on these technologies puts the sector at risk for cyberattacks from anywhere in the world.
Natural Disasters and Extreme Weather (Earthquakes, Hurricanes, Fires, and Floods)
Patterns of more extreme weather increase the geographic magnitude and severity of disasters. These events require huge amounts of resources from the Emergency Services Sector, often for extended periods of time. In addition, disasters present increased hazards to responders and often disrupt critical services needed for an effective response.
Violent Extremist and Terrorist Attacks
There is an increased risk to emergency services personnel from violent extremists. These individuals are characterized by those who support or commit ideologically motivated violence to further political goals. These individuals are increasingly targeting first responders, especially law enforcement, and soft targets. Threats of terrorism persist and are becoming more diversified through a wide range of attack methods. Radicalization on the Internet, and other forms of social media, enable these individuals to connect with others and plan violent attacks.
Chemical, Biological, Radiological, and Nuclear Incidents
Incidents involving chemical, biological, radiological, and nuclear agents present serious risks to emergency services personnel responding to and treating victims of such attacks. Biological agents or infectious diseases can quickly spread through numerous jurisdictions and greatly strain emergency services resources. Responding to HazMat incidents also creates risks as first responders may not be provided with accurate cargo information. All of these agents require extensive training and specialized equipment. Shortfalls in budgets and other resources may limit the capabilities of the sector to rapidly respond to and contain such incidents.

Source: DHS, "Emergency Services Sector-Specific Plan," 2015.

and incidents involving chemical, biological, and nuclear agents.[43] Table 11.10 provides details on these four sector threats.

ESS Goals and Priorities

As part of the 2015 sector-specific plan, the ESS identified goals and priorities to guide the sector's security and resilience efforts over the next four years. Specifically, these four goals are

- **Partnership Engagement**—collaborating with sector partners and encouraging continuous growth and improvement of these partnerships;
- **Situational Awareness**—support an information-sharing environment;
- **Prevention, Preparedness, and Protection**—implement a risk-based approach to improve the preparedness and resilience of the sector's overall capacity to perform its mission through policy initiatives and decision making;
- **Recovery and Reconstitution**—improve the operational capacity, sustainability, and resilience of the sector and increase the speed and efficiency of normal services and activities following an incident.[44]

In support of these goals, sector partners developed 12 priorities to focus their efforts which include:

- Developing and utilizing processes and mechanisms to support sector partnerships and information sharing;
- Sharing best practices;

- Enabling sector partners to implement their missions;
- A focus on technological solutions and cybersecurity;
- Encourage cross-sector and cross-discipline collaborations; and
- Defining current sector risk assessment and information-sharing capabilities and requirements.[45]

ESS: Assessing Risk

The ESS approaches risk with an emphasis on its reliance upon digital technologies and other cyber infrastructures. Two documents released by the sector establish a baseline national-level cyber risk and identify risk mitigation strategies—the 2012 *Emergency Services Sector Cyber Risk Assessment* (ESS-CRA) and the 2014 *Emergency Services Sector Roadmap to Secure Voice and Data Systems* (Roadmap).[46] The ESS-CRA evaluates risk to the sector by focusing on the five ESS disciplines. The ESS-CRA uses the DHS Cyber Assessment Risk Management Approach (CARMA), along with the Cybersecurity Framework, to assess risk in the sector. CARMA provides a methodology for sector stakeholders to define key business functions that must be protected and identifies risks posed to their functional viability.[47]

Risk management in the ESS operates from a point of common understanding and terminology, despite the varying conceptualizations of infrastructure criticality across disciplines and jurisdictions. While the national-level assessments, such as the Strategic National Risk Assessment, are important to an overall understanding of national risk, the ESS conducts risk assessments primarily at the state and local levels.[48] According to the ESS 2015 sector-specific plan, assessments conducted by the ESS adhere to several risk assessment methodology guidelines:

- Documented—information used in the assessment and how it is synthesized to generate risk estimate is documented and transparent to the user and others using the results.
- Reproducible—despite variance in each discipline's facilities, capabilities, and personnel, the results are comparable and repeatable. Subjective judgments are minimized to allow for future policy and value judgments by owners and operators.
- Defensible—assessment components are logically integrated and ESS disciplines are used appropriately for the risk analysis with a parallel effort to accomplish assessment accuracy and transparency.[49]

The ESS uses a number of risk assessment tools to assess the threats, vulnerabilities, and consequences. These assessments are crucial to decision making and policy implementation throughout the sector. Table 11.11 provides a description of these assessment tools.

ESS: Policy and Emerging Issues

Changes in policy, resources, threat types, and public expectations have impacted the ESS risk profile and public policy decision making. The DHS ESS-specific plan outlines eight specific trends and issues, key to security, and resilience efforts.[50] The following list describes these in detail.

- **Increasing public expectations for ESS expertise, rapid response capabilities, and real-time information sharing.** The focus has shifted to an all-hazard incident response planning. The public expects ESS personnel to respond quickly and efficiently to all types of incidents and to communicate information in real time. Public information needs can drain response personnel. Moreover, reduced budgets, increased mandates, training requirements, and high costs associated with equipment maintenance can further hinder resources.[51]

Table 11.11 Emergency Services Sector Risk Assessment Tools

SLTT Hazard Identification and Risk Assessment (HIRA)
SLTT ESS partners conduct HIRAs to understand and examine specific potential or existing circumstances that can generate a disaster or emergency incident at the SLTT level. HIRAs focus on quantitative assessment of hazards and consequences.
Threat and Hazard Identification and Risk Assessment (THIRA)
THIRA expands on HIRA by broadening what is considered through the risk assessment process. SLTT entities use the THIRA process to complete the following qualitative risk assessment steps: identify the threats and hazards of primary concern to the community, develop threat and hazard context, establish targets for each core capability within the National Preparedness Goal, and apply the results to estimate resources required.
Enhanced Critical Infrastructure Protection (ECIP)
This tool facilitates outreach to establish or enhance the relationship between DHS and critical infrastructure owners and operators. Voluntary security surveys are offered as a result of the outreach and are conducted by DHS PSAs to assess the overall security and resilience of the nation's most critical infrastructure sites.
Onsite Assessments
SLTT entities deploy onsite risk assessment teams that focus on high-priority assets within their area of responsibility. Many risk assessment teams concentrate on critical infrastructure vulnerabilities and provide recommendations to the owner/operator to address potential threats and security vulnerabilities. An example of an available Federal onsite assessment is the Cyber Resilience Review—a no-cost voluntary cyber risk assessment facilitated by DHS cybersecurity professionals.
Planning Efforts
Joint committees, such as local emergency planning committees (LEPCs), primarily comprise local emergency services representatives familiar with a variety of jurisdiction-specific risk factors. LEPC meetings are open to all emergency management stakeholders and enable the illumination of risk factors that may be inadvertently excluded from assessments. Some sector partners also rely on capital budgeting plans to determine if the organization's risk management investments are worth pursuing.
Cybersecurity Assessment and Risk Management Approach (CARMA)
Provides a strategic methodology to identify, assess, and manage cyber critical infrastructure risks in the sector.

Source: DHS, "Emergency Services Sector-Specific Plan," 2015.

- **Reduced grant funding constraining state and local resources.** The ability of the ESS to identify risk and make changes to the sector profile may be inhibited by reductions in grant monies and reduced state and local budgets. For example, such restrictions have forced the ESS community to choose between terrorist threat mitigation strategies and additional police activities for drug and gang enforcement. At the same time costs have increased for healthcare, personnel, and fuel and maintenance equipment.[52]
- **Extreme weather events.** In recent years extreme weather events have become more frequent. Overtime these increase response demands which may drain sector personnel, assets, and capabilities. Of particular concern are natural disasters which threaten key services that enable ESS response.[53]
- **Greater dependence on cyber infrastructure.** As with other critical sectors, the ESS has become increasingly dependent on cyber assets, systems, and disciplines to carry out its services. Newer technologies such as next-generation 9-1-1, the transition toward cloud-based information systems, and the usage of geospatial tools have enabled the ESS to expand and improve its operations. The downside is these newer technologies present new risks that further test the sector's ability to swiftly and carefully respond to emergencies.[54]
- **Changing population dynamics.** Changes in population density and characteristics present another set of challenges for the ESS. As people live longer, the average age of the general population increases. Older adults often require more medical response. Language barriers and

threats from global mobility (i.e. risks of biological agents and communicable diseases) are other issues of concern for the sector.[55]

- **Attacks targeting or compromising ESS personnel.** While the nature of the disciplines in the ESS presents an inherent risk, first responders may also become the target of attacks when responding to mass casualty, active shooter, or improvised explosive device (IED) incidents. ESS personnel may also suffer from physical or emotional trauma due to incident response.[56]
- **Aging infrastructure.** The nation's aging infrastructure includes electrical grids, water/wastewater systems, and roads and bridges—all of which increase the risk of failure. These risks create incidents that ultimately require ESS response while also hampering services critical to ESS functions.[57]
- **Loss of workforce expertise.** As the average age of personnel increases, the risk of losing quality people to retirement is a real concern. New recruits tend to be less experienced.[58]

Conclusion

The HPH sector, Transportation sector, and ESS are complex infrastructures. Each requires the use of a variety of methods and tools to assess and manage risk. Specific methods of risk include SHIRA in the Health and Public Health sector, TSSRA in the Transportation Systems sector, and THIRA in the ESS. The increasing link between physical and cyber security in these sectors necessitates the development of policies and initiatives to address evolving threats. It also creates more interdependencies. All three sectors support efforts to implement the Cybersecurity Framework, and a collection of tools is being developed to assess cyber risks. The next chapter presents the communications, information technology, and financial sectors, all of which are also subject to cyber intrusions.

Key Terms

CARMA
Cybersecurity Framework Components
HITRAC
ISACs
Partnership Engagement
SHIRA

Review Questions

1. Discuss the mission statements, goals, and priorities of each of the three sectors presented in the chapter.
2. List and describe the various approaches to managing risk taken by each of the three sectors presented in the chapter. Is there an overlap of methods used?
3. What are some of the key operating characteristics of the Emergency Services sector? What makes these different from the Healthcare and Public Health or Transportation Systems sectors?
4. Explain what is meant by partnership engagement. Which of the three sectors embraces this concept and why? Do you think it is applicable to other sectors? Explain your answer using specific examples.

Notes

1. DHS, "Healthcare and Public Health Sector-Specific Plan," 2010.
2. DHS, "Healthcare and Public Health Sector-Specific Plan," 2010.
3. DHS, "Healthcare and Public Health Sector," https://www.dhs.gov/healthcare-and-public-health-sector

4 DHS, "Healthcare and Public Health Sector," https://www.dhs.gov/healthcare-and-public-health-sector
5 DHS, "Healthcare and Public Health Sector-Specific Plan," 2010.
6 DHS, "Healthcare and Public Health Sector-Specific Plan," 2010.
7 DHS, "Healthcare and Public Health Sector-Specific Plan," 2010.
8 DHS, "Healthcare and Public Health Sector-Specific Plan," 2010.
9 DHS, "Healthcare and Public Health Sector-Specific Plan," 2010.
10 DHS, "Healthcare and Public Health Sector-Specific Plan," 2010.
11 US Department of Health and Human Services, "Assistant Secretary for Preparedness and Response," www.phe.gov.
12 Hanson, C. (2015). "CDRH Schedules January 2016 Cybersecurity Workshop." Inside Medical Devices, Updates on Developments for Medical Devices. www.insidemedicaldevices.com.
13 National Institute of Standards and Technology, "Framework for Improving Critical Infrastructure Cybersecurity," February 12, 2014.
14 National Institute of Standards and Technology, "Framework for Improving Critical Infrastructure Cybersecurity," February 12, 2014.
15 National Institute of Standards and Technology, "Framework for Improving Critical Infrastructure Cybersecurity," February 12, 2014.
16 National Institute of Standards and Technology, "Framework for Improving Critical Infrastructure Cybersecurity," February 12, 2014.
17 National Institute of Standards and Technology, "Framework for Improving Critical Infrastructure Cybersecurity," February 12, 2014.
18 National Institute of Standards and Technology, "Framework for Improving Critical Infrastructure Cybersecurity," February 12, 2014.
19 US Department of Health and Human Services, "Assistant Secretary for Preparedness and Response," www.phe.gov.
20 US Department of Health and Human Services, "Assistant Secretary for Preparedness and Response," www.phe.gov.
21 US Department of Health and Human Services, "Assistant Secretary for Preparedness and Response," www.phe.gov.
22 US Department of Health and Human Services, "Assistant Secretary for Preparedness and Response," www.phe.gov.
23 US Department of Health and Human Services, "Assistant Secretary for Preparedness and Response," www.phe.gov.
24 US Department of Health and Human Services, "Assistant Secretary for Preparedness and Response," www.phe.gov.
25 US Department of Health and Human Services, "Assistant Secretary for Preparedness and Response," www.phe.gov.
26 US Department of Health and Human Services, "Office of Policy and Planning," www.phe.gov.
27 DHS, "Transportation Systems Sector-Specific Plan," 2010.
28 DHS, "Transportation Systems Sector-Specific Plan," 2010.
29 DHS, "Transportation Systems Sector-Specific Plan," 2010, p. 32.
30 DHS, "Transportation Systems Sector-Specific Plan," 2010, p. 35.
31 DHS, "Transportation Systems Sector-Specific Plan," 2010, pp. 34–35.
32 DHS, "Transportation Systems Sector-Specific Plan," 2010, p. 37.
33 DHS, "Transportation Systems Sector-Specific Plan," 2010, p. 36.
34 DHS, "Transportation Systems Sector-Specific Plan," 2010, p. 36.
35 DHS, "Transportation Systems Sector-Specific Plan," 2010, p. 37.
36 DHS, "Transportation Systems Sector-Specific Plan," 2010, p. 37.
37 DHS, "Emergency Services, Sector-Specific Plan," 2015.
38 DHS, "Emergency Services, Sector-Specific Plan," 2015.
39 DHS, "Emergency Services, Sector-Specific Plan," 2015.
40 DHS, "Emergency Services, Sector-Specific Plan," 2015.
41 DHS, "Emergency Services, Sector-Specific Plan," 2015.
42 DHS, "Emergency Services, Sector-Specific Plan," 2015.
43 DHS, "Emergency Services, Sector-Specific Plan," 2015, p. 8.
44 DHS, "Emergency Services, Sector-Specific Plan," 2015.
45 DHS, "Emergency Services, Sector-Specific Plan," 2015.

46 DHS, "Emergency Services, Sector-Specific Plan," 2015.
47 DHS, "Emergency Services Sector, Cyber Risk Assessment," 2012.
48 DHS, "Emergency Services, Sector-Specific Plan," 2015, p. 16.
49 DHS, "Emergency Services, Sector-Specific Plan," 2015, p. 17.
50 DHS, "Emergency Services, Sector-Specific Plan," 2015, p. 7.
51 DHS, "Emergency Services, Sector-Specific Plan," 2015, p. 7.
52 DHS, "Emergency Services, Sector-Specific Plan," 2015, p. 7.
53 DHS, "Emergency Services, Sector-Specific Plan," 2015, p. 7.
54 DHS, "Emergency Services, Sector-Specific Plan," 2015, p. 7.
55 DHS, "Emergency Services, Sector-Specific Plan," 2015, p. 8.
56 DHS, "Emergency Services, Sector-Specific Plan," 2015, p. 7.
57 DHS, "Emergency Services, Sector-Specific Plan," 2015, p. 7.
58 DHS, "Emergency Services, Sector-Specific Plan," 2015, p. 7.

12 Sector-Specific Agencies' Approaches to Risk

Communications Sector, Information Technology Sector, and Financial Sector

Chapter Outline

Introduction

In late 2015, a security researcher discovered multiple vulnerabilities in various name-brand wireless routers. Discovered flaws included path traversals, used to access potentially sensitive Application Programming Interfaces (APIs), and weaknesses that allow unauthenticated attackers to alter the setting of a device, bypass authentication, and remote code executions.[1] In another scenario, Google's Android OS was found to be prone to numerous vulnerabilities that could be exploited through email, web browsing, and MMS when processing media files.

Then, in late 2019, a campaign of cyberattacks breached the computer networks at the Texas-based software company Solar Winds. Over 18,000 customers who applied the company's update, Sunburst, were immediately affected. The impact was widespread and is considered one of the most sophisticated hacking campaigns ever conducted against the federal government—including the U.S. Department of Health, the Treasury, and the U.S. Department of State—and the private sector.

These instances demonstrate a potential risk to the Communications sector via smartphones and the Internet. The Internet has evolved as a critical infrastructure and its protection is shared with the Information Technology sector. In this chapter we will examine both the Communications and Information Technology sectors. Closely interdependent, these two sectors provide essential

DOI: 10.4324/9781003434887-12

operations supporting the U.S. economy, businesses, public safety, and government. The chapter will follow with an examination of the Financial Services sector, which shares some of the same challenges as the Communications and Information Technology sectors. Mobile commerce, social media, and the safeguarding of client and employee information present unique challenges to the Financial Services sector as it does to the Communications and Information Technology sectors. The chapter will conclude with a summary of the remaining seven critical infrastructure sectors not previously discussed.

Communications Sector Profile

America's reliance on a digital infrastructure makes the Communications sector critical. The nation's communications infrastructure is composed of wireline, wireless, satellite, cable, and broadcasting capabilities, as well as those transport networks that support the Internet and other key information systems. This complex system incorporates multiple technologies and services with diverse ownership, much of which comes from the private sector.[2] Many of these companies are interdependent upon each other and, as such, have had a long-standing tradition of cooperation and trust among them. Over time, these companies have included disasters and accidental disruptions into their network resilience designs, business continuity plans, and disaster recovery policies.

The Office of Cybersecurity and Communications (CS&C) is the Sector-Specific Agency for both the Communications and Information Technology Sectors. The CS&C coordinates national-level reporting that is consistent with the National Response Framework and works to prevent or minimize disruptions to critical information infrastructure in order to protect the public, the economy, and government services.[3]

Goals and Priorities of the Communications Sector

The vision and goals for the Communications sector combine a critical reliance on assured communications with the purpose of protecting, enhancing, and improving the sector's national security and emergency preparedness. Table 12.1 presents these statements.

Communications Sector: Assessing Risk

Historically speaking, the Communications sector has been an important infrastructure in need of protection. However, the compelling events of 911, Hurricane Katrina, and Hurricane Sandy in

Table 12.1 Communications Sector: Vision and Goals

Vision Statement for the Communications Sector
The U.S. has a critical reliance on assured communications. The Communications sector strives to ensure that the Nation's communications networks and systems are secure, resilient, and rapidly restored in the event of disruption.
Goals for the Communications Sector
Goal 1: Protect and enhance the overall physical and logical health of communications.
Goal 2: Rapidly reconstitute critical communications services in the event of disruption and mitigate cascading effects.
Goal 3: Improve the sector's national security and emergency preparedness (NS/EP) posture with federal, state, local, tribal, international, and private sector entities to reduce risk.

Source: DHS, "Communications Sector-Specific Plan," 2010, p. 5.

2012 have shown how dramatically the communications infrastructure can be impacted and how significant the sector is in an all-hazards environment. The terrorist attacks on September 11, 2001, severely disrupted communications networks. Moreover, a major telecommunications hub was severely damaged in the collapse of the World Trade Center buildings, affecting more than four million data circuits. Hurricane Katrina impacted a three-state telecommunications infrastructure, leaving three million users without a dial tone and taking 38 911 emergency services centers and 1,000 cell phone towers out of operation.[4] Hurricane Sandy disrupted telecommunications services in several Northeastern states impacting 911 calls and wireless cell towers.

It is clear that the Communications sector faces both natural and manmade threats and that these infrastructure failures can be wide-reaching. When this occurs, it does so in three specific ways: physical destruction of network components, disruption in supporting network infrastructure, and network congestion.[5] Assessing the risk of these elements requires ongoing activity aimed at new asset and network configurations and the analysis of existing assets in light of new threats.[6]

The Communications sector approaches risk using a strategic framework based on the formula of consequence, vulnerability, and threat. As mentioned above, the Communications sector is integrally linked with the Information Technology sector (discussed later in this chapter).

The Communications sector-specific plan (CSSP) Risk Assessment Framework is built on the goals of the CSSP and aims to increase the resilience of the Communications sector. These basic risk assessment goals are:

- **Resilient Infrastructure:** Critical infrastructure and its communications capabilities should be able to withstand natural or manmade hazards with minimal interruption or failure;
- **Diversity:** Facilities should have diverse primary and backup communications capabilities that do not share common points of failure;
- **Redundancy:** Facilities should use multiple communications capabilities to sustain business operations and eliminate single points of failure that could disrupt primary services;
- **Recoverability:** Plans and processes should be in place to restore operations quickly if an interruption or failure occurs.[7]

Table 12.2 summarizes the CSSP risk framework. Note that sector assessments are ongoing and continually updated.

The Communications Sector: Information Sharing Policies

Homeland Security Presidential Directive 7 (HSPD-7) assigns the DHS lead responsibility for coordinating the protection of national infrastructure, including the Communications sector. Initially, the DHS delegated this responsibility to the National Communications System (NCS), an office created in 1962 for crisis communications which was then moved to the DHS when it was created in 2003. In 2012, the NCS was disbanded by Presidential Executive Order 13618, Assignment of National Security and Emergency Preparedness Communications Functions.

As mentioned previously, the Office of Cybersecurity and Communications (CS&C) and since become the sector-specific agency (SSA) for the Communications and Information Technology (IT) sectors. Information sharing is a key part of the DHS mission and the Communications sector. In this capacity, the Communications sector works closely with the DHS's National Cybersecurity and Communications Integration Center (NCCIC) to reduce the likelihood and severity of incidents that could potentially affect the infrastructure.[8] The NCCIC supports cyber situational awareness, incident response, and management 24/7. It is a national center that supports cyber and

Table 12.2 CSSP Risk Framework Summary

Industry Self-Assessments	*Government-Sponsored Assessments*		*Government-Sponsored Cross-Sector Dependency Analysis*
Internal Industry Assessments	National Sector Risk Assessment (NSRA)	Detailed Risk Assessments	Sector Dependency Assessments
Recognition: Industry routinely conducts self-assessments as a part of its business operations **Approach:** Owners and operators conduct self-assessments of their networks voluntarily	**Goal:** Examine Communications Sector architecture to identify national risks **Approach:** Industry and government conduct a high-level qualitative risk analysis of the entire Communications sector	**Goal:** Conduct detailed risk assessments on architecture elements that have been identified as being high risk by NSRA **Approach:** The government works in conjunction with industry to conduct detailed quantitative assessments with respect to mission impact	**Goal:** Assist other sectors in the assessment of communications dependency for high-risk assets **Approach:** Identify high-level critical sector communications dependencies and leverage NCS risk assessment methodologies to identify communications dependencies specific to a facility or function

Source: DHS, "Communications Sector-Specific Plan," 2010, p. 29.

communications integration for the federal government, intelligence community, and law enforcement. The NCCIC shares information among public and private sector partners to build awareness of vulnerabilities, incidents, and mitigations.[9]

Information Technology Sector Profile

The IT sector is central to the nation's security, economy, and public health and safety.[10] A functions-based sector, it comprises not only physical assets but also virtual systems and networks that enable key capabilities and services in both the public and private sectors.[11] The functions of the IT sector are vital to nearly every segment of modern society with businesses, governments, academia, and private citizens increasingly dependent upon its functions. Both virtual and distributed, these functions produce and provide hardware, software, and information technology systems and services, and—in collaboration with the Communications sector—the Internet.[12] IT sector functions are operated by owners and operators and their respective associations. Together, they maintain and reconstitute networks, (i.e. the Internet, local networks, and wide area networks) and their associated services.[13] There are six critical functions of the IT sector:

- Provide IT products and services;
- Provide incident management capabilities;
- Provide domain name resolution services;
- Provide identity management and associated trust support services;
- Provide Internet-based content, information, and communications services; and
- Provide Internet routing, access, and connection services.[14]

Goals and Priorities of the Information Technology Sector

The IT sector has a specific vision and goals to support risk management of its critical functions. These are noted in Table 12.3.

Table 12.3 IT Sector Vision and Goals

IT Sector Vision

To achieve a sustained reduction in the impact of incidents on the sector's critical infrastructure

Goals for the IT Sector

Goal 1: Identify, assess, and manage risks to the IT sector's critical functions and international dependencies.

Goal 2: Improve situational awareness during normal operations, potential or realized threats and disruptions, intentional or unintentional incidents, crippling attacks (cyber or physical) against IT sector infrastructure, technological emergencies and failures, or presidentially declared disasters.

Goal 3: Enhance the capabilities of public and private sector partners to respond to and recover from realized threats and disruptions, intentional or unintentional incidents, crippling attacks (cyber or physical) against IT sector infrastructure, technological emergencies or failures, or presidentially declared disasters and develop mechanisms for reconstitution.

Goal 4: Drive continuous improvement of the IT sector's risk management, situational awareness, and response, recovery, and reconstitution capabilities.

Source: DHS, "Information Sector-Specific Plan," 2010, p. 2.

Information Technology Sector: Assessing Risk

The risk environment of the IT sector consists of varied and complex threats. Aside from man-made unintentional and natural threats, the IT sector faces cyber and physical threats from criminals, hackers, terrorists, and nation-states, all of whom have demonstrated capabilities and intentions to attack IT critical functions.[15] These threats, coupled with a high degree of interdependency, make assessing risk a difficult task. To more fully address this risk environment, the IT sector approaches risk from a top-down and functions-based approach. The IT sector is unique in that it evaluates risk across the sector by focusing on critical functions rather than specific organizations or assets. Specifically, there are two levels of risk management that the IT sector focuses on:

- **The individual enterprise level**: Involves cybersecurity initiatives and practices to maintain the health of information security programs and infrastructures. In the private sector, these enterprise approaches are based on business objectives, such as shareholder values, efficacy, and customer service. On the other hand, the private sector usually bases its enterprise approaches on mission effectiveness or providing a public service.[16]
- **The sector or national level**: At this level, the IT sector manages risk to its six critical functions to promote the assurance and resilience of the IT infrastructure and to protect against cascading consequences based on the sector's interconnectedness and the critical functions' interdependencies.[17]

The IT Sector Baseline Risk Assessment Method

The preferred method of risk in the IT sector is the IT Sector Baseline Risk Assessment (ITSRA) approach. ITSRA serves as a foundation for the sector's national-level risk management activities and was developed in collaboration with public and private sector partners and subject matter experts. The intent of the ITSRA was not to conflict with individual company or organizational risk management activities but, rather, to provide a sector and national-level all-hazards risk profile. Included in this profile is the ability to assess risk from manmade deliberate, manmade unintentional, and natural threats that could affect the ability of the sector to support both the economy and security of our nation.[18] ITSRA leverages existing risk-related definitions, frameworks, and taxonomies from various entities, including public and private IT sector partners and standards and guidance organizations. In doing so, ITSRA reflects current knowledge about risk and adapts it in a way that enables a functions-based risk assessment.[19] The ITSRA methodology consists of five steps:

1. Step 1: Scope Assessment;
2. Step 2: Assess Threats;
3. Step 3: Assess Vulnerabilities;
4. Step 4: Assess Consequences;
5. Step 5: Create Risk Profile.[20]

Inherent within these steps are the threat, vulnerability, and consequence frameworks in the sector's risk assessment methodology. These approaches are described as follows.

Assessing Threats

The IT sector threat analysis approach considers the full spectrum of intentional and unintentional manmade and natural threats. It is interesting to note that traditional threat analysis is not sufficient here. Traditional threat analysis identifies an actor and the actor's intentions, motives, and capabilities to compromise a given target. However, in an IT sector analysis, this approach is not enough because of the unique risk environment where actors are not always easily identifiable or traceable. Furthermore, attacks can move from conception to exploitation in a matter of hours.[21]

This is where the ITSRA becomes a critical piece of the risk assessment process. Because of the difficulty with identifying threat actors, especially those in cyberspace, the IT sector focuses on a threat's capabilities to exploit vulnerabilities before identifying the specific actors.[22] The IT sector defines threat capability as the availability or ease of use of tools or methods that could be used to damage, disrupt, or destroy critical functions.[23] A capability assessment of a natural threat considers those that could have a nationally significant impact. When assessing an intentional manmade threat, the capabilities approach is applied differently using tools or methods that can be easily configured to exploit critical functions. This is challenging, especially, since the IT sector is also vulnerable to unintentional manmade threats due to its high reliance on humans.[24]

Assessing Vulnerabilities

Assessing vulnerabilities in the IT sector considers the people, process, technology, and physical vulnerabilities that, if exploited by a threat, could affect the confidentiality, integrity, or availability of critical functions.[25] More specifically, these vulnerabilities include:

- **People:** includes factors affecting the workforce such as human resource practices (personnel security), demographics (citizenship, qualifications), training and education (quality and quantity of institutions that teach and train the workforce), and market environments (compensation and benefits).
- **Processes:** includes vulnerabilities associated with the sequence and management of operations or activities. For example, manufacturing, logistics, information flow, and efficiency and effectiveness.
- **Technologies:** includes factors like reliance on hardware and software and system dependencies and interdependencies.
- **Physical:** consists of vulnerabilities associated with the physical characteristics of facilities or locations. Geographic locations, weather, and natural vulnerabilities such as earthquakes, floods, and other natural disasters are included.[26]

Assessing Consequences

The IT sector's consequence framework is common to all threat types (deliberate, unintentional, and natural). The IT sector uses HSPD-7 consequence categories and criteria for evaluating events of national significance and the impacts they could have on national security, economic security, public health and safety, and public confidence (if a critical function is disrupted or degraded).[27] Some of the questions considered under the consequence assessment are:

- What is the potential for loss of life, injuries, or adverse impact on public health and safety?
- How many users could be severely affected?
- What are the economic impacts, including asset replacement, business interruption, and remediation costs?
- Will federal, state, or local governments be adversely affected?[28]

Once the threat, vulnerability, and consequence frameworks have been properly assessed, a comprehensive risk profile identifying the risks of concern for the sector can be developed. ITSRA results in a sector-wide risk profile that describes the sector-wide risks as well as the function-specific risks and their associated existing mitigations. Once the profile is in place, it can then be used to develop the risk management strategy for each of the functions and guide mitigation decisions. Table 12.4 shows the IT sector's risks of concern as developed by subject matter experts for the IT sector.

The IT Sector and Policy Initiatives

Since the IT sector comprises a substantial amount of private sector owners and operators, it operates by a slightly different paradigm than the other critical infrastructure and key resource (CIKR) sectors. For example, when it comes to policies of research and development, the government invests a substantial amount of resources in cybersecurity, yet the private sector also makes substantial contributions. Innovation is highly competitive and an ongoing process in the private sector, and as such it fuels new products and capabilities.[29] A collaborative public–private partnership is necessary to achieve as the IT sector advances its R&D agenda. It becomes a balancing act for the IT sector to leverage its private sector R&D investment while sharing information on government R&D initiatives and priorities.

Table 12.4 IT Sector's Risks of Concern Using the Baseline ITSRA

IT Sector Critical Function	*Risks of Concern*
Produce and provide IT products and services	Production or distribution of untrustworthy critical product or service through a successful manmade deliberate attack on a supply chain vulnerability
Provide domain name resolution services	Breakdown of a single interoperable Internet through a manmade attack and resulting failure of governance policy; large-scale manmade denial-of-service attack on the domain name system (DNS) infrastructure
Provide Internet-based content, information, and communications services	Manmade unintentional incident caused by Internet content services results in a significant loss of e-Commerce capabilities.
Provide Internet routing, access, and connection services	Partial or complete loss of routing capabilities through a manmade deliberate attack on the Internet routing infrastructure.
Provide incident management capabilities	Impact detection capabilities because of a lack of data availability resulting from a natural threat.

Source: DHS, "Information Technology Sector-Specific Plan," 2010, p. 26.

As mentioned previously, the IT sector is interdependent and interconnected with other sectors. Therefore, information sharing becomes a critical activity. Currently, the IT sector has means for sharing policy and operational information and information sharing is frequently accomplished on a voluntary basis. Private sector organizations typically are not required or mandated to share information and may even face federal or state government limits on the disclosure of sensitive information.[30] Information sharing can be even further complicated in the public sector by government officials and specific mandates which govern such activities. On the other hand, government agencies might be required to disclose information under the Freedom of Information Actor comparable state disclosure laws.[31]

Because information sharing in the IT sector is so vital to maintaining situational awareness and addressing threats, an enhanced IT sector information-sharing framework has been established. Operational information is exchanged through the Information Technology Information Sharing and Analysis Center (IT-ISAC), which serves as a central repository for security-related information about threats, vulnerabilities, and best practices related to physical and cyber events. The IT-ISAC is the IT sector's focal point for coordinating the sharing and analysis of operational and strategic private sector information between and among members, as well as with other public and private partners.[32]

In addition to IT-ISAC, the IT sector partners with the United States Computer Emergency Readiness Team (U.S.-CERT), which is a partnership between the DHS and the public and private sectors designed to facilitate the protection of cyber infrastructure through cyber analysis, warning, and information sharing; the Multi-State Information Sharing and Analysis Center (MS-ISAC), which serves as a focal point for information sharing with and among state and local governments, and the National Infrastructure Coordinating Center, which is a 24/7 watch operation that maintains operational and situational awareness of the nation's CIKR sectors.[33] In addition to these partnerships, the IT sector's vision for sharing and reporting incidents includes:

- Collect, disseminate, and share information along horizontal and vertical paths of an organization and among organizations;
- Communicate in a regular and predictable manner so that information is passed to all appropriate partners and entities are not inadvertently omitted;
- Establish formal policies or procedures to prescribe the flow of information between and among public and private IT sector partners at all levels;
- Provide intelligence collection and other information requirements to the DHS in accordance with the 2013 NIPP;
- Establish and maintain feedback mechanisms to ensure shared information is useful; and
- Develop formal triggers or incident reporting thresholds to provide consistent guidance to owners and operators for determining when to elevate an event to a higher level or report it to the government.[34]

Financial Services Sector Profile

Recent natural disasters, large-scale power outages, and the increased number of sophisticated cyberattacks have demonstrated the vast array of potential risks facing the Financial Services sector.[35] The DHS describes this vital sector as follows: "The Financial Services Sector includes thousands of depository institutions, providers of investment products, insurance companies, other credit and financing organizations, and the providers of the critical financial utilities and services that support these functions."[36] This complex sector comprises many different types of financial institutions from large global companies to community banks and credit unions. Components at risk in this sector include individual savings accounts, financial derivatives, credit extended to large organizations, or investments made to foreign countries.[37] The Department of Treasury is designated as the SSA for the Financial Services sector.

The Financial Services sector is best understood by examining the services that it provides. These include:

1 Deposit, consumer credit, and payment systems products;
2 Credit and liquidity products;
3 Investment products; and
4 Risk transfer products.[38]

Deposit, Consumer Credit, and Payment Systems Products

These products are provided by depository institutions that offer the bulk of wholesale and retail payment services, such as wire transfers, checking accounts, and credit and debit cards. Large volumes of transfer systems, automated clearinghouses, and automated teller machines (ATM) are the primary point of contact with the sector for many people. Other services such as mortgages, home equity loans, and various lines of credit are also offered through these institutions. Technology has increased the ways in which customers can access these services. Aside from making deposits directly at banking institutions, they can use the Internet or ATM machines. Mobile devices have also become a vehicle through which these services may be accessed, making the average consumer more susceptible to the risks facing this sector.

Credit and Liquidity Products

Individuals seeking a mortgage to purchase a home or a business looking to expand their operations may need a line of credit directly from an institution or indirectly from the liquidity given to a financial service firm. These products allow customers to make purchases that otherwise they might not be able to afford. Laws provide customer protections against fraud involving these products and include federal and state securities laws, banking laws, and laws that are tailored to the specifics of a particular class of lending activity.[39]

Investment Products

Investments include debt securities (bonds and bond mutual funds), equities (stocks, or stock mutual funds), exchange-traded funds, and derivatives (such as options and futures). Individual customers and organizations use securities firms, depository institutions, pension funds, and the like, for investing needs. There is a vast diversity of investment service providers offering these services in this portion of the Financial sector, many competing on a global market.[40]

Risk Transfer Products

There is a market need for services directed toward financial losses due to theft or the destruction of physical or electronic property. Loss events may come from a fire, cybersecurity incident, loss of income due to a death or disability, or other situations. Insurance companies and other organizations offer customers the ability to transfer various types of financial risks under a multitude of circumstances.[41]

The Financial Services Sector Mission and Goals

The Financial Services sector comprises closely interconnected private companies working together with the government to improve security and resilience. The mission and vision of the Financial Services Sector (FSS) combines security and reliance. Table 12.5 illustrates the FSS sector mission vision, and goals.

The FSS sector promotes four primary goals to improve its security and resilience. These provide a framework for identifying and prioritizing collaborative programs and initiatives. These goals consist of information sharing, best practices, incident responses, and recovery, and policy support. Table 12.6 details these goals and shared sector priorities, which help to guide the daily operations of the FSS sector.

Table 12.5 Mission, Vision Statement, and Goals of the Financial Services Sector

Mission Statement for the Financial Services Sector
Continuously enhance security and resilience within the Financial Services sector through a strong community of private companies, government agencies, and international partners that establishes shared awareness of threats and vulnerabilities, continuously enhances baseline security levels, and coordinates rapid response to and recovery from significant incidents as they occur.
Vision Statement for the Financial Services Sector
A secure and stable financial system operating environment that maintains confidence in the integrity of global financial transactions, assets, and data.

Source: DHS, "Financial Services Sector-Specific Plan," 2015, p. 13.

Table 12.6 Financial Services Sector Goals and Priorities

Information Sharing	
Goal 1	*Implement and maintain structured routines for sharing timely and actionable information related to cybersecurity and physical threats and vulnerabilities among firms, across sectors of industry, and between the private sector and government.*
Priority	1 Improve the timeliness, quality, and reach of threat and trend information shared within the sector, across sectors, and between the sector and government. 2 Address interdependencies by expanding information sharing with other sectors of critical infrastructures and international partners. 3 Accelerate the sharing of information through structured information-sharing processes and routines.
Best Practices	
Goal 2	*Improve risk management capabilities and the security posture of firms across the Financial Services sector and the service providers they rely on by encouraging the development and use of common approaches and best practices.*
Priority	1 Promote sector-wide usage of the National Institute of Standards and Technology (NIST) Cybersecurity Framework, including among smaller and medium-sized institutions. 2 Encourage the development and use of best practices for managing third-party risk.
Incident Response and Recovery	
Goal 3	*Collaborate with the homeland security, law enforcement, and intelligence communities; financial regulatory authorities; other sectors of industry; and international partners to respond to and recover from significant incidents.*
Priority	1 Streamline, socialize, and enhance the mechanisms and processes for responding to incidents that require a coordinated response. 2 Routinely exercise government and private sector incident response processes.
Policy Support	
Goal 4	*Discuss policy and regulatory initiatives that advance infrastructure security and resilience priorities through robust coordination between government and industry.*
Priority	1 Identify, prioritize, and support government research and development funding for critical financial infrastructure protection. 2 Identify and support policies that enhance critical financial infrastructure security and resilience, including a more secure and resilient Internet. 3 Encourage close coordination among firms, financial regulators, and executive branch agencies to inform policy development efforts.

Source: DHS, "Financial-Services Sector-Specific Plan," 2015, p. 14.

The Financial Services Sector: Assessing Risk

U.S. banking and financial system institutions face an evolving and dynamic set of risks. As noted above, these include operational, liquidity, credit, legal, and reputational risks.[42] The organizations that support our nation's financial system are also a vital component of the global economy. Furthermore, these organizations are tied together through a network of electronic systems with countless points of entry.[43] As technology has made sector services more accessible, the risk factors have increased. Cybersecurity has emerged as a key concern for this sector, and one in which ongoing risks are consistently being assessed. Incidents in the Financial Services sector, manmade or natural, can have a devastating impact on the U.S. economy. Recent events have illustrated how physical disruptions can have significant outcomes for this sector. The following are some examples:

- Securities markets and several futures exchanges were closed in lower Manhattan after the September 11, 2001, terrorist attacks until communications and other services were either transferred to other sites or restored.
- A series of coordinated distributed denial-of-service attacks against financial institutions began in the summer of 2012. These attacks impacted customers access to banking information but avoided the core systems and processes.
- Superstorm Sandy hit the East Coast on October 29, 2012, causing a two-day closure of major equities exchanges while fixed-income markets were closed for one day.
- Major data breaches to retailers and other networks by cybercriminals have resulted in stolen credit card information and other financial data.[44]

Risk assessment is a long-standing activity in the Financial Services sector. There is both a regulatory and individual institution component. The U.S. Department of Treasury, financial regulators, the U.S. Department of Homeland Security (DHS), law enforcement, and other government partners coordinate with financial institutions to share information about current and emerging threats to the sector.[45] These entities work together by exchanging data, developing threat mitigation information, and meeting to collaborate on specific actions and regulatory processes.

As discussed previously, the sector's cybersecurity and physical risks are critical component that must be assessed. This is accomplished by identifying critical processes and their dependence on information technology and supporting operations for the delivery of financial products and services. As technology improves the sector's services, newer risks may emerge. This is one of the major challenges to the Financial Services sector—staying ahead of the next cyber intrusion or attack. Furthermore, financial institutions and technology service providers are tightly interconnected—an incident impacting one firm has the potential to have cascading impacts on other firms or even sectors.[46]

Intensifying the risk are the interdependencies the Financial Services sector has with other sectors for critical services such as electricity, communications, and transportation. Managing risk requires the cooperation of many sector partners and other levels of government. A good example of this can be found in Section 9 of Executive Order (EO) 13636, *Improving Critical Infrastructure Cybersecurity*, which requires that the DHS identify critical infrastructure against which a cybersecurity incident could result in catastrophic regional or national effects on public health or safety, economic security, or national security. EO 1336 allows owners and operators of identified critical infrastructure whose business and operations are cyber dependent (rely extensively on network and communications technology) to be eligible for expedited processing of clearance through the DHS Private Sector Clearance Program. This program provides access to classified government cybersecurity threat information that may be crucial for these organizations.[47]

Financial Services Sector: Policy Initiatives

Risk in the FSS sector is evolving and as mentioned above, cybersecurity has become increasingly significant. Sector partners continue to make progress in building private–public partnerships in numerous ways including:

- Creating a public–private cybersecurity exercise program to test and improve incident response processes;
- Significantly expanding the sector's cybersecurity information sharing capabilities, including through the rapid growth of the Financial Services Information Sharing and Analysis Center (FSISAC) and the establishment of the Treasury's Financial Sector Cyber Intelligence Group (CIG);
- Establishing a formalized structure of joint working groups to advance specific tasks;
- Formalizing processes for coordinating technical assistance activities; and
- Expanding collaboration with cross-sector international partners.[48]

A key component of the FSS sector's efforts toward security and resilience is the effective public policy framework it embraces. Partnerships with public and private sector organizations allow for the development and implementation of public policy proposals in support of keeping our financial systems secure. One example is the resources government provides for research and development (R&D) of new technologies. This, along with the expertise from private sector owners and operators, has resulted in R&D resource allocation decision-making by government agencies such as the DHS Office of Science and Technology Policy and the National Science Foundation.[49] Finally, these collaborative efforts not only inform public policy decisions but provide a way for continued progress in securing this vital sector.

Summary of Remaining Sectors

We have covered in some detail the sector profiles and risk assessment methods for 9 of the 16 critical sectors. Those discussed were not deemed as more critical than others but rather were selected because they illustrate substantial differences in their approaches to managing risk. Table 12.7 offers a snapshot of the remaining seven sectors, a brief profile, and their stated approach to risk. These sectors are summarized here because they represent methods previously discussed in other sectors. For a deeper explanation of these sectors and their individual approaches to risk management, you can access the sector-specific plans at the official website of the DHS at https://www.dhs.gov/critical-infrastructure-sectors.

Conclusion

The Communications sector, Information Technology sector, and Financial Services sector not only illustrate the complexities of critical infrastructure but show the interconnectedness of risks. Each of these three sectors relies on digital technologies and supports critical services that depend upon each other in some significant way. The remaining seven sectors were presented in a brief overview, which shows the sector profile and individual approach to risk management. We conclude that, while individual sectors offer unique methods and tools of risk assessment, all are continually working through the SSAs to improve upon SSPs. These SSPs direct and combine not only the sectors' efforts to secure and strengthen the resilience of the respective sector but also the sectors' contributions to national infrastructure security and resilience as set forth in Presidential

Table 12.7 Summary Remaining Critical Infrastructure Sectors Approach to Risk Management

Sector	*Profile*	*Approach to Risk Management*
Commercial Facilities Sector	Facilities in this sector operate on the principle of open public access. Includes a diverse range of sites that draw large crowds of people for shopping, business, and entertainment, or lodging. The majority are privately owned. Consists of eight subsectors: • Entertainment and Media • Gaming • Lodging • Outdoor Events • Public Assembly • Real Estate • Retail • Sports Leagues	Uses the basic DHS framework for risk = threat, vulnerability, and consequences. The DHS has provided strategic coordination and field operations support to assist owners and operators with risk assessments, such as the Computer-Based Assessment Tool. Uses information-sharing programs such as the Classified Intelligence Forum. Manages cyber risks through the Real Estate Information Sharing and Analysis Center (RE-ISAC), fusion centers, and the Industrial Control Systems Cyber Emergency Response Team (ICS-CERT).
Critical Manufacturing Sector	The Critical Manufacturing Sector is crucial to the economic prosperity of the U.S. The core of the sector includes four industries: 1 Primary Metal Manufacturing 2 Machinery Manufacturing 3 Electrical Equipment, Appliance, and Component Manufacturing 4 Transportation Equipment Manufacturing	Uses the DHS framework for threat, vulnerability, and consequences. Assesses consequences based on four general categories as set forth in the NIPP: • Public health and safety impact • Economic impact • Psychological impact • Impact on government Historically the Critical Manufacturing Sector has utilized assessment methodologies such as Fault Tree Analyses, Process Hazard Analyses, and others to identify vulnerabilities. Works closely with the DHS Homeland Infrastructure Threat and Risk Analysis Center (HITRAC) to share information.
Dams Sector	There are more than 87,000 dams throughout the U.S., about 60% are privately owned. This sector delivers critical water retention and control services in the U.S. including hydroelectric power generation, municipal and industrial water supplies, agricultural irrigation, sediment and flood control, river navigation for inland bulk shipping, industrial waste management, and recreation. The Dams sector irrigates at least 10% of U.S. cropland, helps protect more than 43% of the population from flooding, and generates about 60% of electricity in the Pacific Northwest.	Uses the basic DHS framework for risk = threat, vulnerability, and consequences at the facility and sector level. USACE, the Bureau of Reclamation, and Tennessee Valley Authority (TVA) conduct comprehensive risk assessments at federally owned dams and levees under their self-regulating authorities. Private and municipal hydroelectric utilities under the jurisdiction of Federal Energy Regulatory Commission (FERC) also complete mandatory vulnerability and security assessments. Aside from regulatory aspects, the DHS has conducted voluntary security assessments of 70% of critical Dams Sector assets. Conducts information sharing through the HSIN-CI Dams Portal and USACE.

(*Continued*)

Table 12.7 (Continued)

Sector	*Profile*	*Approach to Risk Management*
Chemical Sector	The Chemical Sector is an integral part of the U.S. economy. It relies on and supports a wide range of other critical infrastructure sectors. The majority of the facilities are privately owned, thus requiring the DHS to work closely with private sector companies and associations. The sector is divided into five main segments: • Basic chemicals • Specialty chemicals • Agricultural chemicals • Pharmaceuticals • Consumer products.	Uses the basic DHS framework for risk = threat, vulnerability, and consequences. Conducts information sharing through the Homeland Security Information Network-Critical Infrastructure (HSIN-CI). Manages cyber risks through: • Cyber-Dependent Infrastructure Identification • Critical Infrastructure Cyber Community Voluntary Program • Partnership-Developed Cybersecurity Resources (where owners and operators have worked with the DHS and other partners to develop a DVD containing sector-specific tools and resources).
Nuclear Reactors, Materials, and Waste Sector	There are 99 commercial nuclear power plants in the U.S. which provide about 20% of the nation's electrical generated power. The sector includes: • Nuclear power plants • Non-power nuclear reactors used for research, testing, and training. • Manufacturers of nuclear reactors or components. • Radioactive materials used primarily in medical, industrial, and academic settings. • Nuclear fuel cycle facilities. • Decommissioned nuclear power reactors. • Transportation, storage, and disposal of nuclear and radioactive waste.	Uses the basic DHS framework for risk = threat, vulnerability, and consequences at the facility and sector level. Has used Probabilistic Risk Assessments (PRAs) for more than 30 years to analyze risk. PRA is a process for examining how engineered systems, such as nuclear power plants and human interactions with these systems work together to ensure safety and security. Uses DHS Radiological/Nuclear Terrorism Risk Assessment to analyze information from the intelligence, law enforcement, scientific, medical, and public health communities to estimate human casualty and economic consequences of radiological and nuclear terrorism. The sector also employs force-on-force exercises, continuous security enhancements, and integrated planning, training, and exercises to address specific risks.
Defense Industrial Base	This sector is the worldwide industrial complex that enables research and development, as well as design, production, delivery, and maintenance of military weapons systems, subsystems, and components, or parts to meet U.S. military requirements. Included are more than 100,000 Defense Industrial Base companies and their subcontractors—both domestic and foreign entities. This sector provides products and services that are essential to mobilize, deploy, and sustain military operations.	There are currently no regulatory requirements for the Defense Industrial Base (DIB) companies to conduct risk assessments. The Department of Defense (DOD) works with DIB partners to assess those risks to DOD missions resulting from disruption or degradation of DIB critical infrastructure and key resources. Uses CIP-MAA to evaluate plausible threats or hazards. Manages cyber threats by working with HITRAC on information sharing. Also uses the DIB-DOD Collaborative Information Sharing Environment.
Government facilities	Includes a wide variety of government buildings, located in the U.S. and overseas, that are owned or leased by federal, state, local, and tribal governments. These include general-use office buildings and special-use military installations, embassies, courthouses, national laboratories, and structures that may house critical equipment, systems, networks, and functions. Cyber elements are also included.	Uses the basic DHS framework for risk = threat, vulnerability, and consequences. Sector includes two subsectors: • Education Facilities • National Monuments and Icons.

Source: DHS, "Critical Infrastructure Sector Profiles and Sector-Specific Plans," www.dhs.gov

Policy Directive 21. Collectively, these 16 critical infrastructure sectors and their individual approaches to managing risk offer a unique policy perspective. Understanding critical infrastructure protection as an emergent process that is continuously evolving is vital to the goal of a secure and resilient nation.

Key Terms

CSSP Risk Framework
FSISAC
ITSRA
Office of Cybersecurity and Communications (CS&C)
Resilient Infrastructure
Risk Transfer Products
U.S.-CERT

Review Questions

1. Discuss the Communications and Information Technology sectors. What is the function of each, and how are they different?
2. What are the interdependencies that all three sectors presented in this chapter share?
3. Explain the emerging issue of cybersecurity and the impact it has on the Communications, Information Technology, and Financial Services sectors.
4. Discuss the various approaches to managing risk that each sector uses. Is there an overlap in methods? Why or why not?

Notes

1 Kovacs, E. (2015). "Wireless Routers Plagued by Unpatched Flaws," *Security Week*, December 14.
2 DHS, "Communications Sector-Specific Plan," 2010, p. 2.
3 DHS, "Communications Sector-Specific Plan," 2010, p. 2.
4 DHS, "Communications Sector-Specific Plan," 2010, p. 6.
5 Townsend, A.M., and Moss, M.L. (2015). "Telecommunications Infrastructure in Disasters: Preparing Cities for Crisis Communications," Center for Catastrophe and Preparedness Response, New York University.
6 DHS, "Communications Sector-Specific Plan," 2010, p. 27.
7 DHS, "Communications Sector-Specific Plan," 2010, p. 28.
8 DHS, "National Cybersecurity and Communications Integration Center," https://www.dhs.gov/national-cybersecurity-and-communications-integration-center.
9 DHS, "National Cybersecurity and Communications Integration Center," https://www.dhs.gov/national-cybersecurity-and-communications-integration-center.
10 DHS, "Information Technology Sector-Specific Plan," 2010.
11 DHS, "Information Technology Sector-Specific Plan," 2010, p. 1.
12 DHS, "Information Technology Sector-Specific Plan," 2010.
13 DHS, "Information Technology Sector-Specific Plan," 2010.
14 DHS, "Information Technology Sector-Specific Plan," 2010, p. 1.
15 DHS, "Information Technology Sector-Specific Plan," 2010, p. 17.
16 DHS, "Information Technology Sector-Specific Plan," 2010, p. 2.
17 DHS, "Information Technology Sector-Specific Plan," 2010, p. 2.
18 DHS, "Information Technology Sector-Specific Plan," 2010.
19 DHS, "Information Technology Sector-Specific Plan," 2010, p. 22.
20 DHS, "Information Technology Sector-Specific Plan," 2010, p. 18.
21 DHS, "Information Technology Sector-Specific Plan," 2010, p. 22.
22 DHS, "Information Technology Sector-Specific Plan," 2010, p. 22.
23 DHS, "Information Technology Sector-Specific Plan," 2010, p. 22.

24 DHS, "Information Technology Sector-Specific Plan," 2010, p. 22.
25 DHS, "Information Technology Sector-Specific Plan," 2010, p. 23.
26 DHS, "Information Technology Sector-Specific Plan," 2010, p. 24.
27 DHS, "Information Technology Sector-Specific Plan," 2010, p. 25.
28 DHS, "Information Technology Sector-Specific Plan," 2010, p. 25.
29 DHS, "Information Technology Sector-Specific Plan," 2010, p. 41.
30 DHS, "Information Technology Sector-Specific Plan," 2010, p. 56.
31 DHS, "Information Technology Sector-Specific Plan," 2010, p. 56.
32 DHS, "Information Technology Sector-Specific Plan," 2010, pp. 55–58.
33 DHS, "Information Technology Sector-Specific Plan," 2010, p. 58.
34 DHS, "Information Technology Sector-Specific Plan," 2010, pp. 58–59.
35 DHS, "Financial Services Sector-Specific Plan," 2015.
36 DHS, "Financial Services Sector-Specific Plan," 2015.
37 DHS, "Financial Services Sector-Specific Plan," 2015.
38 DHS, "Financial Services Sector-Specific Plan," 2015.
39 DHS, "Financial Services Sector-Specific Plan," 2015.
40 DHS, "Financial Services Sector-Specific Plan," 2015.
41 DHS, "Financial Services Sector-Specific Plan," 2015.
42 DHS, "Financial Services Sector-Specific Plan," 2015.
43 DHS, "Financial Services Sector-Specific Plan," 2015.
44 DHS, "Financial Services Sector-Specific Plan," 2015.
45 DHS, "Financial Services Sector-Specific Plan," 2015.
46 DHS, "Financial Services Sector-Specific Plan," 2015.
47 DHS, "Financial Services Sector-Specific Plan," 2015.
48 DHS, "Financial Services Sector-Specific Plan," 2015.
49 DHS, "Financial Services Sector-Specific Plan," 2015.

13 The Future of Critical Infrastructure Protection

Risk, Resilience, and Policy

Chapter Outline

Introduction

Studying critical infrastructures from a policy perspective provides a comprehensive examination of the executive orders (EOs), national strategies, presidential policy directives, and methods that have evolved in the homeland security enterprise over the past 22 years. Sector-specific agencies (SSAs), along with the individual approaches each sector takes toward managing risk, are also important to consider as the protection of these 16 critical infrastructures continues to evolve. As we look to the future of protecting our critical infrastructures, several key issues emerge. This chapter examines those issues that continue to challenge and shape our responses to critical infrastructure protection, risk management, and resilience efforts.

Increased Nexus between Cyber and Physical Security

A March 2016 cyberattack on MedStar Health, one of the largest healthcare systems in the Baltimore, Washington D.C. area, shows how rapidly the digitization of the Healthcare and Public Health sector is creating new threats. Pressure to put patient healthcare records, test results, and other medical systems online has made the healthcare industry an easy target, and its security systems tend to be less mature than those in other sectors. As discussed in Chapter 11, this reliance on technology for medical purposes has opened a new set of vulnerabilities where devices can malfunction, treatments can be interrupted, and breaches of patient information can occur. The MedStar data breach began with a virus that infiltrated its computer systems and forced the shutdown of its entire network. As a result, email, patient records, and other medical systems were shut down. Patients were turned away, surgeries were postponed, and paper records of visits became the norm. Some employees reported seeing pop-up messages indicating the attack was "ransomware"—a

DOI: 10.4324/9781003434887-13

kind of software that can lock people out of systems until they make a Bitcoin payment.[1] Clearly the push for digitizing the healthcare industry is rapidly making it a target for hackers.

An emerging issue for critical infrastructure protection is the increased nexus between cyber and physical security. Executive Order 13636 Improving Critical Infrastructure Cybersecurity and Presidential Policy Directive 21—Critical Infrastructure Security and Resilience are two of the first official acknowledgments of the complex connection between physical and cyber security.[2] These federal policies, along with public–private plans, establish the roles and responsibilities of federal agencies working with the private sector and other entities to enhance the cyber and physical security of public and private critical infrastructures.[3] With increased technologies, there is a new linkage between physical and cyber infrastructures.

We now rely on cyber systems to run just about everything from mass transit to pipelines, electricity, and, as the MedStar example illustrates, hospital networks.[4] This connection now means that both cyber and physical security measures are needed to protect critical infrastructures against potential threats. Physical security measures prevent unauthorized access to servers and other technologies that carry sensitive information. Cyber security measures can thwart an attack, which may have physical consequences. For example, an attack on a water treatment control system could have damaging effects on human lives as well as the environment and the economy.[5] For these reasons, the federal government and the Department of Homeland Security (DHS) have taken an integrated approach to critical infrastructure protection by including the evolving risk and increased role of cybersecurity in securing physical assets.

In a November 2015 Government Accountability Office (GAO) Report, it was concluded that SSAs need to better measure cybersecurity progress.[6] In the study, the GAO's objectives were to determine the extent to which SSAs have (1) identified the significance of cyber risks to their respective sectors' networks and industrial control systems, (2) taken actions to mitigate cyber risks within their respective sectors, (3) collaborated across sectors to improve cybersecurity, and (4) established performance metrics to monitor improvements in their respective sectors.[7] The GAO analyzed policy, plans, and other documentation and interviewed public and private sector officials for 8 of the 9 SSAs who have responsibility for 15 of the 16 sectors. It was found that the Departments of Defense, Energy, and Health and Human Services established these performance metrics for their three sectors but that the SSAs for the other 12 sectors had not yet developed them and were unable to report on the effectiveness of cyber risk mitigation activities in their sectors. One of the reasons reported was the SSA's reliance on private sector partners to voluntarily share information needed to measure these efforts. In response, the GAO recommended that collaboration with sector partners needs to be more prevalent in certain SSAs and that performance metrics need to be established.[8]

Interdependence

As we have learned, critical infrastructures do not operate alone. They interact with other sectors and help them function by providing essential resources and services. While each has unique functions, they are also dependent upon and interdependent with other critical infrastructures. Dependency may be defined as a relationship between two infrastructures in a single direction or one infrastructure influencing the state of another. For example, the reliance on electric power and fuels by a multitude of industries means that all critical infrastructure sectors are dependent upon the Energy sector. The Food and Agriculture sector is dependent upon the Water and Wastewater Systems sector for clean irrigation and processed water, the Transportation System for movement of products and livestock, and the Chemical sector for fertilizers and pesticides used in the production of crops. Interdependency is bidirectional and multidirectional with two or more infrastructures influencing each other. The Water and Wastewater Systems sector is a good example of a critical infrastructure that contains numerous interdependencies with other critical sectors. A disruption in water service, posed by a natural hazard, terrorism, or accident, threatens other sectors such as emergency services, healthcare, and transportation. Furthermore, interdependencies are the centerpiece of the National Infrastructure Protection Plan (NIPP).[9]

Consideration of dependencies and interdependencies of critical infrastructures is vital in assessing risk and resiliency. A case study of the interconnectedness of risks posed by Hurricane Sandy for New York shows how a single disaster can cause enormous economic damage because of the interdependent infrastructure systems. Hurricane Sandy had a significant impact on the energy sector with the most damage found in power outages. More than two million people were impacted by the loss of power. In the energy sector, regional refineries were shut down and more than eight million customers lost power in 21 affected states. The water and wastewater sector was affected by crippled treatment plants and raw sewage that flowed into the waterways of New York and New Jersey, continuing a month after the storm. The Healthcare sector took a significant hit with five acute care hospitals and one psychiatric hospital closed. As a result, nearly 2,000 patients had to be evacuated. The transportation sector was also impacted with tunnels, subways, and railroad tunnels shut down due to flooding and damage to equipment. It has been estimated that approximately 8.6 million daily public transit riders, 4.2 million drivers, and 1 million airport passengers were impacted.[10]

One of the challenges for critical infrastructure protection and risk management is the potential for cascading failures. As illustrated by Hurricane Sandy, a wider network of risks must be considered. From a policy perspective, that means that risk assessment and plans for resilience must consider the interconnectedness of all critical infrastructures and their various functions. In addition, there have been efforts to develop models that accurately simulate critical infrastructure behavior and identify interdependencies and vulnerabilities.[11] Over the years, several taxonomies have been presented to describe the various types of dependencies between sectors. Rinaldi, Peerenboom, and Kelly (2001) describe four general categories:

1. Physical—a physical reliance on material flow from one infrastructure to another;
2. Cyber—a reliance on information transfer between infrastructures;
3. Geographic—a local environmental event affects components across multiple infrastructures due to physical proximity;
4. Logical—a dependency that exists between infrastructures that does not fall into one of the above categories.[12]

The complexity of these interdependencies calls for the use of modeling and simulation capabilities. Modeling and analysis of interdependencies between critical infrastructures may be used to understand infrastructure systems and may be used to support vulnerability and risk assessments, training exercises, and performance measurement. The challenges are similar to any modeling domain: data accessibility, model development, and model validation.[13] The National Infrastructure Simulation and Analysis Center is responsible for developing these modeling capabilities for our nation's infrastructures. Expert analysts study the details of all 16 critical infrastructures, from asset level to systems level, the interactions between sectors, and how various sectors respond to natural disasters, cyber threats, or terrorist attacks.[14]

Risks Associated with Climate Change

Extreme weather events continue to cause undue damage throughout the U.S. In 2022, 18 separate weather and climate disasters totaled at least $1 billion. That number puts 2022 into a three-way tie with 2017 and 2011 for the third-highest number of billion-dollar disasters in a calendar year, all behind 22 events in 2020 and 20 events in 2021.[15]

Bizarre weather patterns are becoming the norm. In December 2021, a four-day outbreak of tornados impacted five states. One of the hardest hit areas was in western Kentucky. The city of Mayfield, Kentucky, suffered extreme damages with over 3,778 residences and 183 commercial properties impacted, including the county courthouse and the Mayfield Consumer Products candle factory (NWS). In recent years, the impact of climate change on severe weather events has received a lot of attention. Although no specific weather event has been directly credited to climate change, there is awareness in the scientific community that it can worsen the impact, frequency, and intensity of such events.[16]

Typically, these events result in significant damage to infrastructures we rely upon every day such as water, energy, transportation, communications, and emergency services.[17] Researchers maintain that to better assess risk from weather events it is important to determine how future climates might impact our critical infrastructure systems. Presently, there are areas in the U.S. that are at risk from the impact of climate change, such as in the Gulf Coast where several of the largest seaports in the U.S. are located. The combination of relative sea level rise and more intense hurricanes and tropical storms in this area could lead to significant disruptions and damage. Another example of climate risk locations can be found in the Tri-State area of New York, New Jersey, and Connecticut where many transportation infrastructure facilities are located, all within the range of current and projected 50-year coastal storm surges.[18]

As discussed above, the interdependent nature of our critical infrastructure also creates new vulnerabilities and opportunities for disruption across supply chains. Extreme weather events associated with climate change can also impact these interconnected systems. The Energy sector is particularly vulnerable. For example, in September 2011 high temperatures and a high demand for electricity tripped a transformer in Yuma, Arizona. This triggered a chain of events that shut down the San Onofre nuclear power plant and resulted in a large-scale power outage across the entire San Diego distribution system. Approximately 2.7 million customers were without power with outages lasting as long as 12 hours.[19] In another example, three Browns Ferry Reactors in Alabama automatically shut down when strong storms knocked out off-site power. Emergency diesel generators had to be used for five days.

The increased number of billion-dollar natural disasters has strained the federal government both in dollars and in resources. In response, the DHS has developed a Climate Change Adaptation Roadmap and Climate Action Plan, which aligns with the President's Climate Action Plan and Executive Order 13653, *Preparing the United States for the Impacts of Climate Change.*[20] This plan directs federal agencies to:

- **Modernize federal programs to support climate-resilient investments**—the EO directs agencies to review their policies and programs to find better ways to create stronger standards for building infrastructure.
- **Manage lands and waters for climate preparedness and resilience**—the EO directs agencies to identify changes that must be made to land, and water-related policies, programs, and regulations to strengthen the climate resilience of our watersheds, natural resources, and ecosystems.
- **Provide information, data, and tools for climate change preparedness and resilience**—the EO directs agencies to work together and with information users to develop new climate preparedness tools.
- **Plan for climate change-related risk**—the EO directs federal agencies to develop and implement strategies to evaluate and address their most significant climate change-related risks.[21]

The challenges that climate change presents for critical infrastructure protection are not without controversy. On the one hand, creating stronger infrastructure to withstand the impacts of climate change and extreme weather events is an important undertaking. So much so that the DHS and the White House have established a plan for this risk environment. Additionally, the DHS National Protection and Programs Directorate works to manage risk to critical infrastructure by supporting climate preparedness, adaptation, and resilience efforts locally.[22] However, critics argue that money spent on improving structures, resilience technologies, and other target-hardening efforts is unwise. These investments are costly, and one extreme weather event could easily destroy an entire critical sector.

Mitigation efforts are costly. For example, the St. John's Regional Medical Center in Joplin, Missouri, was wiped out by a 2011 tornado. It has since been rebuilt with newer technologies that include windows that can resist up to 250 miles per hour at a cost of $170 per square foot—$70 more than standard windows.[23] In Edna, Texas, a coastal city at risk for hurricanes, a $2.5 million dome shelter was built for the town's 5,500 residents. The shelter doubles as a gymnasium and was built to withstand winds up to 300 miles per hour. The Federal Emergency Management Agency (FEMA) provided 75% of the cost with plans to invest $683 million in similar shelters in 18 additional states.[24]

At the local level, many city leaders are developing innovative ideas to address extreme weather due to climate change. There is a recognition that increasing a city's resilience to climate change keeps people and businesses safe—a key for economic growth and stability. Some of these initiatives include upgraded public transportation systems, providing cleaner and more reliable energy, and improving air quality.[25] In Cleveland, Ohio, outbreaks of winter cold are becoming more severe as are more heat waves. Located in the Great Lakes Region, the city of Cleveland, Ohio, is not subject to the most dramatic evidence of climate change that the coastal areas are experiencing. While there is no concern for rising sea levels or hurricanes, there is plenty of evidence that climate change is impacting severe weather in this area. More severe weather equates to higher costs. It is estimated that between 2010 and 2015, 36% of U.S. extreme weather events that caused more than $1 billion in damage occurred in the Great Lakes Region.[26] The Center for American Progress, a Washington-based think tank, recently published a report entitled, "Resilient Midwestern Cities: Improving Equity in a Changing Climate."[27] In it, the city of Cleveland was one of five cities praised for their climate resilience efforts. These include energy efficiency and stormwater management programs that target some of the city's most vulnerable neighborhoods. Poverty in Cleveland is rivaled only by Detroit, Michigan, with more than one-third of its total residents and half of its children living in poverty.[28] As we saw with Hurricane Katrina, poor and low-income residents are impacted the most by a severe weather event because of existing economic and social hardships.

Addressing the issue of critical infrastructure protection and the stresses of climate change is an evolving issue. Efforts are being focused on programs and initiatives at both the local and regional levels. While climate change may be controversial, the intensity of storms and increase in weather events is clearly on the rise. Critical infrastructures are vulnerable to these weather and climate changes, and our policies should reflect a stronger effort to mitigate them.

An Aging and Outdated Infrastructure

A serious issue facing the future of critical infrastructure protection in the U.S. is the poor preservation and maintenance of our infrastructure systems. Old and deteriorating bridges, highways, transportation systems, and the power grid all pose significant risks to the economy and our ability to recover from natural or manmade disasters, the effects of climate change, and terrorist events.

Of particular concern are the Energy and Transportation Systems sectors. In April 2015, the first installment of the Energy Department's Quadrennial Energy Review was published offering a grim picture of the U.S. electrical grid, power transmission lines, natural gas and oil, pipelines, ports, railways, and other critical pieces of national infrastructure.[29]

The report concludes that the U.S. energy landscape is changing and that policy discussions have shifted from concerns about rising oil imports and high gasoline prices to debates about how much and what kinds of U.S. energy should be exported.[30] Additional concerns include the transportation of large amounts of crude oil by rail and how to meet the demands for future energy supplies. The report found that threats to the energy sector are growing, especially with respect to the electrical grid that is badly in need of modernization. The risk of terrorist attacks and severe weather events caused by climate change are on the rise while the reliability and safety of the grid have remained stagnant or, in some cases gone backward.[31]

Similarly, the oil and gas infrastructures have not kept pace with changes in the size and geography of oil and gas production. "Our ports, waterways, and rail systems are congested," the report concludes,

> with the growing demands for handling energy commodities increasingly in competition with transport needs for food and other non-energy freight. Although improvements are being made, much of the relevant infrastructure—pipelines, rail systems, ports, and waterways alike—is long overdue for repairs and modernization.[32]

The report describes our pipeline system as one compelling example. Approximately 50% of the nation's gas transmission and gathering pipelines were constructed in the 1950s and 1960s. These were built out of the interstate pipeline network in response to the booming post-World War II economy. The cost to modernize this network has been estimated between $2.6 billion and $3.5 billion per year between 2015 and 2030, depending upon the demand for natural gas. Additionally, the total cost of replacing cast iron and bare steel pipes in gas distribution systems has been estimated at $270 billion.[33]

Aligning with the previous discussion on climate change, the report found the most important environmental factor affecting the Energy sector now and going forward is global climate change. Rising sea levels, thawing permafrost, and frequency of extreme storms are already impacting our aging infrastructures.[34] In response to these issues, the Obama administration announced a Climate Action Plan in June 2013. This plan included the following three pillars:

1 Reducing U.S. emissions of GHGs;
2 Increasing domestic preparedness for and resilience against the changes in climate that no longer can be avoided;
3 Engaging internationally to encourage and assist other countries in taking similar steps.[35]

Specific infrastructure projects that must be addressed in the Energy sector are also outlined in the report; some require Congressional appropriations. These include a $2.5 billion Energy Department initiative to improve natural gas distribution; a $3.5 billion plan to modernize the U.S. electric grid; and at least $1.5 billion to improve and extend the life of the crucial Strategic Petroleum Reserve.[36]

The U.S. Transportation System sector is also vulnerable to an aging and deteriorating infrastructure. For instance, more than 63,000 highway bridges have been classified as structurally deficient due to age, erosion, and other structural problems. As conditions on these bridges worsen, restrictions are put in place, or they are taken out of service. Extreme weather and natural disasters also contribute to deterioration and risk. The impact on other critical sectors is undeniable. Most sectors are dependent upon highway bridges to conduct daily operations such as the delivery of products and goods.[37] Furthermore, the failure of a bridge can directly impact the Energy and Communications sectors as power lines, fiber-optic cables, and other utility lines can be collocated underneath a bridge.[38] The implications of an aging critical infrastructure are particularly important for the Emergency Services sector. The ability to quickly and efficiently respond to a disaster can become less reliable if a bridge collapses or the ability to communicate is compromised by a failure in the power grid. Some of these aging infrastructures can also become a hazard just like the I-35 bridge collapse in Minnesota in 2007.[39]

The consequences of a decaying urban infrastructure can be devastating, as found with the Flint, Michigan Water Crisis, which began in April 2014. Flint, Michigan, is a city 70 miles north of Detroit with over 40% of its residents living below the poverty level. The State of Michigan took over the finances of the city in 2011 after an audit projected a $25 million deficit.[40] A shortfall in the funding for the water supply prompted a change from the treated Detroit water system to the Flint River, a system in which officials had failed to apply corrosion inhibitors. Beginning in April 2014, the drinking water in Flint had a series of problems that culminated with lead contamination. The corrosive Flint River water caused lead from aging pipes to leak into the water supply, resulting in extremely high levels of lead in the drinking water.

It has been estimated that between 6,000 and 12,000 children in the city of Flint have been exposed to toxic levels and as a result may suffer from serious health issues. Research shows that the percentage of children under 5 with elevated blood lead levels has more than doubled since the change in the water supply.[41] In addition, cases of Legionnaire's disease spiked in the county and numerous residents were complaining of hair loss, rashes, as well as vision and memory problems. On January 5, 2016, Governor Synder declared a state of emergency for Genesee County, home to Flint. This was followed by a state of emergency declaration by President Barack Obama. National Guard Troops and the Red Cross responded to Flint, distributing bottled water and water filters to homes. Criminal charges were filed against two states and one city employee in April 2016. The accusations include misleading federal regulatory officials, manipulating water sampling, and tampering with reports.[42]

The Flint Water Crisis not only sheds light on poor decision-making and critical infrastructure problems in Michigan but throws a harsh spotlight on urban infrastructure issues across the country.[43] The infrastructure crisis has been long unfolding inside city halls. Mayors are often the last to receive money for infrastructure projects such as water quality, pipe stability, buses, transit systems, and crumbling bridges. Local governments are struggling to find dollars from both their governors and their states to support these projects that seem to take a backseat to other policy priorities.

In November 2021, the Bipartisan Infrastructure Law was passed under the Biden administration. This law marked a new era in correcting and addressing some of the major issues in our aging and outdated infrastructure. The law allocates funding for over 350 infrastructure projects and programs. One of the critical infrastructures slated for these monies includes the transportation sector, with $40 billion dedicated to improving and building bridges, and over $8 billion slated for passenger, freight, and highway projects.

Additionally, the information technology and communications sectors will both benefit from the funding of broadband Internet. Through the Bipartisan Infrastructure Law, approximately $65 billion has been invested to ensure all citizens have access to affordable and high-speed Internet (a problem that came to light during the COVID-19 pandemic). Other sectors scheduled for project updates include the water and wastewater sector and energy sectors.[44]

The Internet of Things

The Internet of Things (IoT) has emerged in recent years as a major area of concern for cybersecurity and critical infrastructure protection. Non-computer devices ("things") that possess Internet capabilities are becoming increasingly common. Everything from baby monitors to doorbells, thermostats, fitness trackers, and medical devices are now able to be connected to the Internet. These connected devices are responsible for the rapid growth of vulnerabilities, threats, and attacks on the nation. As more things become connected, the more ways there are to breach them. Typically, weak passwords are the most common ways that hackers access these types of devices. Furthermore, the convergence of IoT with operational technologies (OTs) creates a wide-open area for nation state actors and cyber criminals. OT attacks often serve as gateways into a critical infrastructure facility's IT network.[45]

An interesting example of exploiting IoT vulnerabilities occurred in 2015, when two independent hackers-turned-researchers conducted a study into the ability to hack into and gain remote control over a 2014 Jeep Cherokee.[46] The hackers exploited a security vulnerability in the vehicle's Uconnect system—a touchscreen entertainment system with Wi-Fi, navigation, apps, and cellular communication capabilities—after using the Sprint cellular network to determine the vehicle's IP address. Control over the Uconnect system allowed the hackers to fiddle with the radio, including changing stations and volume, and mess with the heating, ventilation, and air conditioning (HVAC) system to blast the AC or heat. In both of these cases, the driver could not regain control over the radio and HVAC system.

The Jeep's Uconnect system was just the tip of the iceberg. Many modern-day vehicles now come equipped with driver-assist capabilities, such as adaptive cruise control, lane departure, park assist, and forward-collision warning. Through the Uconnect system, the hackers could reach a chip with the ability to communicate with the vehicle's central computer. They were then granted full control over the car's braking, steering, and windshield wipers, and could even completely shut off the brakes and car itself. After their findings, the researchers shared their information with Chrysler, the parent company of Jeep, who then patched the vulnerabilities. This study is just one harrowing example of the complexities in today's growing threats regarding IoTs.

How to Manage Risk Across a Disparate Set of Threats

As a nation, we continue to be unprepared for the unthinkable. Since the first edition of this textbook, the U.S. has continued to experience unprecedented natural disasters, a global pandemic, and a divided political environment, all amid an increasingly unsettled world stage. COVID-19 is a prime example of how challenging it is to manage risk to our critical infrastructure. The pandemic impacted all 16 critical sectors in many different ways. From supply chain shortages in the transportation sector to a lack of personal protective equipment (PPE) in our supply chains, and the strain on the healthcare system, each sector's vulnerabilities were exposed. In March 2020, Cybersecurity and Infrastructure Security Agency (CISA) began to monitor ongoing risks from COVID-19 to the 55 National Critical Functions (NCFs).[47] While there are both challenges and opportunities in using the NCF risk assessment framework, overall, it provides a step forward in managing risks across an increasingly complex threat environment.

Information Sharing

The NIPP 2013 emphasizes the importance of integrating information sharing as an essential component of the risk management framework. Specifically, the NIPP directs the sharing of actionable and relevant information across the critical infrastructure community to build awareness and risk-informed decision-making.[48] The NIPP also directs that this information sharing be of a voluntary nature. "Voluntary collaboration between private sector owners and operators (including their partner associations, vendors, and others) and their government counterparts has been and will remain the primary mechanism for advancing collective action toward national critical infrastructure security and resilience."[49]

The key word here is voluntary. While most would agree that information sharing is crucial for Homeland Security and critical infrastructure protection, the execution of shared information is not seamless. Getting sector organizations out of their functional, jurisdictional, and competitive silos continues to be a major challenge. For example, a recent GAO report found that a lack of information sharing from the private sector was responsible for some sector-specific agencies' inability to implement proper cybersecurity metrics.[50] These findings fueled a debate among some lawmakers that critical infrastructure industries should be required to report more cybersecurity data to the government. While an attempt to include such a mandate in the recently passed Cybersecurity Information Sharing Act of 2015 (CISA) failed, the controversy surrounding information sharing continues.

On December 18, 2015, President Barack Obama signed into law the CISA of 2015, which encourages the private sector and the federal government to share information on cyber threats. Essentially, it allows private businesses to share cyber information with the federal government (specifically the Department of Defense) without the risk of being sued. Critics have argued that the provisions are too broad and that it opens the door to other, unrelated information being shared. This is but one example where neither the public nor private sector appears to be satisfied with the information it receives from the other. This "expectations gap" in information sharing comes from a mutually acknowledged reluctance to exchange sensitive information.[51] It has been argued that multiple variables conspire to hinder effective cross-sector information sharing including an unsettled organizational landscape, questions of trust, and the fact that information sharing rarely provides an immediate payoff for businesses.[52]

The benefits of sharing critical infrastructure information cannot be overstated, and the flow of it needs to be both horizontal and vertical. Various levels of government need to communicate with each other up and down while jurisdictions need to share information among all stakeholders—public and private. This two-way approach to information sharing is essential for critical infrastructure protection and risk management. Problems occur when information is held back or filtered before it is passed on. Because information is not mandated but rather encouraged, some critical sectors will have better success at it than others. Since 80% of the nation's critical infrastructure is owned by the private sector, collaboration between government and private business owners becomes vital. The DHS has established several operations and tools to support information sharing within and among critical infrastructure sectors. These include

- Homeland Security Information Network—Critical Infrastructure (HSIN-CI);
- Infrastructure Protection Gateway (IP Gateway);
- National Infrastructure Coordinating Center (NCIC);
- Office of Cyber & Infrastructure Analysis (OCIA);
- Protected Critical Infrastructure Information (PCII) program;
- Protective Security Advisors (PSAs);
- TRIP*wire* (Technical Resource for Incident Prevention);
- DHS Daily Open Source Infrastructure Report.[53]

The DHS also partners with other organizations to provide additional information sharing support to its security partners. These include the following:

- SSAs;
- Information Sharing Environment;
- National Explosives Task Force;
- Fusion Centers.[54]

Public–Private Partnerships

There is an underlying tension between public and private sector views when it comes to critical infrastructure protection. With over 80% of our nation's critical infrastructure under private sector control, public–private sector collaboration is essential. While the DHS and its' Office of Infrastructure Protection provide structure for these coordination efforts, obstacles to these partnerships still exist. In *Realizing the promise of public-private partnerships in US critical infrastructure protection*, Austen D. Givens and Nathan E. Busch (2013) argue that

> Challenges result from imprecise contracts that create a mismatch in expectations, a lack of centralized mechanisms for coordinating integrated actions, a tendency on the part of actors in a partnership to act out of self-interest, and the prospect of public and private sector actors relying on the other to bear the costs of the partnership.[55]

An example of unmet expectations and cost overruns can be found in the DHS Secure Border Initiative known as "Virtual Fence."[56] Virtual Fence was a program designed to monitor the U.S.–Mexico border through a series of surveillance radars, cameras, and sensors. The initiative failed because the harsh terrain caused malfunctions and different technologies that made up Virtual Fence were difficult to integrate. Additionally, the program became too expensive. The project, carried out by the Boeing Corporation under a contract initially signed by President George W. Bush in 2005, was plagued by delays and cost overruns. Originally, the Virtual Fence Project was estimated to cost $7 billion for the fence to cover the entire 2,000-mile U.S. southern border. However, a pilot test discovered the actual cost to be $1 billion to cover only 53 miles of the border—just 2% of the total project.[57] The DHS cancelled the Virtual Fence Project in January 2011, noting that it "did not meet the current standards for viability and cost effectiveness."[58] The failure of the Virtual Fence Project illustrates the complexities of public–private partnerships and how unmet expectations, poor execution, and out-of-control costs can ruin an initiative.[59]

When it comes to costs, businesses need incentives to spend on their own protection measures. There is a difference between the private sector spending money on protecting their own operations versus developing critical infrastructure technologies. The DHS has instituted a number of initiatives to encourage businesses to participate in critical infrastructure protection measures such as adopting the 2014 Cybersecurity Framework. These incentives include:

- Cybersecurity Insurance;
- Grants;
- Process Preference;
- Liability Limitation;
- Streamline Regulations;
- Public Recognition;
- Rate Recovery for Price Regulated Industries;
- Cybersecurity Research.[60]

The private sector also has a built-in incentive to collaborate on critical infrastructure protection measures. The continuity of operations in the aftermath of a crisis is critical and not doing so could force a business into bankruptcy, as evident in the wake of Hurricane Katrina.[61] Enhancing the effectiveness of public–private partnerships can be a challenge, and it will take considerable efforts by both sides to improve these collaborations.

Conclusion

Critical infrastructure protection, risk, and resilience are dynamic and ever-changing components of the Homeland Security Enterprise (HSE). The threats described throughout this text are daunting, and the challenges of addressing these threats require sophisticated tools. It is clear that the DHS, CISA, FEMA, and other partners throughout the HSE must continue to support these efforts in an environment of increasing cyber threats and diminished resources.

Current challenges include the increased nexus between cyber and physical security, interdependencies between and across sectors, the effects of climate change, and an aging infrastructure. Newer threats such as the IoT and how to manage risk through information sharing and public–private partnerships only add to these challenges. Addressing these challenges and the priorities and concerns of all critical infrastructure stakeholders is an essential step toward developing Homeland Security strategic plans that reflect the whole community and provide for a safe and secure nation.

Key Terms

Cybersecurity
Information sharing
Interdependencies
Physical Security
Public–Private Partnerships

Review Questions

1 Describe the nexus between cyber and physical security. What is the importance of it for the current climate of critical infrastructure protection?
2 Why is information sharing an issue between public and private sector partners? What suggestions do you have to improve upon quality and frequency of information being shared?
3 How does climate change impact critical infrastructure protection? Cite two examples of how it is changing the way we perceive extreme weather events.
4 Discuss the key challenges facing critical infrastructure protection in the future. How do these challenges impact the concepts of risk and resilience? What suggestions do you have to address these challenges?
5 Define the Internet of Things (IoT), and think of an example that you or those in your community use every day. What kinds of threats pose a risk to this IoT? How would it impact you or your community?

Notes

1 Johnson, C., and Zapotosky, M. (2016). "Under Pressure to Digitize Everything, Hospitals Are Hackers' Biggest New Target." *The Washington Post*, April 1.
2 Berger, V. (2013). "Converging Physical and Cyber Security." *FCW Magazines*, December 19.
3 GAO Report to the Committee on Homeland Security, House of Representatives, "Critical Infrastructure Protection Sector-Specific Agencies Need to Better Measure Cybersecurity Progress," December 2015.
4 DHS, "Written Testimony of National Protection and Programs Directorate Under Secretary Rand Beers for a House Committee on Appropriations, Subcommittee on Homeland Security oversight hearing titled, Cybersecurity and Critical Infrastructure," March 20, 2013. www.dhs.gov

5 DHS, "Written Testimony of National Protection and Programs Directorate Under Secretary Rand Beers for a House Committee on Appropriations, Subcommittee on Homeland Security Oversight Hearing Titled, Cybersecurity and Critical Infrastructure," March 20, 2013. www.dhs.gov
6 GAO Report to the Committee on Homeland Security, House of Representatives, "Critical Infrastructure Protection Sector-Specific Agencies Need to Better Measure Cybersecurity Progress," December 2015.
7 GAO Report to the Committee on Homeland Security, House of Representatives, "Critical Infrastructure Protection Sector-Specific Agencies Need to Better Measure Cybersecurity Progress," December 2015.
8 GAO Report to the Committee on Homeland Security, House of Representatives, "Critical Infrastructure Protection Sector-Specific Agencies Need to Better Measure Cybersecurity Progress," December 2015.
9 Zimmerman, R. (2009). "Understanding the Implications of Critical Infrastructure Interdependencies for Water." *Published Articles & Papers. Paper 7.*
10 Haragucki, M., and Soojun, K. (2015). "Critical Infrastructure Systems: A Case Study of the Interconnectedness of Risk Posed by Hurricane Sandy for New York City." Input Paper Prepared for the Global Assessment Report on Disaster Reduction.
11 Pederson, P., Dudenhoffer, D., Hartley, S., and Permann, M. (2006). "Critical Infrastructure Interdependency Modeling: A Survey of US and International Research." Technical Support Working Group, Idaho National Laboratory, Washington, DC.
12 Rinaldi, S., Peerenboom, J., and Kelly, T. (2001). "Identifying, Understanding, and Analyzing Critical Infrastructure Interdependencies." *IEEE Control Systems Magazine*, December, pp. 11–25.
13 "Modeling and Simulation of Critical Infrastructure Systems for Homeland Security Applications." DRAFT for discussion at DHS/NIST Workshop on Homeland Security Modeling & Simulation, June 14–15, 2011.
14 National Infrastructure Simulation and Analysis Center. http://www.sandia.gov/nisac/overview/
15 https://www.climate.gov/news-features/blogs/beyond-data/2022-us-billion-dollar-weather-and-climate-disasters-historical#:~:text=2022%20Highlights&text=1%20winter%20storm%2Fcold%20wave,(in%20Missouri%20and%20Kentucky).
16 https://www.weather.gov/pah/December-10th-11th-2021-Tornado#:~:text=Over%203%2C778%20residences%2C%20183%20commercial,least%20200%20people%20were%20injured
17 Barr, L., and Nider, S. (2015). "Critical Infrastructure & Climate Adaptation." George Mason University Center for Infrastructure Protection & Homeland Security *CIP Report*, August 20.
18 Barr, L., and Nider, S. (2015). "Critical Infrastructure & Climate Adaptation." George Mason University Center for Infrastructure Protection & Homeland Security *CIP Report*, August 20.
19 Barr, L., and Nider, S. (2015). "Critical Infrastructure & Climate Adaptation." George Mason University Center for Infrastructure Protection & Homeland Security *CIP Report*, August 20.
20 Barr, L., and Nider, S. (2015). "Critical Infrastructure & Climate Adaptation." George Mason University Center for Infrastructure Protection & Homeland Security *CIP Report*, August 20.
21 FACT SHEET: Executive Order on Climate Preparedness, November 1, 2013. https://www.whitehouse.gov/the-press-office/2013/11/01/fact-sheet-executive-order-climate-preparedness.
22 Barr, L., and Nider, S. (2015). "Critical Infrastructure & Climate Adaptation." George Mason University Center for Infrastructure Protection & Homeland Security *CIP Report*, August 20.
23 "Pound Foolish: Federal Community-Resilience Investments Swamped by Disaster Damages." Center for American Progress. https://www.americanprogress.org/issues/green/report/2013/06/19/67045/pound-foolish/
24 "Pound Foolish: Federal Community-Resilience Investments Swamped by Disaster Damages." Center for American Progress. https://www.americanprogress.org/issues/green/report/2013/06/19/67045/pound-foolish/
25 "Storm Ready Cities: How Climate Resilience Boosts Metro Areas and the Economy." Center for American Progress. https://www.americanprogress.org/issues/green/report/2013/06/19/67045/pound-foolish/
26 Litt, S. (2016). "Cleveland Praised for Climate Change Resilience Efforts by Center for American Progress." *The Cleveland Plain Dealer*, April 16.
27 Center for American Progress, Report, "Resilient Midwestern Cities: Improving Equity in Changing Climate," April 2016.
28 Litt, S. (2016). "Cleveland Praised for Climate Change Resilience Efforts by Center for American Progress." *The Cleveland Plain Dealer*, April 16.
29 Wolfgang, B. (2015). "Billions of Dollars Needed to Fix Aging Vulnerable US Energy Infrastructure, Report Says." *The Washington Times*, April 21.
30 QER Report: "Energy Transmission, Storage, and Distribution Infrastructure," April 2015.

31 QER Report: "Energy Transmission, Storage, and Distribution Infrastructure," April 2015.
32 QER Report: "Energy Transmission, Storage, and Distribution Infrastructure," April 2015, p. S-5.
33 QER Report: "Energy Transmission, Storage, and Distribution Infrastructure," April 2015, p.S-5.
34 QER Report: "Energy Transmission, Storage, and Distribution Infrastructure," April 2015, p. S-6.
35 QER Report: "Energy Transmission, Storage, and Distribution Infrastructure," April 2015, p. S-7.
36 Wolfgang, B. (2015). "Billions of Dollars Needed to Fix Aging Vulnerable US Energy Infrastructure, Report Says." *The Washington Times*, April 21.
37 DHS National Protection and Programs Directorate (OCIA) Critical Infrastructure Security and Resilience Note. "Aging and Failing Infrastructure Systems: Highway Bridges." www.dhs.gov
38 DHS National Protection and Programs Directorate (OCIA) Critical Infrastructure Security and Resilience Note. "Aging and Failing Infrastructure Systems: Highway Bridges." www.dhs.gov
39 FEMA Strategic Foresight Initiative, "Critical Infrastructure: Long-term Trends and Drivers and Their Implications for Emergency Management," June 2011.
40 "Flint Water Crisis Fast Facts." *CNN*. http://www.cnn.com/2016/03/04/us/flint-water-crisis-fast-facts/
41 "Flint Water Crisis Fast Facts." *CNN*. http://www.cnn.com/2016/03/04/us/flint-water-crisis-fast-facts/
42 McLaughlin, E., and Shoichet, C. (2016). "Charges Against 3 in Flint Water Crisis, 'only the beginning'." *CNN*, April 20. http://www.cnn.com/2016/04/20/health/flint-water-crisis-charges/
43 McLaughlin, E., and Shoichet, C. (2016). "Charges against 3 in Flint Water Crisis, 'only the beginning'." *CNN*, April 20. http://www.cnn.com/2016/04/20/health/flint-water-crisis-charges/
44 https://www.whitehouse.gov/wp-content/uploads/2022/05/BUILDING-A-BETTER-AMERICA-V2.pdf
45 Kass, D. (2022). "Critical Infrastructure Attacks: Convergence of IoT and OT Gives Attackers a Huge Attack Surface." https://www.msspalert.com/news/critical-infrastructure-attacks-convergence-of-iot-and-ot-gives-hackers-a-huge-attack-surface
46 https://ioactive.com/pdfs/IOActive_Remote_Car_Hacking.pdf
47 https://www.rand.org/pubs/research_reports/RRA210-1.html
48 DHS, NIPP, "Partnering for Critical Infrastructure Security and Resilience," 2013. www.dhs.gov
49 DHS, NIPP, "Partnering for Critical Infrastructure Security and Resilience," 2013, p. 10. www.dhs.gov
50 GAO Report to the Committee on Homeland Security, House of Representatives, "Critical Infrastructure Protection Sector-Specific Agencies Need to Better Measure Cybersecurity Progress," December 2015.
51 Givens, A., and Busch, N. (2013). "Realizing the Promise of Public-Private Partnerships in US Critical Infrastructure Protection." *International Journal of Critical Infrastructure Protection*, Vol. 6, Issue 1. pp. 43–44
52 Givens, A., and Busch, N. (2013). "Realizing the Promise of Public-Private Partnerships in US Critical Infrastructure Protection." *International Journal of Critical Infrastructure Protection*, Vol. 6, Issue 1.
53 DHS, "Information Sharing: A Vital Resource for Critical Infrastructure Security." www.dhs.gov.
54 DHS, "Information Sharing: A Vital Resource for Critical Infrastructure Security." www.dhs.gov.
55 Givens, A., and Busch, N. (2013). "Realizing the Promise of Public-Private Partnerships in US Critical Infrastructure Protection." *International Journal of Critical Infrastructure Protection*, Vol. 6, Issue 1.
56 Busch, N., and Givens, A. (2012). "Public-Private Partnerships in Homeland Security: Opportunities and Challenges." *Homeland Security Affairs*. Vol. VIII, pp. 43–44.
57 Busch, N., and Givens, A. (2012). "Public-Private Partnerships in Homeland Security: Opportunities and Challenges." *Homeland Security Affairs*.
58 Busch, N., and Givens, A. (2012). "Public-Private Partnerships in Homeland Security: Opportunities and Challenges." *Homeland Security Affairs*.
59 Busch, N., and Givens, A. (2012). "Public-Private Partnerships in Homeland Security: Opportunities and Challenges." *Homeland Security Affairs*.
60 Daniel, M. (2013). "Incentives to Support Adoption of the Cybersecurity Framework." DHS, White-House blog.
61 Givens, A., and Busch, N. (2013). "Realizing the Promise of Public-Private Partnerships in US Critical Infrastructure Protection." *International Journal of Critical Infrastructure Protection*, Vol. 6, Issue 1.

Index

For Product Safety Concerns and Information please contact our EU
representative GPSR@taylorandfrancis.com
Taylor & Francis Verlag GmbH, Kaufingerstraße 24, 80331 München, Germany

www.ingramcontent.com/pod-product-compliance
Ingram Content Group UK Ltd.
Pitfield, Milton Keynes, MK11 3LW, UK
UKHW051828180425
457613UK00007B/245

* 9 7 8 1 0 3 2 5 6 3 0 5 3 *